ELEMENTARY REAL ANALYSIS

ELEMENTARY REAL ANALYSIS

BY

H. G. EGGLESTON
Professor of Mathematics in the University of London

CAMBRIDGE
AT THE UNIVERSITY PRESS
1962

CAMBRIDGE UNIVERSITY PRESS
Cambridge, New York, Melbourne, Madrid, Cape Town, Singapore, São Paulo, Delhi

Cambridge University Press
The Edinburgh Building, Cambridge CB2 8RU, UK

Published in the United States of America by Cambridge University Press, New York

www.cambridge.org
Information on this title: www.cambridge.org/9780521048767

First published 1962
This digitally printed version 2008

A catalogue record for this publication is available from the British Library

ISBN 978-0-521-04876-7 hardback
ISBN 978-0-521-09868-7 paperback

CONTENTS

PREFACE

The aim of this book is to provide for undergraduates, who are specializing in mathematics, a text of sufficient scope to cover all the theoretical aspects of real variable analysis required for the first year or possibly the first and second year of an honours course. Complex numbers are used but the complex variable calculus is not included. The text is based on lectures given at various times in the Universities of Wales, of Cambridge and of London.

There are included many worked examples and unworked exercises with which the reader may test his (or her) understanding of the text. Hints for the solution of many of the exercises are given at the end of the book. Answers to those questions which require a specific answer are also given.

It was my intention to write a book in which no assumptions were made except those stated as such, in which definitions were made as explicit as possible, in which there were no forward references to results to be proved later and in which the development of the subject was logically self-contained. With these restrictions I found it advisable to defer the introduction of the elementary functions, cos, sin, exp, log until a comparatively late stage, with the result that no exercises involving these functions could be included until after chapter 19.

Only those parts of analysis that any analyst will be certain to need have been included. Anything in the nature of 'applied analysis' such as differential equations, Fourier series, etc., has been omitted. Many of the exercises are taken from Cambridge University Examination papers and reproduced here by kind permission of the Cambridge University Press.

I am particularly indebted to Mr J. G. Godwin of Swansea University College for a careful and critical reading of the first draft of this book, and to Dr J. C. Burkill and Dr W. L. C. Sargent for a number of suggestions made at a later stage. Mr H. Kestleman has made many valuable comments and the Appendix is largely based on his ideas.

H. G. EGGLESTON

BEDFORD COLLEGE
REGENT'S PARK
LONDON

NOTATION AND CONVENTIONS

Notation

$\mathbf{R}$	the set of real numbers
$\mathbf{Z}$	the set of complex numbers
P	the set of positive integers
I	the set of integers
ϕ	the empty set
$\mathscr{R}$	real part
$\mathscr{I}$	imaginary part
$\in$	belongs to, $\notin$ does not belong to
$\subset$	contained in
$\exists$	there exists
$\forall$	for all
$\mathfrak{f}, \mathfrak{g}, \ldots$	German letters denote functions (but explicitly defined functions are not necessarily denoted by German letters)
$\mathbf{d}(\mathfrak{f})$	the domain of $\mathfrak{f}$
$\mathbf{r}(\mathfrak{f})$	the range of $\mathfrak{f}$
$\mathsf{B}(a, b)$	the set of functions of bounded variation over (a, b)
$\mathsf{R}(a, b)$	the set of functions Riemann integrable over (a, b)
$\mathsf{RS}_{\mathfrak{g}}(a, b)$	the set of functions Riemann–Stieltjes integrable with respect to $\mathfrak{g}$ over (a, b)

Conventions

$\epsilon > 0$	implies that ϵ is both real and positive	
n	is used for a positive integer	unless the contrary is expressly stated
$\mathfrak{f}$	is single-valued	unless the contrary is expressly stated
if	when introducing new notation or symbols means *if and only if*. This is always the meaning in a *definition*	

PRELIMINARIES

(a) Sets and functions

1. The concept of a set

It is assumed that the reader has some notion of what is meant by 'a set'. We may say that a set is an aggregate distinguished by possessing a certain property and we shall consider only those sets for which the question as to whether a given entity possesses the property or not can be answered. Those entities which possess the property form the set. They are said to 'belong to the set' or to be 'members of the set'.

We shall use the notation $x \in X$ to mean 'x is a member of the set X' and $x \notin X$ to mean 'x does not belong to X'. We shall also refer to a member of X as being an 'element' of X or a 'point' of X. When the members of X are themselves sets then we refer to X as being a 'collection of sets' or a 'class of sets' or an 'aggregate of sets'.

The fact that X is a set defined by a certain property P will be denoted by the symbolism

$$X = \{x|x \text{ has property } P\}$$

and this should be read 'X is the set of those elements x which have the property P'.

The concept of a set will be used primarily for the classification of knowledge. The logical difficulties raised by such expressions as 'the set of all sets' are beyond the scope of this book and we shall not consider them.

2. The idea of a function

A *function* is a correspondence between two sets.

Given two sets X and Y and a rule (or method or formula) by means of which to each member of X we can associate one or more members of Y, then we say that this rule is a *function* and denote it by say $\mathfrak{f}$. We shall use $\mathfrak{f}$ for the function and $\mathfrak{f}(x)$ for the member or members of Y that correspond to x, $x \in X$, by means of the function $\mathfrak{f}$. Functions will always be denoted by German letters except for particular functions when the formula can be given explicitly. X is called the *domain* of $\mathfrak{f}$ and is denoted by $\mathbf{d}(\mathfrak{f})$.

The set Z defined by

$$Z = \{y | y = \mathfrak{f}(x) \text{ for some } x \in X\}$$

is called the *range* or *set of values* of $\mathfrak{f}$ and is denoted by $\mathbf{r}(\mathfrak{f})$. We say that $\mathfrak{f}$ is 'defined over X' and has 'values in Y'. We can distinguish between different functions defined over X and with values in Y by the use of affixes, but it is clear that this notation has limitations if we have to consider functions with different domains and ranges.

For this reason we shall sometimes use $\mathfrak{f}(x)$ in place of the symbol $\mathfrak{f}$ for the function itself.

If $\mathfrak{f}(x)$ is a single member of Y for each $x \in X$ then we say that $\mathfrak{f}$ is *single-valued*. In what follows *all functions will be single-valued unless the contrary is expressly stated*. Such a function whose range is Z and which is such that every value in Z is taken by $\mathfrak{f}$ for exactly one member x of X is said to define a 1-1 *correspondence* between X and Z.

If $\mathfrak{f}$ is a 1-1 correspondence with $\mathbf{d}(\mathfrak{f}) = X$ and $\mathbf{r}(\mathfrak{f}) = Z$ and we wish to pay special attention to Z as opposed to X, then instead of distinguishing the members of Z by $z = \mathfrak{f}(x)$, we shall write z_x, and different suffixes x will denote different members of Z. Then Z is said to be *indexed by* X and X is said to be an *index set*.

If $\mathbf{d}(\mathfrak{f}) = X$ and $\mathbf{r}(\mathfrak{f})$ is a single point then $\mathfrak{f}$ is said to be *constant* over X. A constant function and the value taken by a constant function are two quite distinct ideas.

3. Subsets, unions and intersections of sets

A *subset* of a set X is a set formed by some but not necessarily all of the members of X. If X_1 is a subset of X then we say that X_1 is *contained* in X and write $X_1 \subset X$ or $X \supset X_1$. If two sets X_1, X_2 are such that $X_1 \subset X_2$ and $X_1 \supset X_2$ then they are *identical* and we shall write $X_1 = X_2$ and say that X_1 is X_2. If X_1 and X_2 are two non-identical sets then we write $X_1 \neq X_2$.

The *union* of the sets X_1, X_2 is that set X whose members belong either to X_1 or to X_2 or to both X_1 and X_2. In symbols we write $X = X_1 \cup X_2$ and the definition is equivalent to

$$X_1 \cup X_2 = \{x | x \in X_1 \text{ or } x \in X_2\}.$$

More generally if X_i is a set for every $i \in I$ where I is an index set then the union of the sets X_i, $i \in I$ is denoted by $\bigcup_{i \in I} X_i$ and is defined by

$$\bigcup_{i \in I} X_i = \{x | x \in X_i \text{ for some } i \in I\}.$$

The *intersection* of the sets X_1 and X_2 is that set Y whose members belong both to X_1 and to X_2. We write $Y = X_1 \cap X_2$ and the definition is equivalent to

$$X_1 \cap X_2 = \{x \mid x \in X_1 \text{ and } x \in X_2\}.$$

More generally the intersection of the sets X_i, $i \in I$, is denoted by $\bigcap_{i \in I} X_i$ and defined by

$$\bigcap_{i \in I} X_i = \{x \mid x \in X_i \text{ for all } i \in I\}.$$

For any two sets X_1, X_2

$$X_1 \cup X_2 \supset X_1 \supset X_1 \cap X_2$$

and for any class of sets X_i, $i \in I$,

$$\bigcup_{i \in I} X_i \supset X_i \supset \bigcap_{i \in I} X_i.$$

The set which has no members is called the *empty* set or the *void* set. It is denoted by ϕ and is regarded as being a subset of every set.

Every subset of X apart from X itself and the empty set ϕ is called a *proper subset* of X.

4. Composite and inverse functions

If $\mathfrak{f}$ and $\mathfrak{g}$ are functions such that the domain of $\mathfrak{g}$ contains the range of $\mathfrak{f}$, then we use $\mathfrak{g}(\mathfrak{f})$ or $\mathfrak{g}\mathfrak{f}$ to denote the *composite* function whose domain coincides with that of $\mathfrak{f}$ and whose range is contained in that of $\mathfrak{g}$ and whose value for $x \in \mathbf{d}(\mathfrak{f})$ is given by $\mathfrak{g}\mathfrak{f}(x) = \mathfrak{g}(\mathfrak{f}(x))$.

If $X_1 \subset \mathbf{d}(\mathfrak{f})$ then we use $\mathfrak{f}(X_1)$ to denote the set of values taken by $\mathfrak{f}(x)$ for $x \in X_1$, i.e.

$$\mathfrak{f}(X_1) = \{y \mid y = \mathfrak{f}(x) \text{ for some } x \in X_1\}.$$

We now have an ambiguous notation since for $x \in X$, $\mathfrak{f}(x)$ is a member of Y whilst for $X_1 \subset X$, $\mathfrak{f}(X_1)$ as defined above is a subset of Y. We make the convention of regarding each member of a set as being also a subset of that set and we shall use the same symbol for a member of a set and for the subset consisting of just this one member. With this convention we have either $x \in X$ or $x \subset X$ and there is now no ambiguity in the notations $\mathfrak{f}(x)$, $\mathfrak{f}(X_1)$.

If $X_1 = \phi$ we define $\mathfrak{f}(X_1)$ to be ϕ.

Similarly if $Y_1 \subset \mathbf{r}(\mathfrak{f})$ we define $\mathfrak{f}^{-1}(Y_1)$ to be a subset of $\mathbf{d}(\mathfrak{f})$ as follows

$$\mathfrak{f}^{-1}(Y_1) = \{x \mid x \in X \text{ and } \mathfrak{f}(x) \in Y_1\},$$

$$\mathfrak{f}^{-1}(\phi) = \phi.$$

We extend the definition of $\mathfrak{f}^{-1}(Y_1)$ to the case when Y_1 is not contained in $\mathbf{r}(\mathfrak{f})$ by defining $\mathfrak{f}^{-1}(Y_1) = \mathfrak{f}^{-1}(Y_1 \cap \mathbf{r}(\mathfrak{f}))$. For example, if there is no $x \in \mathbf{d}(\mathfrak{f})$ for which $\mathfrak{f}(x) \in Y_1$ then $\mathfrak{f}^{-1}(Y_1) = \phi$.

If $\mathbf{d}(\mathfrak{f}) = X$ and $\mathbf{r}(\mathfrak{f}) = Z$ then $\mathfrak{f}^{-1}$ is a function, possibly many-valued for which $\mathbf{d}(\mathfrak{f}^{-1}) = Z$ and $\mathbf{r}(\mathfrak{f}^{-1}) = X$, $\mathfrak{f}^{-1}$ is called the *inverse* function of $\mathfrak{f}$ whether single-valued or not.

(*b*) Numbers

1. Real numbers

It is assumed that the reader is familiar with the properties of integers and the idea of mathematical induction, with the properties of rational numbers and the definition of real numbers (for those not familiar with one of the possible definitions a brief account is given in the Appendix). The Binomial theorem for positive integral exponents is also assumed.

A real number which is not rational is called *irrational* and a real number which is the root of an equation whose coefficients are integers is called *algebraic.*

The set of all real numbers will be denoted by $\mathbf{R}$. Two basic properties are:

(i) If L and U are subsets of $\mathbf{R}$ such that

(*a*) $\mathbf{R} = L \cup U$,

(*b*) if $x \in L$ and $y \in U$ then $x < y$,

(*c*) $L \neq \phi$, $U \neq \phi$,

then either L has a greatest member or U has a least member. The pair of sets (L, U) is called a *Dedekind section* and the number which is either the greatest member of L or the least member of U is said to be *defined by the section.* If it is denoted by x and $y \in \mathbf{R}$, $y \neq x$ then we have $y < x$ or $y > x$ according as $y \in L$ or $y \in U$.

(ii) If a, b are two real numbers such that $a > 0$ then there exists an integer n such that $na > b$.

Of these two properties the first is referred to as the *completeness* of the real number system and the second is called *Archimedes Axiom.*

A set of real numbers will be called a *real set.*

We use $\boldsymbol{P}$ for the set of positive integers and $\boldsymbol{I}$ for the set of all integers.

A function of basic importance is the modulus of x, $|x|$ defined for every real number x by

$$|x| = x \quad \text{if} \quad x \geqslant 0, \qquad |x| = -x \quad \text{if} \quad x < 0.$$

2. Complex numbers

A complex number is an ordered pair of real numbers (x_1, x_2) where $x_1 \in \mathbf{R}$, $x_2 \in \mathbf{R}$. $(x_1, x_2) = (x_2, x_1)$ if and only if $x_1 = x_2$. The addition and multiplication of complex numbers are defined by

$$(x_1, x_2) + (y_1, y_2) = (x_1 + y_1, x_2 + y_2),$$

$$(x_1, x_2)(y_1, y_2) = (x_1 y_1 - x_2 y_2, x_1 y_2 + x_2 y_1).$$

We shall use $\mathbf{Z}$ for the set of all complex numbers and often use a single symbol z for a member of $\mathbf{Z}$. If $z = (x_1, x_2)$ then x_1 is the *real part* of z denoted by $\mathscr{R}(z)$ and x_2 is the *imaginary part* of z denoted by $\mathscr{I}(z)$. Both $\mathscr{R}(z)$ and $\mathscr{I}(z)$ are real numbers.

3. Real and complex numbers

We shall assume that the reader has already met the ideas of adding, multiplying, subtracting and dividing complex numbers, but we do not assume the properties of powers of numbers apart from

(*a*) positive integral powers of complex numbers z defined inductively by $z^1 = z$, $z^n = z . z^{n-1}$ $(n > 1)$,

(*b*) negative integral powers of z defined by

$$z^{-n} = 1/z^n \quad (n = 1, 2, \ldots, z \neq 0),$$

(*c*) the zero power of a complex number $z \neq 0$ defined by $z^0 = 1$ and finally

(*d*) rational powers of positive real numbers.

The square root of a real positive number is always taken to be positive unless the contrary is expressly stated. A positive number is one which is *greater than zero.* The number zero itself is neither positive nor negative.

The *modulus* of the complex number $z = (x_1, x_2)$ is defined to be $(x_1^2 + x_2^2)^{\frac{1}{2}}$ and is denoted by $|z|$. It is clear that $|(x_1, 0)| = |x_1|$.

The operations of addition, multiplication, subtraction and division can be used to define functions both over $\mathbf{R}$ and over $\mathbf{Z}$. For example, if a is a fixed real number then $a + x$ is defined for all $x \in \mathbf{R}$ and the rule which associates with x the number $a + x$ is a function whose range and domain are both $\mathbf{R}$. It is customary to denote this function by $a + x$. Similarly we can regard $x + x$ or $2x$ as another function and $x . x$ or x^2 as a third. Any function defined by means of the operations of addition and multiplication used a finite number of times is called a *polynomial.* Any polynomial $\mathfrak{f}(x)$ can be written in the form $\mathfrak{f}(x) = a_0 + a_1 x + \ldots + a_n x^n$ where n

is a positive or zero integer. The numbers $a_0, a_1, \ldots, a_n$ are called *coefficients*.

We make the convention that in what follows n is a positive integer unless the contrary is expressly stated.

The sum of a set of numbers $a_1, a_2, \ldots, a_n$ is denoted by $\sum_{i=1}^{n} a_i$ and the product of these numbers is denoted by $\prod_{i=1}^{n} a_i$. In particular we write $n!$ for $\prod_{i=1}^{n} i$ and define $0!$ to be 1.

The complex number $(x_1, -x_2)$ is said to be *conjugate* to the complex number (x_1, x_2) and we denote the conjugate of z by $\bar{z}$. Thus

$$\overline{(\bar{z})} = z, \quad z.\bar{z} = (|z|^2, 0), \quad z+\bar{z} = (2\mathscr{R}(z), 0).$$

The product of z with $\bar{z}$ is particularly useful in dealing with $|z|^2$. For example we may prove the fundamental inequality

$$|z_1+z_2| \leqslant |z_1| + |z_2|$$

as follows: $(z_1+z_2)\overline{(z_1+z_2)} = (z_1+z_2)(\bar{z}_1+\bar{z}_2)$

$$= z_1\bar{z}_1 + z_2\bar{z}_2 + z_1\bar{z}_2 + z_1\bar{z}_2$$

and taking the real parts of each side

$$|z_1+z_2|^2 = |z_1|^2 + |z_2|^2 + 2\mathscr{R}(z_1\bar{z}_2).$$

Now $\mathscr{R}(z) \leqslant |z|$ and $|\bar{z}| = |z|$. Thus $\mathscr{R}(z_1\bar{z}_2) \leqslant |z_1\bar{z}_2| = |z_1|.|\bar{z}_2|$ and $|z_1+z_2|^2 \leqslant (|z_1|+|z_2|)^2$.

The required inequality now follows on taking square roots since we know that $|z_1|+|z_2|$ is not negative. We shall have need to refer to this inequality frequently so we call it the *modulus inequality*.

The real numbers **R** do not form a subset of the complex numbers **Z** but the complex numbers z for which $\mathscr{I}(z) = 0$ have exactly the same algebraic properties as the real numbers. In many circumstances it is convenient to replace the real numbers x by the complex numbers $(x, 0)$ and to regard **R** as a subset of **Z**. Apart from certain special situations we shall always do this.

In what follows the range of every function is contained in **R** or **Z** unless the contrary is expressly stated. If $\mathbf{r}(\mathfrak{f}) \subset \mathbf{R}$ we say that $\mathfrak{f}$ is a *real function*, if $\mathbf{r}(\mathfrak{f}) \subset \mathbf{Z}$ then $\mathfrak{f}$ is a *complex function* and we shall use the terms positive function or negative function according as every value of $\mathfrak{f}$ is positive or is negative.

If $\mathfrak{f}$ and $\mathfrak{g}$ are both real or both complex functions defined over the same set X then we define the sum of $\mathfrak{f}$ and $\mathfrak{g}$, $\mathfrak{f}+\mathfrak{g}$ by

$$\mathfrak{f}+\mathfrak{g}\,(x) = \mathfrak{f}(x)+\mathfrak{g}(x) \quad (x \in X)$$

and the product $\mathfrak{f}.\mathfrak{g}$ by

$$\mathfrak{f}.\mathfrak{g}\,(x) = \mathfrak{f}(x).\mathfrak{g}(x) \quad (x \in X).$$

If $\mathfrak{f}(x) \neq 0$ we also define the 'inverse' $1/\mathfrak{f}$ by

$$1/\mathfrak{f}\,(x) = 1/(\mathfrak{f}(x)) \quad (x \in X).$$

This inverse should not be confused with that defined earlier.

One of the techniques that we shall use repeatedly is the following. Suppose that we have to show that the two real numbers a, b satisfy $a \leqslant b$. In practice it is often fairly easy to show that if ϵ is a given positive real number then $a \leqslant b+\epsilon$ and yet difficult to prove that $a \leqslant b$ directly. In this case we try to establish that $a \leqslant b+\epsilon$ *by an argument that does not depend upon the particular* ϵ *involved.* Then it will follow that $a \leqslant b+\epsilon$ for $\epsilon > 0$ and hence that $a \leqslant b$ (for if $a > b$ then there exists $\epsilon > 0$ such that $a > b+\epsilon$ for example $\epsilon = \frac{1}{2}(a-b)$).

1

ENUMERABILITY AND SEQUENCES

Remark. Much of analysis is concerned with the extension of ideas that are easily defined for finite sets of numbers to infinite sets of numbers. It is a fact that in order to extend ideas of addition we need to be able to 'order' the set in some way. For an enumerable set this may be done by arranging the set in a sequence and the precise definitions and methods are explained in this chapter.

A set X is said to be *finite* if for some integer N, $N > 0$, $\exists\,\mathfrak{f}$ for which $\mathbf{d}(\mathfrak{f})$ is the set of integers $1, 2, \ldots, N$ and $\mathbf{r}(\mathfrak{f}) = X$ in particular if we regard the integers $1, 2, \ldots, N$ as an index set and if $\mathfrak{f}$ does not take the same value twice then we can denote the members of X by $x_1, x_2, \ldots, x_N$. If there is no such function for any positive integer N then the set is said to be *infinite.* If X is itself a set of positive integers then X will always contain a least member. X will contain a greatest member if and only if it is a finite set.

A set X is said to be *enumerable* if and only if $\exists\,\mathfrak{f}$ for which $\mathbf{d}(\mathfrak{f})$ is the set of positive integers and $\mathbf{r}(\mathfrak{f}) = X$. Any finite set is enumerable and we make the convention of regarding ϕ as an enumerable set. A set which is not enumerable is said to be non-enumerable.

A function $\mathfrak{f}$ such that $\mathbf{d}(\mathfrak{f})$ is the set of positive integers is called a *sequence*.

We may indicate such a sequence by writing $x_1, x_2, \ldots, x_n, \ldots$, where $x_n = \mathfrak{f}(n)$, and we shall also use the notation $\{x_n\}$ or $\{\mathfrak{f}(n)\}$ to indicate the sequence.

Examples of sequences are

$$1, 2, 3, \ldots, n, \ldots, \quad \text{where} \quad \mathfrak{f}(n) = n,$$

$$2, 4, 6, \ldots, 2n, \ldots, \quad \text{where} \quad \mathfrak{f}(n) = 2n,$$

$$1, -1, 2, -2, \ldots, n, -n, \ldots, \quad \text{where} \quad \mathfrak{f}(n) = \tfrac{1}{2}(n+1),\ n \text{ odd}; \quad -\tfrac{1}{2}n,\ n \text{ even},$$

$$z, z^2, z^3, \ldots, z^n, \ldots, \quad \text{where} \quad \mathfrak{f}(n) = z^n.$$

The order of the integers 1, 2, 3, ... induces an order among the elements of the set. x_n is called the nth *term* of the sequence and we say that x_i precedes or follows x_j according as $i < j$ or $i > j$. The values x_i need not be distinct. The sequence is changed if we alter the order in which the distinct elements occur.

A sequence of positive integers $\mathfrak{f}$, i.e. a function $\mathfrak{f}$ such that $\mathbf{r}(\mathfrak{f}) \subset P$, $\mathbf{d}(\mathfrak{f}) = P$ is said to be *strictly increasing* if and only if $\mathfrak{f}(n) > \mathfrak{f}(m)$ whenever $n > m$.

A *subsequence* is obtained from a given sequence by omitting certain of its terms and writing down those that are left in precisely the same order as that in which they occur in the original sequence. More precisely a sequence $\mathfrak{h}$ is said to be a subsequence of $\mathfrak{g}$ if and only if, for some strictly increasing sequence of positive integers $\mathfrak{f}$, $\mathfrak{h}$ is the composite function $\mathfrak{g}\mathfrak{f}$. If $\mathfrak{g}$ is $\{x_n\}$ then $\mathfrak{g}\mathfrak{f}$ is $\{x_{n_i}\}$ where $\mathfrak{f}$ is defined by $\mathfrak{f}(i) = n_i$. For example, if $\mathfrak{g}$ is the sequence $1, -1, +2, -2, +3, -3, \ldots$, then $1, 2, 3, \ldots$ is the subsequence defined by $\mathfrak{h} = \mathfrak{g}\mathfrak{f}$ where $\mathfrak{f}(n) = 2n-1$; and $-1, -4, -9, -16, \ldots$ is the subsequence defined by $\mathfrak{g}\mathfrak{f}$ with $\mathfrak{f}(n) = 2n^2$.

Worked example

For each positive integer i X_i is an enumerable set. Show that $\bigcup_{i \in P} X_i$ is also enumerable.

We have to show that there is a function defined over the positive integers whose values include all of the set $\bigcup_{i \in P} X_i$.

Since X_i is enumerable $\exists\, \mathfrak{f}_i(n)$ defined for $n = 1, 2, 3, \ldots$ and such that $X_i = \mathbf{r}(\mathfrak{f}_i)$. Define a new function $\mathfrak{g}$ as follows. If n is a product of a power of 2 by a power of 3 say $n = 2^p 3^q$ where p, q are positive integers then $\mathfrak{g}(n) = \mathfrak{f}_p(q)$. Otherwise $\mathfrak{g}(n) = \mathfrak{f}_1(1)$. Then $\mathbf{d}(\mathfrak{g}) = \boldsymbol{P}$, $\mathbf{r}(\mathfrak{g}) = \bigcup_{i \in \boldsymbol{P}} X_i$ for if x belongs to this union then $x \in X_i$ for some i and therefore $x = \mathfrak{f}_i(j)$ for some j where i, j are both positive integers. Hence $x = \mathfrak{g}(n)$ where $n = 2^i 3^j$.

Thus $\bigcup_{i \in \boldsymbol{P}} X_i$ is enumerable.

Exercises

1. Show that every subset of an enumerable set is enumerable.

2. Show that the set of rational numbers is enumerable.

3. Show that the set of all polynomials with integer coefficients is enumerable. Deduce that the set of all algebraic numbers is enumerable.

4. X is an enumerable set and U is the set whose members are those subsets of X which are finite. Show that U is enumerable.

5. $\boldsymbol{P}$ is the set of positive integers and V is the set of all subsets of $\boldsymbol{P}$. Show that if $\{v_n\}$ is a sequence of members of V then there is another member of V which does not occur in the sequence.

6. X is an infinite enumerable set. Show that it is possible to establish a 1-1 correspondence between the members of X and the positive integers.

7. An enumerable set X is such that if X_1, X_2 are any two subsets of X for which $X = X_1 \cup X_2$ and $X_1 \cap X_2 = \phi$, then at least one of the sets X_1, X_2 is a finite set. Show that X is a finite set.

8. X is an enumerable set and Y a set such that a function $\mathfrak{f}$ can be defined with $\mathbf{d}(\mathfrak{f}) \subset X$, $\mathbf{r}(\mathfrak{f}) = Y$. Show that Y is enumerable.

9. $\{\mathfrak{f}(n)\}$ is the sequence of positive integers defined by $\mathfrak{f}(n) = 2n$ and $\{\mathfrak{g}(n)\}$ is a second strictly increasing sequence of positive integers with $\mathfrak{g}(1) = 1$. If the sequences $\{\mathfrak{f}\mathfrak{g}(n)\}$ and $\{\mathfrak{g}\mathfrak{f}(n)\}$ are identical show that $\mathfrak{g}(n) = n$.

10. $\mathfrak{f}$ is a sequence and $\mathfrak{g}$ is a subsequence of $\mathfrak{f}$. Show that every subsequence of $\mathfrak{g}$ is a subsequence of $\mathfrak{f}$.

11. X is an enumerable set and $\{x_n\}$ is a sequence formed from distinct elements of X (i.e. x_n is x_m if and only if $n = m$). $\{\mathfrak{f}(n)\}$ and $\{\mathfrak{g}(n)\}$ are two strictly increasing sequences of positive integers. Show that $\{x_{\mathfrak{f}(n)}\}$ is a subsequence of $\{x_{\mathfrak{g}(n)}\}$ if and only if $\mathbf{r}(\mathfrak{g}) \supset \mathbf{r}(\mathfrak{f})$.

2

BOUNDS FOR SETS OF NUMBERS

Remark. Of a finite set of real numbers it is often important to know the values of the largest or least. For infinite sets there may be no largest member of the set even when it is known that the set does not contain arbitrarily large numbers. For example $\{x | x \in \mathbf{R}, x < 1\}$ contains no largest member. The concepts of least upper bound and greatest lower bound introduced in this chapter are valid for infinite sets. These bounds have similar properties to those of the largest and least members of a finite set.

In the following definitions it is assumed that the sets A and X are non-void.

Definition 1. *If $A \subset \mathbf{R}$ and $\exists\, k \in \mathbf{R}$ such that for all $a \in A$, $a \leqslant k$, then k is called an upper bound of A and A is said to be bounded above.*

Definition 2. *If $A \subset \mathbf{R}$ and $\exists\, l \in \mathbf{R}$ such that for all $a \in A$, $a \geqslant l$, then l is called a lower bound of A and A is said to be bounded below.*

Definition 3. *If $A \subset \mathbf{R}$ and A is bounded above and bounded below then we say that A is bounded.*

Definition 4. *If $X \subset \mathbf{Z}$ and $Y \subset \mathbf{R}$ is defined by*

$$Y = \{y \mid y = |x|,\ x \in X\}$$

then we say that X is bounded if and only if Y is bounded.

Theorem 1. (*a*) *If $A \subset \mathbf{R}$ is bounded above then there is amongst its upper bounds one which is least.*

(*b*) *If $A \subset \mathbf{R}$ is bounded below then there is amongst its lower bounds one which is greatest.*

(*a*) Define the sets L, U by $L = \{x \mid x \text{ is not an upper bound of } A\}$ $U = \{x \mid x \text{ is an upper bound of } A\}$. It may be verified that (L, U) is a Dedekind section. Let y be the number defined by this section. y is an upper bound for A, for otherwise $\exists\, a \in A$ and $a > y$ then $\frac{1}{2}(a-y)+y > y$. Thus $\frac{1}{2}(a-y)+y \in U$ since y is the number defined by (L, U), but this means from the definition of U that $\frac{1}{2}(a-y)+y$ is an upper bound for A. But this is not so since $\frac{1}{2}(a-y)+y < a$.

This contradiction shows that y is an upper bound of A. Also if x is an upper bound of A then $x \in U$ and since y is defined by (L, U) $y \leqslant x$. Thus y is the least of all the upper bounds.

(*b*) The second part of the theorem may be proved similarly or may be deduced from (*a*) by considering the set B defined by $B = \{x \mid -x \in A\}$. The least upper bound of this set exists and its negative can be shown to be the greatest lower bound of A.

The theorem is proved.

The least upper bound and the greatest lower bound of a bounded real set A are denoted by $\sup\limits_{x \in A} x$ or $\sup A$ and $\inf\limits_{x \in A} x$ or $\inf A$, respectively. If A is a finite set say, $x_1, \ldots, x_k$ then $\sup A$ is the largest x_i and $\inf A$ is the least x_i. In this case we also use $\max(x_1, \ldots, x_k)$ and $\min(x_1, \ldots, x_k)$, respectively.

If the real set A is unbounded above then we write

$$\sup_{x \in A} x = \sup A = \infty$$

and if A is unbounded below then we write $\inf\limits_{x \in A} x = \inf A = -\infty$.

Neither ∞ nor $-\infty$ is a number and we introduce these symbols here in order to avoid exceptional cases in some of the statements of later theorems. They are called *infinity* and *minus infinity*. A bound which is neither ∞ nor $-\infty$ is referred to as a *finite bound*. We sometimes use $+\infty$ instead of ∞.

If A is bounded above the number $\sup A$ is characterized by the two properties:

(*a*) $a \in A$ implies that $a \leqslant \sup A$;

(*b*) $\forall \epsilon > 0 \, \exists \, a \in A$ such that $a > \sup A - \epsilon$.

For (*a*) is the statement that $\sup A$ is an upper bound of A and (*b*) is the statement that no number less than $\sup A$ is an upper bound of A, i.e. $\sup A$ is the least upper bound of A.

The number $\inf A$ can be characterized similarly.

Worked examples

(i) *A is a bounded real set and B is a non-void subset of A. Show that*

$$\inf A \leqslant \inf B \leqslant \sup B \leqslant \sup A.$$

If $a \in B$ then $a \in A$ and thus

$$\inf A \leqslant a \leqslant \sup A.$$

This is true for all members of B. Thus $\sup A$ is an upper bound and $\inf A$ is a lower bound of B. Hence $\sup B$, the *least* upper bound of B, satisfies $\sup B \leqslant \sup A$ and $\inf B$, the *greatest* lower bound of B, satisfies $\inf B \geqslant \inf A$. Finally for any member $a \in B$

$$\inf B \leqslant a, \quad a \leqslant \sup B.$$

Thus $\inf B \leqslant \sup B$ and all the stated inequalities have been proved.

(ii) *A and B are two bounded real sets and C is defined by*

$$C = \{x | x = a + b \text{ where } a \in A, \, b \in B\}.$$

Show that C is bounded above and that

$$\sup C = \sup A + \sup B.$$

$\forall x \in C$, $x = a + b$ where $a \in A$ and $b \in B$. Now $a \leqslant \sup A$ and $b \leqslant \sup B$. Thus $x \leqslant \sup A + \sup B$. This is true for all $x \in C$: thus

$\sup A+\sup B$ is an upper bound for C, C is bounded above and $\sup C \leqslant \sup A+\sup B$.

$\forall \epsilon > 0 \, \exists a \in A$ such that $a > \sup A - \frac{1}{2}\epsilon$, $\exists b \in B$ such that $b > \sup B - \frac{1}{2}\epsilon$. But $a+b \in C$. Hence

$$\sup C \geqslant a+b > \sup A+\sup B-\epsilon.$$

This is true for all $\epsilon > 0$ and thus $\sup C \geqslant \sup A+\sup B$.

The reverse inequality has been proved above. Hence

$$\sup C = \sup A+\sup B.$$

Exercises

1. Determine for each of the following sets whether they are bounded above or below and in each case find the least upper bound and the greatest lower bound.

(i) $X = \{x \mid x \in \mathbf{R},\ x < 1\}$. (ii) $X = \{x \mid x = 1/n,\ n = 1, 2, 3, \ldots\}$.
(iii) $X = \{x \mid x$ an irrational real number with $x^2 < 2\}$.
(iv) $X = \{x \mid x$ a positive integer$\}$. (v) $X = \{x \mid x = 0\}$.

2. A is a bounded real set. Show that $\inf A$ is characterized by the two properties:
(a) $a \in A$ implies that $a \geqslant \inf A$;
(b) $\forall \epsilon > 0 \, \exists a \in A$ such that $a < \inf A+\epsilon$.

3. With the notation of worked example (ii) above show that

$$\inf C = \inf A+\inf B.$$

4. A is a bounded real set l, m real numbers with $m > 0$. B is defined by $B = \{l+mx \mid x \in A\}$. Show that B is bounded and

$$\sup B = l+m\sup A, \quad \inf B = l+m\inf A.$$

5. A, B are two bounded real sets. Show that

$$\sup (A \cup B) = \max(\sup A, \sup B), \quad \inf (A \cup B) = \min(\inf A, \inf B).$$

6. Z is a set of complex numbers and the real sets A, B, C are defined by

$$A = \{x \mid x = |z|,\ z \in Z\}, \quad B = \{x \mid x = \mathscr{R}(z),\ z \in Z\},$$

$$C = \{x \mid x = \mathscr{I}(z),\ z \in Z\}.$$

Show that A is bounded above if and only if both B and C are bounded above and that in this case

$$\sup A \leqslant \max(|\sup B|, |\inf B|)+\max(|\sup C|, |\inf C|).$$

3

BOUNDS FOR FUNCTIONS AND SEQUENCES; THE O NOTATION

Remark. The functions that arise in applications of analysis are rarely defined in an explicit numerical manner. They are usually given in terms of one or more properties and we have to deduce as much as we can about the function. A useful property of functions is that of boundedness, i.e. the boundedness of their range of values.

Definition 5. *$\mathfrak{f}$ is said to be bounded above, bounded below or bounded according as $\mathbf{r}(\mathfrak{f})$ is bounded above, bounded below or bounded respectively.*

Thus if x is real the functions x and x^2 are bounded over $0 < x \leqslant 1$ but $1/x$ is not bounded over $0 < x \leqslant 1$ even though it is defined at each point of this set. The function $1/x$ is not defined at $x = 0$ but we could define a function $\mathfrak{f}$ by $\mathfrak{f}(0) = 0$, $\mathfrak{f}(x) = 1/x$ $0 < x \leqslant 1$. Then $\mathfrak{f}$ is defined at all points of $0 \leqslant x \leqslant 1$ but is not bounded over this set.

Definition 6. *The least upper bound and the greatest lower bound of the real function $\mathfrak{f}$ are defined to be the least upper bound and the greatest lower bound of $\mathbf{r}(\mathfrak{f})$ respectively.*

We use the notation $\sup\limits_{x \in Y} \mathfrak{f}(x)$ or $\sup \mathfrak{f}(Y)$ and $\inf\limits_{x \in Y} \mathfrak{f}(x)$ and $\inf \mathfrak{f}(Y)$ where $Y = \mathbf{d}(\mathfrak{f})$. If $\mathfrak{f}$ is a sequence $\{x_n\}$ then we write $\sup\limits_{n} x_n$ and $\inf\limits_{n} x_n$. For example $\{1/n\}$ is bounded and $\sup\limits_{n} 1/n = 1$, $\inf\limits_{n} 1/n = 0$. When it is clear what sequences are involved we write simply $\sup x_n$ and $\inf x_n$.

The O notation

It is often more important to know whether or not a sequence $\{x_n\}$ is bounded than it is to know the values of $\sup\limits_{n} x_n$ and $\inf\limits_{n} x_n$ if x_n is real or of $\sup\limits_{n} |x_n|$ if x_n is complex. We write $x_n = O(1)$ to indicate that x_n is bounded, i.e. that $\exists\, k$ such that $|x_n| < k$ for $n = 1, 2, 3, \ldots$. Further if $\{x_n\}$, $\{y_n\}$ are two sequences and each y_n is a non-negative real number then we write $x_n = O(y_n)$ if and only if $\exists\, k$ such that $|x_n| < k y_n$ $(n = 1, 2, 3, \ldots)$.

Remark. If $y_n > 0$ all n then $x_n = O(y_n)$ is equivalent to $x_n/y_n = O(1)$. The first notation will be used when we compare the sequence $\{x_n\}$ with some simpler sequence $\{y_n\}$ in the course of some calculation and having the relation $x_n = O(y_n)$ we can replace x_n by y_n in the later stages of the problem. Whereas $x_n/y_n = O(1)$ will be used when the observation of this fact concludes whatever arguments are in hand.

The non-enumerability of the real numbers

We show next that the real numbers satisfying $a < x < b$ where $a < b$ are non-enumerable.

If x_n is a sequence such that $a < x_n < b$ then $\exists\, y$ such that $a < y < b$ and $x_n \neq y$ for $n = 1, 2, \ldots$.

The argument is based upon the fact that if $X = \{x \mid a < x < b\}$ and $r \in \mathbf{R}$ we can choose $c, d \in \mathbf{R}$ such that $a < c < d < b$ and either $r < c$ or $d < r$, i.e. $r \notin \{x \mid c < x < d\}$.

Define p_1, q_1 real numbers such that

$$a < p_1 < q_1 < b \quad \text{and} \quad x_1 \notin \{x \mid p_1 < x < q_1\}.$$

Generally when $p_1, q_1;\ p_2, q_2;\ \ldots;\ p_{n-1}, q_{n-1}$ have been defined we define p_n, q_n such that

$$p_{n-1} < p_n < q_n < q_{n-1} \quad \text{and} \quad x_n \notin \{x \mid p_n < x < q_n\}.$$

Write $y = \sup p_n$. Then:

(*a*) Since y is an upper bound of all the p_n $y \geqslant p_{n+1}$. But $p_{n+1} > p_n$. Thus $y > p_n$ $(n = 1, 2, \ldots)$.

(*b*) By a simple induction argument $p_r < q_{n+1}$ $(r = 1, 2, \ldots)$. Thus $y \leqslant q_{n+1}$. But $q_{n+1} < q_n$ and hence $y < q_n$ $(n = 1, 2, \ldots)$.

By (*a*) and (*b*) $p_n < y < q_n$ and thus $a < y < b$ and $y \notin x_n$ for any n (since $x_n \notin \{x \mid p_n < x < q_n\}$).

Worked examples

(i) *Prove that if $\{a_n\}$ and $\{b_n\}$ are bounded real sequences then*

$$\sup a_n + \sup b_n \geqslant \sup (a_n + b_n) \geqslant \sup a_n + \inf b_n.$$

Since $\sup a_n \geqslant a_m$, $\sup b_n \geqslant b_m$ we have

$$\sup a_n + \sup b_n \geqslant a_m + b_m \quad (m = 1, 2, 3, \ldots).$$

Thus the number $\sup a_n + \sup b_n$ is an upper bound for the sequence $\{a_n + b_n\}$ and must be greater than or equal to its least upper bound $\sup (a_n + b_n)$. Thus

$$\sup a_n + \sup b_n \geqslant \sup (a_n + b_n).$$

Next for any $m = 1, 2, \ldots, b_m \geqslant \inf b_n$ and $\forall \epsilon > 0 \,\exists$ an integer N such that $a_N > \sup a_n - \epsilon$. Hence

$$a_N + b_N > \inf b_n + \sup a_n - \epsilon.$$

Thus $$\sup(a_n + b_n) > \inf b_n + \sup a_n - \epsilon.$$

Thus this is true for every $\epsilon > 0$, hence finally

$$\sup(a_n + b_n) \geqslant \inf b_n + \sup a_n.$$

(ii) *The real function* $\mathfrak{f}$ *is defined for all real* x *and satisfies*

$$\mathfrak{f}(x+y) = \mathfrak{f}(x) . \mathfrak{f}(y), \quad \mathfrak{f}(x) \leqslant k.$$

Show that $\mathfrak{f}$ *is a constant function.*

If $\mathfrak{f}$ is not the constant function with value $0 \,\exists\, x_0$ such that $\mathfrak{f}(x_0) \neq 0$. But by hypothesis

$$\mathfrak{f}(x_0 + 0) = \mathfrak{f}(x_0) . \mathfrak{f}(0)$$

and since $x_0 + 0$ is x_0 it follows on division by $\mathfrak{f}(x_0)$ that $\mathfrak{f}(0) = 1$. Further for any x

$$\mathfrak{f}(x) . \mathfrak{f}(-x) = \mathfrak{f}(x - x) = \mathfrak{f}(0) = 1$$

and thus $$\mathfrak{f}(x) = 1/\mathfrak{f}(-x).$$

Again we have $\mathfrak{f}(x) = \mathfrak{f}(\tfrac{1}{2}x + \tfrac{1}{2}x) = [\mathfrak{f}(\tfrac{1}{2}x)]^2 \geqslant 0.$

Thus either

(*a*) $\mathfrak{f}(x)$ is the constant function 0, or

(*b*) $\mathfrak{f}(0) = 1$, $\mathfrak{f}(x) \geqslant 0$ all x and $\mathfrak{f}(x) = 1/\mathfrak{f}(-x)$.

In case (*b*) either $\mathfrak{f}(x) = 1$ for all x or $\exists\, x_1$ such that $\mathfrak{f}(x_1) > 1$ (from the third condition of (*b*)). If $\mathfrak{f}(x_1) = 1 + y$, $y > 0$ then by the definition of $\sup \mathfrak{f} \,\exists\, x_2$ such that $\mathfrak{f}(x_2) > (\sup \mathfrak{f})/\mathfrak{f}(x_1)$. Hence

$$\mathfrak{f}(x_1 + x_2) = \mathfrak{f}(x_1) . \mathfrak{f}(x_2) > \sup \mathfrak{f}.$$

This is impossible and there is no x_1 such that $\mathfrak{f}(x_1) > 1$.

An alternative method is to consider the sequence $\{\mathfrak{f}(nx_1)\}$. If $\mathfrak{f}(x_1) > 1$ then since $\mathfrak{f}(nx_1) = (\mathfrak{f}(x_1))^n$ this sequence is unbounded above.

Thus $\mathfrak{f}$ is a constant function either with value 0 or with value 1.

Exercises

1. Prove that the polynomial $a_0 + a_1 x + \ldots + a_n x^n$ is bounded over $\{x \mid 0 \leqslant x \leqslant 1\}$ and is not bounded over $\{x \mid 0 \leqslant x\}$, if $a_n \neq 0$ $(n > 0)$.

2. For $x \in X \subset \mathbf{R}$, $M > |\mathfrak{f}(x)| > \delta > 0$. Show that $\mathfrak{f} . \mathfrak{g}$ is bounded over X if and only if $\mathfrak{g}$ is bounded over X.

3. $\mathfrak{f}$ is bounded over each of the sets $X_1, X_2, \ldots, X_k$. Show that $\mathfrak{f}$ is bounded over $\bigcup_{n=1}^{k} X_n$.

4. A is a bounded real set. Show that there is an enumerable subset of A, say B, such that

$$\sup B = \sup A, \quad \inf B = \inf A.$$

5. Prove that $\sup_n 1/(1+n+n^2) = \frac{1}{3}$, $\inf_n 1/(1+n+n^2) = 0$.

6. Give an example of a sequence $\{a_n\}$ such that if B is any finite subset of the elements of the range set of the sequence then

$$\inf a_n < \inf B \leqslant \sup B < \sup a_n.$$

7. $\{a_n\}$ and $\{b_n\}$ are bounded real sequences. Show that

$$\sup a_n + \sup b_n \geqslant \sup(a_n+b_n) \geqslant \begin{pmatrix} \sup a_n + \inf b_n \\ \inf a_n + \sup b_n \end{pmatrix}$$
$$\geqslant \inf(a_n+b_n) \geqslant \inf a_n + \inf b_n.$$

Give an example of two sequences $\{a_n\}$, $\{b_n\}$ for which strict inequality occurs in each of these above inequalities.

8. $\{a_n\}$, $\{b_n\}$ are bounded non-negative real sequences. Prove that

$$\sup(a_n b_n) \leqslant \sup a_n . \sup b_n,$$
$$\inf(a_n b_n) \geqslant \inf a_n . \inf b_n.$$

9. $\{a_{n_i}\}$ is a subsequence of the bounded real sequence $\{a_n\}$. Show that

$$\inf a_n \leqslant \inf_i a_{n_i} \leqslant \sup_i a_{n_i} \leqslant \sup a_n.$$

10. $\{a_n\}$ is a bounded real sequence such that $\inf a_n > 0$. Show that

$$\sup(1/a_n) = 1/(\inf a_n),$$
$$\inf(1/a_n) = 1/(\sup a_n).$$

11. $\{a_n\}$ is a bounded real sequence. Show that

$$\sup((a_1+\ldots+a_n)/n) \leqslant \sup a_n, \quad \inf((a_1+\ldots+a_n)/n) \geqslant \inf a_n,$$

and if $a_n \geqslant 0$ $(n = 1, 2, \ldots)$, then

$$\sup(a_1 a_2 \ldots a_n)^{1/n} \leqslant \sup a_n, \quad \inf(a_1 a_2 \ldots a_n)^{1/n} \geqslant \inf a_n.$$

12. Prove that

(i) $7n^2+n = O(n^2)$, (ii) $n^2 = O(7n^2+n)$, (iii) $n = O(n^2)$.

13. $a_n = O(b_n)$ and $c_n \geqslant 0$. Show that $a_n c_n = O(b_n c_n)$.

14. $a_n = O(b_n)$. Show that $(a_1+a_2+\ldots+a_n)/n = O(b_1+b_2+\ldots+b_n)/n$ and that if also $a_n \geqslant 0$ then $(a_1 a_2 \ldots a_n)^{1/n} = O((b_1 b_2 \ldots b_n)^{1/n})$.

Give two examples of pairs of sequences $\{a_n\}$, $\{b_n\}$ of which the first satisfies

$$a_n = O(b_n), \quad a_1 a_2 \ldots a_n \neq O(b_1 b_2 \ldots b_n)$$

and the second satisfies

$$a_1 a_2 \ldots a_n = O(b_1 b_2 \ldots b_n) \quad \text{and} \quad a_n \neq O(b_n), \quad b_n > 0.$$

15. $a_n = O(b_n)$ and $b_n = O(c_n)$. Prove that

$$\text{(i)} \;\; -a_n = O(b_n), \qquad \text{(ii)} \;\; a_n = O(c_n),$$
$$\text{(iii)} \;\; a_n + b_n = O(c_n), \qquad \text{(iv)} \;\; a_n^2 = O(b_n^2).$$

4

LIMITS OF SEQUENCES; THE o, $\sim$ NOTATION

Remark. It is important to be able to consider properties of sequences which depend upon the sequence as a whole and particularly those properties that are unchanged if we alter a finite number of members of the sequence. The most important property is whether or not the sequence tends to a limit as n tends to infinity. The meaning of these phrases is given below.

***Definition* 7 (*a*).** *$\{a_n\}$ is said to converge or tend to the limit l as n tends to ∞ if and only if $\forall \epsilon > 0 \, \exists$ an integer N such that $|a_n - l| < \epsilon$ for all $n > N$.*

N may depend on ϵ. If there exists any such integer N with the above properties then, given ϵ, there will be a least possible integer N which we denote by $\mathfrak{N}(\epsilon)$. We shall write

$$a_n \to l \quad \text{as} \quad n \to \infty \quad \text{or} \quad \lim a_n = l.$$

if the conditions of the definition hold. If $\{a_n\}$ is a real sequence then the definition is equivalent to $\forall \epsilon > 0 \, \exists N$ such that

$$l - \epsilon < a_n < l + \epsilon \quad \text{for all } n > N.$$

***Definition* 7 (*b*).** *The real sequence $\{a_n\}$ diverges to ∞ if and only if for any real number $k \, \exists$ an integer N such that $a_n > k$ for all $n > N$.*

We write in this case $\lim a_n = \infty$ or $a_n \to \infty$ as $n \to \infty$.

***Definition* 7 (*c*).** *The real sequence $\{a_n\}$ diverges to $-\infty$ if and only if for any real number $k \, \exists$ an integer N such that $a_n < k$ for any $n > N$.*

In this case we write $a_n \to -\infty$ as $n \to \infty$ or $\lim a_n = -\infty$.

Any sequence satisfying the conditions of definition 7*a* is said to *converge*. Any other sequence not satisfying these conditions is said to *diverge*. If $a_n \to l$ as $n \to \infty$ is false, we write $a_n \not\to l$.

A sequence whose limit is zero is a *null sequence* and one whose terms are positive real numbers is a *positive sequence*.

Remark. In testing a sequence for convergence we need to know the limit l to which it converges and then establish the required inequalities by using arguments that do not depend upon the particular n involved but are true for all $n > N$. We shall later see how to decide whether a sequence converges or not even when we do not know its limit.

Exercises. (1) Prove that $1/n \to 0$ as $n \to \infty$; $n^2 \to \infty$ as $n \to \infty$; $-n \to -\infty$ as $n \to \infty$.

(2) $a_n \to l$ as $n \to \infty$. Prove that $a_{n+1} \to l$ as $n \to \infty$.

The o, $\sim$ notation

If $\{a_n\}$ is a null sequence then we write $a_n = o(1)$. If $\{b_n\}$ is a positive sequence and $a_n/b_n = o(1)$ then we write $a_n = o(b_n)$.

If $\{a_n\}$, $\{b_n\}$ are such that $a_n/b_n \to 1$ as $n \to \infty$ then we write $a_n \sim b_n$.

For example: $n = o(n^2)$; $1/n = o(1)$; $n^2 + n \sim n^2$.

Statement

It will be convenient to use the definition of convergence in a more general form. Let $\mathfrak{f}$ be any positive function defined for positive x. Then if $\{a_n\}$ converges to l as $n \to \infty$ $\forall \epsilon > 0 \, \exists$ an integer N such that $|a_n - l| < \mathfrak{f}(\epsilon)$ for all $n > N$. In fact since $\mathfrak{f}(\epsilon) > 0$ $\mathfrak{N}(\mathfrak{f}(\epsilon))$ is defined and will do for N. $\mathfrak{f}(\epsilon)$ is usually some elementary function of ϵ such as $\frac{1}{2}\epsilon$ or ϵ^2.

For certain functions $\mathfrak{f}$ there is a converse of the above. Suppose that $\forall \epsilon > 0 \, \exists \delta > 0$ such that $0 < \mathfrak{f}(x) < \epsilon$ if $0 < x < \delta$. Then the above condition implies the convergence of a_n to l as $n \to \infty$. For $\forall \epsilon > 0 \, \exists y$ such that $0 < \mathfrak{f}(y) < \epsilon$ (by the condition on $\mathfrak{f}$) and $\exists$ an integer N such that $|a_n - l| < \mathfrak{f}(y)$ for all $n > N$ (by the condition on $\{a_n\}$). Thus $|a_n - l| < \epsilon$ for all $n > N$, i.e. $a_n \to l$ as $n \to \infty$. The most useful function for $\mathfrak{f}$ is $\mathfrak{f}(x) = kx$ where k is a positive constant.

Elementary properties of limits

Throughout the remainder of this chapter all given limits of sequences are to be taken to be numbers, i.e. not ∞ or $-\infty$.

(i) $\lim a_n = l$ *implies that* $\lim |a_n| = |l|$ *and* $\{a_n\}$ *is bounded.*

$\forall \epsilon > 0 \, \exists$ an integer N such that $|a_n - l| < \epsilon$ for all $n > N$. By the modulus inequality

$$|a_n| \leqslant |l| + |a_n - l| \quad \text{and} \quad |l| \leqslant |a_n| + |a_n - l|.$$

Thus
$$||a_n| - |l|| \leqslant |a_n - l|.$$

Hence $$||a_n|-|l|| < \epsilon \quad \text{for all } n > N$$
and $$|a_n| \to |l| \quad \text{as} \quad n \to \infty.$$

Further $$|a_n| \leqslant \max\{|a_1|, |a_2|, \ldots, |a_N|, |l|+\epsilon\}$$
and thus $\{a_n\}$ is bounded.

(ii) *If* $\lim a_n = l$, $\lim b_n = m$, *then*

$$(a)\ \lim(a_n+b_n) = l+m, \quad (b)\ \lim a_n b_n = lm,$$

$$(c)\ \textit{if } l \neq 0,\ a_n \neq 0\ (n = 1, 2, 3, \ldots) \textit{ then } \lim(1/a_n) = 1/l.$$

(*a*) Since $a_n \to l$ $\forall \epsilon > 0 \exists$ an integer N_1 such that $|a_n - l| < \epsilon$ for $n > N_1$. Since $b_n \to m \exists$ an integer N_2 such that $|b_n - l| < \epsilon$ for $n > N_2$. Thus if $n > \max(N_1, N_2)$ from the modulus inequality we have
$$|a_n + b_n - l - m| \leqslant |a_n - l| + |b_n - m| < 2\epsilon$$
and by the statement above this means that $a_n + b_n \to l + m$ as $n \to \infty$.

(*b*) By (i) we see $\exists M$ such that $|b_n| < M$ $(n = 1, 2, \ldots)$. Suppose first that $l = 0$. Then $\forall \epsilon > 0 \exists N$ such that $|a_n| < \epsilon$ for all $n > N$. Hence $|a_n b_n| < M\epsilon$ for all $n > N$, and by the remark made above this means that $a_n b_n \to 0$ as $n \to \infty$.

Next we reduce the general case to the special case as follows. In definition 7*a* the inequalities required to establish that $a_n \to l$ as $n \to \infty$ are precisely the same as those to establish that $a_n - l \to 0$ as $n \to \infty$. Thus we are given that $a_n - l \to 0$ as $n \to \infty$ and $b_n - m \to 0$ as $n \to \infty$. Moreover, a constant sequence converges to its constant value. Thus $l \to l$ as $n \to \infty$ and $lm \to lm$ as $n \to \infty$.

By the above case $(a_n - l)\,b_n \to 0$ as $n \to \infty$ and $(b_n - m)\,l \to 0$ as $n \to \infty$. By (*a*) above $(a_n - l)\,b_n + (b_n - m)\,l \to 0$ as $n \to \infty$ and by another application of (*a*) $(a_n - l)\,b_n + (b_n - m)\,l + lm \to lm$ as $n \to \infty$, i.e. $a_n b_n \to lm$ as $n \to \infty$.

(*c*) Since $a_n \to l$ as $n \to \infty$ $\forall \epsilon > 0 \exists N$ such that $|a_n - l| < \epsilon$ for $n > N$. Now $l \neq 0$ thus we could take $\frac{1}{2}|l|$ for ϵ. Then $\exists N_1$ such that $|a_n - l| < \frac{1}{2}|l|$ for $n > N_1$. From the modulus inequality
$$|a_n| \geqslant |l| - |a_n - l| > \tfrac{1}{2}|l|, \quad n > N_1.$$
Thus for $n > \max(N, N_1)$ dividing $|a_n - l| < \epsilon$ by $|a_n| \cdot |l|$ we have
$$\left|\frac{1}{l} - \frac{1}{a_n}\right| < \frac{\epsilon}{|l|\,|a_n|} < \frac{2\epsilon}{|l|^2} \quad \text{all} \quad n > \max(N, N_1).$$
By the statement above this implies that $1/a_n \to 1/l$ as $n \to \infty$.

Remark. A convergent sequence *must* tend to a limit. It is true that $1/n - 1/n^2 \to 0$ as $n \to \infty$ *but it is nonsense to write* $1/n \to 1/n^2$ *as* $n \to \infty$. Such a symbol has no meaning. We may, however, be able to use the $\sim$ notation in such a case. For example $1/n - 1/(n+1) \to 0$ or $1/n \sim 1/(n+1)$.

One of the most useful devices for proving the convergence of a given real sequence $\{a_n\}$ to a limit l is to compare it with two simpler sequences $\{b_n\}$, $\{c_n\}$ for which

$$b_n \leqslant a_n \leqslant c_n \quad \text{and} \quad b_n \to l, \quad c_n \to l.$$

For example, if

$$a_n = \frac{7n+1}{n+7} \quad \text{take} \quad c_n = \frac{7n+1}{n}, \quad b_n = \frac{7n}{n+7}.$$

Then $$b_n \leqslant a_n \leqslant c_n, \quad c_n = 7 + \frac{1}{n} \to 7 \quad \text{as} \quad n \to \infty,$$

$$b_n = \frac{7}{1+(7/n)} \to 7 \quad \text{as} \quad n \to \infty.$$

Thus $a_n \to 7$ as $n \to \infty$.

(iii) *If the real sequence* $\{a_n\}$ *converges to* l *and* $a_n \geqslant a$ *for* $n = 1, 2, \ldots$ *then* $l \geqslant a$.

For if $l < a$ then $\frac{1}{2}(a-l) > 0$. $\forall \epsilon > 0 \, \exists N$ such that $|a_n - l| < \epsilon$ for $n > N$, i.e. $a_n < l + \epsilon$ $(n > N)$. In particular if $\epsilon = \frac{1}{2}(a-l)$ then $a_n < l + \frac{1}{2}(a-l) < a$. This is impossible as $a_n \geqslant a$ for all n. Thus the assumption that $l < a$ is false and $l \geqslant a$.

Worked examples

(i) $\{a_n\}$ *converges to* l *and* $\{a_{n_i}\}$ *is a subsequence of* $\{a_n\}$. *Show that* $a_{n_i} \to l$ *as* $i \to \infty$.

Since $a_n \to l$ as $n \to \infty \, \forall \epsilon > 0 \, \exists N$ an integer such that $|a_n - l| < \epsilon$ for all $n > N$. Since $\{a_{n_i}\}$ is a subsequence of $\{a_n\}$, $n_i > n_j$ for $i > j$ and this by a simple induction argument implies that $n_i \geqslant i$. Thus if $i > N$ then $n_i > N$ and $|a_{n_i} - l| < \epsilon$ for all $i > N$. Thus $a_{n_i} \to l$ as $i \to \infty$.

(ii) *Discuss the convergence or divergence of the sequence* $(-1)^n$.

Consider the two subsequences $\{(-1)^{2n}\}$ and $\{(-1)^{2n-1}\}$. These are respectively $1, 1, 1, \ldots$ and $-1, -1, -1, \ldots$. The first converges to 1 and the second to -1. Now if $\{(-1)^n\}$ converges to l then by (i) above $l = 1$ and $l = -1$. This contradiction shows that $\{(-1)^n\}$ diverges.

(iii) $\{a_n\}$ *converges to* l. *Show that* $(a_1 + a_2 + \ldots + a_n)/n \to l$ *as* $n \to \infty$.

Remark. If $a_n = l$ then $(a_1+a_2+\ldots+a_n)/n = l$ and the idea is that if $a_n \to l$ then only a 'few' of the a_n can differ substantially from l.

Write $b_n = a_n - l$. Then $b_n \to 0$ as $n \to \infty$ and

$$(b_1+\ldots+b_n)/n = (a_1+\ldots+a_n)/n - l.$$

Thus it is sufficient to show that $(b_1+\ldots+b_n)/n \to 0$ as $n \to \infty$. Now $\forall \epsilon > 0 \; \exists$ an integer N such that $|b_n| < \epsilon$ if $n > N$. Then for $n > N$ by the modulus inequality

$$\left|\frac{b_1+\ldots+b_n}{n}\right| \leqslant \frac{\left|\sum_1^N b_r\right|}{n} + \sum_{N+1}^{n} \frac{|b_r|}{n} < \frac{\left|\sum_1^N b_r\right|}{n} + \frac{n-N}{n}\epsilon.$$

As $n \to \infty \left|\sum_1^N b_r\right|/n \to 0$ and thus $\exists \, N_1$ such that if $n > N_1 \left|\sum_1^N b_r\right|/n < \epsilon$. Then for all $n > \max(N, N_1)$

$$|(b_1+b_2+\ldots+b_n)|/n \leqslant \epsilon + \frac{n-N}{n}\epsilon < 2\epsilon.$$

By the statement on p. 18 it follows that $(b_1+\ldots+b_n)/n \to 0$ as $n \to \infty$.

(iv) $\{a_n\}$ *converges to* l *and* $\{b_n\}$ *converges to* m. *Show that*

$$\lim\,(a_1b_n + a_2b_{n-1} + \ldots + a_nb_1)/n = lm.$$

Write $c_n = a_n - l$. Then $c_n \to 0$ as $n \to \infty$ and

$$(c_1b_n + c_2b_{n-1} + \ldots + c_nb_1)/n = (a_1b_n + \ldots + a_nb_1)/n - l(b_1+\ldots+b_n)/n.$$

By example (iii) above $(b_1+\ldots+b_n)/n \to m$ as $n \to \infty$ and thus it is sufficient to know that $(c_1b_n + c_2b_{n-1} + \ldots + c_nb_1)/n \to 0$ as $n \to \infty$. Now $c_n \to 0$ and therefore $|c_n| \to 0$ (p. 18(i)), $b_n \to l$ and therefore $|b_n|$ is bounded (p. 18(i)). Suppose that $|b_n| < M$ all n. Then

$$\left|\frac{c_1b_n+\ldots+c_nb_1}{n}\right| \leqslant \frac{|c_1b_n|+\ldots+|c_nb_1|}{n} \leqslant \frac{M(|c_1|+\ldots+|c_n|)}{n}$$

by the modulus inequality. By example (iii) $(|c_1|+\ldots+|c_n|)/n \to 0$. Thus $\forall \epsilon > 0 \; \exists$ an integer N such that

$$\left|\frac{c_1b_n+\ldots+c_nb_1}{n}\right| \leqslant M\epsilon \quad \text{for all } n > N.$$

Hence $$\frac{c_1b_n+\ldots+c_nb_1}{n} \to 0 \quad \text{as} \quad n \to \infty.$$

Exercises

1. Prove that

$$\lim (n+1)/(n^2-n+1) = 0, \quad \lim (n^4-n)/(n^4+n^2) = 1,$$
$$\lim ((n+1)^{\frac{1}{2}}-n^{\frac{1}{2}}) = 0.$$

2. $\{a_n\}$ is a complex sequence. Show that $\{a_n\}$ converges to l if and only if $\mathscr{R}(a_n) \to \mathscr{R}(l)$ and $\mathscr{I}(a_n) \to \mathscr{I}(l)$ as $n \to \infty$.

3. $\{a_n\}$ converges to l. Show that ka_n converges to kl.

4. $\{a_n\}$ and $\{b_n\}$ are real sequences. $\{a_n\}$ diverges to ∞ and $\{b_n\}$ converges to l. Show that (i) $\lim (a_n+b_n) = \infty$, (ii) if $l > 0$ then $\lim a_n b_n = \infty$, (iii) if $l < 0$ then $\lim a_n b_n = -\infty$.

5. $a_n > 0$ and $\lim (1/a_n) = 0$. Show that $\lim a_n = \infty$.

6. $\{a_n\}$ and $\{b_n\}$ are real sequences such that $a_n > 0 > b_n$ and $\lim (a_n - b_n) = 0$. Show that $\lim a_n = \lim b_n = 0$.

7. $\{a_n\}$ is a real sequence bounded above and for each integer n, $a_n < \sup a_n$. Show that there is a subsequence $\{a_{n_i}\}$ of $\{a_n\}$ such that $a_{n_i} \to \sup a_n$ as $i \to \infty$.

8. $\{a_n\}$ is a positive null sequence. Show that there is an infinite set of integers J such that if $N \in J$ and $n \geqslant N$ then $a_n \leqslant a_N$.

9. $\{a_n\}$ is a sequence (possibly complex) such that $\lim (a_{n+1} - a_n) = l$. Show that $\lim a_n/n = l$.

10. $\{a_n\}$ and $\{b_n\}$ converge to l and to m, respectively. Show that

$$(a_1 b_1 + a_2 b_2 + \ldots + a_n b_n)/n \to lm \quad \text{as} \quad n \to \infty.$$

11. Show that

(i) if $a_n = o(1)$ and $b_n = O(1)$ then $a_n b_n = o(1)$,
(ii) if $a_n = o(b_n)$ and $b_n = O(c_n)$ then $a_n = o(c_n)$,
(iii) if $a_n = o(b_n)$ and $b_n \sim c_n$ then $a_n = o(c_n)$,
(iv) if $a_n \sim b_n$ and $b_n \sim c_n$ then $a_n \sim c_n$.

12. Show that if $a_n = o(b_n)$ and $a_n > 0$, $b_n > 0$ then

$$(a_1 a_2 \ldots a_n)^{1/n} = o((b_1 b_2 \ldots b_n)^{1/n}).$$

13. $a_n \sim b_n$ and $b_n > 0$, $a_n > 0$. Show that $(a_1 a_2 \ldots a_n)^{1/n} \sim (b_1 b_2 \ldots b_n)^{1/n}$.

14. $a_n \sim b_n$ and $b_n = o(1)$. Show that $a_n = o(1)$.

15. $a_n - b_n = o(1)$ and $1/b_n = O(1)$. Show that $a_n \sim b_n$. Give an example of sequences $\{a_n\}$, $\{b_n\}$ for which the relation $a_n - b_n = o(1)$ is true but the relation $a_n \sim b_n$ is false.

16. $a_n \sim b_n$ and $b_n = O(1)$. Show that $a_n - b_n = o(1)$. Give an example of sequences $\{a_n\}$, $\{b_n\}$ for which the relation $a_n \sim b_n$ is true but the relation $a_n - b_n = o(1)$ is false.

17. $\{a_n\}$ is such that every subsequence of $\{a_n\}$ contains a null sub-subsequence. Show that $\{a_n\}$ is a null sequence.

18. $x_n \to x$ as $n \to \infty$ and for each integer n $x_{n,i} \to x_n$ as $i \to \infty$. Show that there exists a sequence of integers $\{p(n)\}$ such that $x_{n,p(n)} \to x$ as $n \to \infty$.

19. $\{a_n\}$ is such that $\lim a_n^2 = l^2$ and $\sup |a_n - a_{n+1}| < 2|l|$ where $l \neq 0$. Show that either $\lim a_n = l$ or $\lim a_n = -l$.

20. $\{a_n\}$ is such that $\lim (a_{n+2} - 2a_{n+1} + a_n) = k$. Show that

$$\lim (a_{n+1} - a_n)/(n+1) = k$$

and that if $a_n = o(n^2)$ then $k = 0$.

5

MONOTONIC SEQUENCES; TWO IMPORTANT EXAMPLES; IRRATIONAL POWERS OF POSITIVE REAL NUMBERS

Remark. In this chapter we introduce a class of sequences with simple convergence properties and illustrate their importance by a number of examples. It is possible to develop the theory by taking theorem 2 below as an axiom, and deducing theorem 1 from it (see worked example (i)).

Definition 8. *A real sequence $\{a_n\}$ is said to be:*

(i) *increasing, written*

$$\{a_n\}\uparrow \quad \text{if} \quad a_{n+1} \geqslant a_n \quad (n = 1, 2, \ldots),$$

(ii) *decreasing, written*

$$\{a_n\}\downarrow \quad \text{if} \quad a_{n+1} \leqslant a_n \quad (n = 1, 2, \ldots),$$

(iii) *strictly increasing, written*

$$\{a_n\}\Uparrow \quad \text{if} \quad a_{n+1} > a_n \quad (n = 1, 2, \ldots),$$

(iv) *strictly decreasing, written*

$$\{a_n\}\Downarrow \quad \text{if} \quad a_{n+1} < a_n \quad (n = 1, 2, \ldots).$$

Definition 9. *A sequence that is either increasing or decreasing is said to be monotonic. A sequence that is either strictly increasing or strictly decreasing is said to be strictly monotonic.*

If it is asserted that $\{a_n\}$ is monotonic then it is always implied that $\{a_n\}$ is a real sequence.

Theorem 2 (i) *If* $\{a_n\}\uparrow$ *then either* $\lim a_n = \infty$ *or* $\{a_n\}$ *is bounded above and* $\lim a_n = \sup a_n$.

(ii) *If* $\{a_n\}\downarrow$ *then either* $\lim a_n = -\infty$ *or* $\{a_n\}$ *is bounded below and* $\lim a_n = \inf a_n$.

(i) If $\{a_n\}$ is bounded above then $\forall \epsilon > 0 \; \exists$ an integer N such that $a_N > \sup a_n - \epsilon$. But $\{a_n\}\uparrow$ and thus

$$a_n \geqslant a_N > \sup a_n - \epsilon \quad \text{for} \quad n \geqslant N.$$

For all n, $a_n \leqslant \sup a_n$. Thus $|a_n - \sup a_n| < \epsilon$ for all $n \geqslant N$. Thus $\lim a_n$ exists and is equal to $\sup a_n$.

If $\{a_n\}$ is not bounded above then $\forall K \; \exists$ an integer N such that $a_N > K$. But $\{a_n\}\uparrow$ and thus $a_n \geqslant a_N > K$ for all $n \geqslant N$. Hence $\lim a_n = \infty$.

(ii) The proof is similar to that of (i). Alternatively (ii) may be deduced from (i) by considering the sequence $\{-a_n\}$. The details are left to the reader.

An important example. $\{(1+1/n)^n\}$ *is increasing*, $\{(1+1/n)^{n+1}\}$ *is decreasing. They both converge to the same limit as* $n \to \infty$.

We need the following lemma.

Lemma. *If* $x \geqslant -1$ *and* n *is a positive integer then*

$$(1+x)^n \geqslant 1+nx.$$

Equality occurs only if either $n = 1$ *or* $x = 0$.

If $n = 1$ then $(1+x)^n$ is $1+nx$. If $n > 1$ suppose that

$$(1+x)^{n-1} \geqslant 1+(n-1)x \quad \text{for} \quad x \geqslant -1.$$

Then

$$(1+x)(1+x)^{n-1} \geqslant (1+x)(1+(n-1)x) = 1+nx+(n-1)x^2.$$

Thus $(1+x)^n \geqslant 1+nx$ and moreover $(1+x)^n > 1+nx$ unless $(n-1)x^2 = 0$, i.e. $n = 1$ or $x = 0$. By induction it follows that the inequality is true for all positive integers n.

The lemma is proved.

To show that $\{(1+1/n)^n\}\uparrow$ we have to show that

$$(1+1/n)^n \leqslant (1+1/(n+1))^{n+1} \quad (n = 1, 2, \ldots)$$

which can be reduced to $n/(n+1) \leqslant (1-1/(n+1)^2)^{n+1}$ and this last inequality follows from the lemma with $n+1$ in place of n and $-1/(n+1)^2$ in place of x.

To show that $\{(1+1/n)^{n+1}\}\downarrow$ we have to show that

$$(1+1/n)^{n+1} \geqslant (1+1/(n+1))^{n+2} \quad (n = 1, 2, \ldots),$$

i.e. that
$$\left(\frac{n+1}{n}\right)^{n+1} \geqslant \left(\frac{n+2}{n+1}\right)^{n+2}.$$

This last inequality is equivalent to

$$\left(\frac{(n+1)^2}{n(n+2)}\right)^{n+1} \geqslant \frac{n+2}{n+1},$$

i.e. to
$$\left(1+\frac{1}{n(n+2)}\right)^{n+1} \geqslant 1+1/(n+1).$$

Now by the lemma with $n+1$ in place of n and $1/(n+2)\,n$ in place of x we have

$$\left(1+\frac{1}{n(n+2)}\right)^{n+1} \geqslant 1+\frac{n+1}{n(n+2)}.$$

Since $(n+1)/n(n+2) > 1/(n+1)$ we have the required inequality proved and $\{(1+1/n)^{n+1}\}\downarrow$.

Further since $(1+1/n)^n < (1+1/n)^{n+1}$ and, from the monotoneity proved above, $2^1 \leqslant (1+1/n)^n$, $(1+1/n)^{n+1} \leqslant 2^2$ it follows that the sequences $\{(1+1/n)^n\}$, $\{(1+1/n)^{n+1}\}$ are both bounded.

By theorem 2 the sequences tend to limits say l, m and

$$(1+1/n)^n \leqslant l \leqslant m \leqslant (1+1/n)^{n+1}.$$

Thus $1 \leqslant m/l \leqslant 1+1/n$. This is true for every positive integer n. Hence $l = m$. *We denote this common limiting value by* e.

The function x^α where $x > 0$ and α is real and irrational

Consider first the case $x > 1$. By the lemma above, for $x > 1$ and n a positive integer $(1+x/n)^n \geqslant 1+x > x$. Thus $1+x/n > x^{1/n} > 1$. Let $\{\alpha_n\}$ be a null sequence of positive rational numbers. Then $\forall$ a positive integer $M \, \exists$ an integer N such that $\alpha_n < 1/M$ for all $n > N$. Hence $1 < x^{\alpha_n} < 1+x/M$ for all $n > N$. Thus $x^{\alpha_n} \to 1$ as $n \to \infty$.

From the property of limits (ii) (c) on p. 19 $(1/x)^{\alpha_n} \to 1$ as $n \to \infty$. For any rational null sequence $\{\alpha_n\}$

$$(1/x)^{|\alpha_n|} = x^{-|\alpha_n|} \leqslant x^{\alpha_n} \leqslant x^{|\alpha_n|}.$$

But $\{|\alpha_n|\}$ is a positive rational null sequence. Thus $x^{|\alpha_n|} \to 1$ and $(1/x)^{|\alpha_n|} \to 1$ and $n \to \infty$. Thus $x^{\alpha_n} \to 1$ as $n \to \infty$.

Now given any real number α there is an increasing rational sequence $\{\alpha_n\}$ such that $\alpha_n \to \alpha$ and $n \to \infty$. For example, write $\alpha_n = p/2^n$ where p is an integer (positive negative or zero) such that $p/2^n$ is the largest number of this form less than α. Then $\{x^{\alpha_n}\}$ is an increasing sequence bounded above by say x^β where β is any rational greater than α. $\{x^{\alpha_n}\}$ therefore tends to a limit say k as $n \to \infty$. If $\{\beta_n\}$ is another rational sequence convergent to α then $\{(\alpha_n - \beta_n)\}$ is a null rational sequence and thus

$$x^{\beta_n} = x^{\alpha_n}.x^{\beta_n - \alpha_n} \to k.1 \quad \text{as} \quad n \to \infty.$$

(by (ii) (b), p. 19). Thus the limit k is the same for any rational sequence convergent to α. Also if α is rational then $\{(\alpha_n - \alpha)\}$ is a null rational sequence and

$$x^{\alpha_n - \alpha} = x^{\alpha_n}.x^{-\alpha} \to k.x^{-\alpha} \quad \text{as} \quad n \to \infty.$$

But $x^{\alpha_n - \alpha} \to 1$ as $n \to \infty$. Thus $k = x^\alpha$ if α is rational. *If α is irrational we define x^α to be the number k.*

If $x = 1$ we define $x^\alpha = 1$ for all real α. If $0 < x < 1$ we define x^α by $x^\alpha = (1/x)^{-\alpha}$.

A second example. If $-1 < x < 1$ and α is a real number then $n^\alpha x^n \to 0$ as $n \to \infty$.

Since $|n^\alpha x^n| = n^\alpha |x|^n$ it is sufficient to consider the case $0 \leqslant x < 1$. Write $x = 1/y = 1/(1+\delta)$, $\delta > 0$. By the Binomial expansion $(1+\delta)^n > \binom{n}{k} \delta^k$ for any integer k and $n \geqslant k \geqslant 0$. Thus if $n > \alpha$ and $k > \alpha$

$$0 \leqslant n^\alpha x^n < n^\alpha \bigg/ \binom{n}{k} \delta^k.$$

Now
$$\binom{n}{k} = \frac{n!}{k!\,(n-k)!} = \frac{n(n-1)\dots(n-k+1)}{k!}.$$

Thus since $k > \alpha$, $n^\alpha \bigg/ \binom{n}{k} \to 0$ as $n \to \infty$. Hence $n^\alpha x^n \to 0$ as $n \to \infty$.

Worked examples

(i) *Deduce theorem* 1 *from theorem* 2.

We assume that every bounded monotonic real sequence is convergent and deduce from this that every real set bounded above has a least upper bound.

Let A be a real non-void set bounded above and let a_1 be the least integer which is an upper bound for A (a_1 may of course be a negative integer). Generally let a_n be the least number $p/2^n$ where p is an integer that is an upper bound for A. There must be such a least number since if $a \in A$ and k is an upper bound of A there are at most a finite set of numbers of the form $p/2^n$ being between a and k. The least of these is a_n. Then $\{a_n\}$ is a decreasing sequence. For any $a \in A$, $a_n \geqslant a$. Thus $\{a_n\}$ is bounded below. Hence $\{a_n\}$ converges say to l. We show that l is the least upper bound of A.

First, $a_n \geqslant a$ for $n = 1, 2, \ldots$ and $a \in A$. Therefore $\lim a_n \geqslant a$ and thus l is an upper bound of A. Next if k is any upper bound of A and $k < l$ we can find n an integer so large that $l-k > 1/2^n$. Then $\exists$ an integer p such that $k < p/2^n < l$. Now k is an upper bound of A. Thus so is $p/2^n$. Hence $a_n \leqslant p/2^n < l$. But $\{a_n\}\downarrow$ and $a_n \to l$ thus $a_n \geqslant l$ for all n. This contradiction shows that for any upper bound k of A $k \geqslant l$.

Thus we have established the existence of a least upper bound of A.

(ii) *Evaluate* $\lim n^{1/n}$.

$n \geqslant 1$ and thus $n^{1/n} \geqslant 1$.

If $n^{1/n} \geqslant 1+x$ where $x > 0$ then by the Binomial expansion $n \geqslant (1+x)^n > \frac{1}{2}n(n-1)x^2$ for $n \geqslant 2$. Thus $n-1 \leqslant 2/x^2$. Hence if $n > 1+2/x^2$ then $n^{1/n} < 1+x$. That is to say $\forall \epsilon > 0 \, \exists N$ (namely any integer greater than $1+2/\epsilon^2$) such that $|1-n^{1/n}| < \epsilon$ for all $n > N$, i.e. $\lim n^{1/n} = 1$.

(iii) *The real numbers a and x_1 satisfy $a \geqslant 1$, $0 < x_1 < a^2$ and $\{x_n\}$ is defined by $x_{n+1} = a-(a^2-x_n)^{\frac{1}{2}}$, $n = 1, 2, \ldots$. Show that*

(*a*) $\{x_n\}\downarrow$ *and* $\lim x_n = 0$,

(*b*) $x_{n+1}/x_n \to 1/2a$ *as* $n \to \infty$.

We have $x_{n+1} = x_n/(a+(a^2-x_n)^{\frac{1}{2}})$.

(*a*) Since $a \geqslant 1$, $x_{n+1} \leqslant x_n$. Also if $x_n \geqslant 0$ then $x_{n+1} \geqslant 0$. Thus $x_n \geqslant 0$ for all n (by induction). By theorem 2 $\{x_n\}$ decreases to a limit l as $n \to \infty$. Further l must satisfy the equation

$$l = l/(a+(a^2-l)^{\frac{1}{2}}).$$

Thus $l = 0$ or $a+(a^2-l)^{\frac{1}{2}} = 1$. The second alternative cannot occur since $a > 1$, $l \leqslant a^2$ and thus $a+(a^2-l)^{\frac{1}{2}} > 1$. Hence $l = 0$.

$$(b) \quad \frac{x_{n+1}}{x_n} = \frac{1}{a+(a^2-x_n)^{\frac{1}{2}}} \to \frac{1}{2a} \quad \text{as} \quad n \to \infty.$$

Remark. We have used the fact that if $b_n \to l$, $b_n > 0$, $l > 0$, then $b_n^{\frac{1}{2}} \to l^{\frac{1}{2}}$ as $n \to \infty$. This follows from the relation

$$|b_n^{\frac{1}{2}} - l^{\frac{1}{2}}| = \frac{|b_n - l|}{|b_n^{\frac{1}{2}} + l^{\frac{1}{2}}|} < \frac{|b_n - l|}{l^{\frac{1}{2}}}.$$

(iv) *Show that if $x > 0$ $(1+(x/n))^n \to e^x$ as $n \to \infty$. (It may be assumed that if $a_n \to l$, $a_n > 0$, $l > 0$, then $a_n^\alpha \to l^\alpha$.)*

We show first that $\exists L$ independent of n and r such that

$$\left|\left(1+\frac{x}{n}\right)^n - \left(1+\frac{x}{n+r}\right)^{n+r}\right| \leqslant \frac{Lr}{n} \quad (r = 0, 1, 2, \ldots).$$

Let K be an integer greater than x. Then for $0 \leqslant p \leqslant n$, p an integer

$$0 < \left(1+\frac{x}{n}\right)^p < \left(1+\frac{K}{n}\right)^n < \left(1+\frac{1}{n}\right)^{Kn} \leqslant e^K,$$

where we have used both the lemma of p. 24 and the fact that $(1+1/n)^n \uparrow$ and tends to e. It is sufficient to prove the stated inequality when $r = 1$ for the general case then follows by addition.

$$\left|\left(1+\frac{x}{n}\right)^n - \left(1+\frac{x}{n+1}\right)^{n+1}\right|$$

$$= \left|\left(1+\frac{x}{n}\right)^n - \left(1+\frac{x}{n+1}\right)^n - \left(1+\frac{x}{n+1}\right)^n \frac{x}{n+1}\right|$$

$$\leqslant \left|\left\{\left(1+\frac{x}{n}\right) - \left(1+\frac{x}{n+1}\right)\right\}\right.$$

$$\left.\times\left\{\left(1+\frac{x}{n}\right)^{n-1} + \left(1+\frac{x}{n}\right)^{n-2}\left(1+\frac{x}{n+1}\right) + \ldots + \left(1+\frac{x}{n+1}\right)^{n-1}\right\}\right|$$

$$+ \left|\left(1+\frac{x}{n+1}\right)^n \frac{x}{n+1}\right|$$

$$\leqslant \frac{x}{n(n+1)} n\, e^{2K} + \frac{x}{n+1} e^K$$

$$\leqslant \frac{1}{n}\{x(e^{2K} + e^K)\}$$

and this is the required inequality with L equal to $x(e^{2K}+e^K)$.

Suppose next that x is rational, say $x = p/q$ and for each integer n let kp be the largest multiple of p less than n, then as $n \to \infty$, $k \to \infty$ and by the inequality above

$$\left(1+\frac{x}{n}\right)^n - \left(1+\frac{x}{kp}\right)^{kp} \to 0.$$

But $$\left(1+\frac{x}{kp}\right)^{kp} = \left\{\left(1+\frac{1}{kq}\right)^{kq}\right\}^{p/q} \to e^{p/q}.$$

Thus $$\left(1+\frac{x}{n}\right)^n \to e^{p/q} = e^x \quad \text{as} \quad n \to \infty.$$

If x is irrational then $\forall \epsilon > 0$ we can find rational numbers y and z such that $y < x < z$ and

$$e^y > e^x - \epsilon, \quad e^z < e^x + \epsilon.$$

Then, for all $n \geqslant N$ where N is chosen appropriately, depending upon the convergence of

$$\left(1+\frac{y}{n}\right)^n \quad \text{to} \quad e^y \quad \text{and of} \quad \left(1+\frac{z}{n}\right)^n \quad \text{to} \quad e^z,$$

$$e^x - 2\epsilon < e^y - \epsilon < \left(1+\frac{y}{n}\right)^n < \left(1+\frac{x}{n}\right)^n,$$

since $y < x$, and

$$e^x + 2\epsilon > e^z + \epsilon > \left(1+\frac{z}{n}\right)^n > \left(1+\frac{x}{n}\right)^n,$$

since $z > x$, i.e.

$$\left|\left(1+\frac{x}{n}\right)^n - e^x\right| < 2\epsilon \quad \text{for all } n > N.$$

Thus
$$\left(1+\frac{x}{n}\right)^n \to e^x \quad \text{as} \quad n \to \infty.$$

Exercises

1. $\{a_n\}$ is monotonic. Show that $\{(a_1+a_2+\ldots+a_n)/n\}$ is monotonic and increases or decreases according as $\{a_n\}\uparrow$ or $\{a_n\}\downarrow$.

2. $\{a_n\}$ and $\{b_n\}$ are real sequences with $a_1 > b_1 > 0$. Show that in each of the three following cases both sequences converge to the same limit.

(i) $a_{n+1} = \frac{1}{2}(a_n+b_n), \quad b_{n+1} = (a_n b_n)^{\frac{1}{2}}.$

(ii) $a_{n+1} = \frac{1}{2}(a_n+b_n), \quad b_{n+1} = (a_{n+1} b_n)^{\frac{1}{2}}.$

(iii) $a_{n+1} = \frac{1}{2}(a_n+b_n), \quad b_{n+1} = (a_n b_n)/a_{n+1}.$

3. $\{a_n\}$ is a real bounded sequence and $2a_n \leqslant a_{n-1}+a_{n+1}$. Show that $\lim(a_{n+1}-a_n) = 0$.

4. Prove that

$$\lim\left(\frac{n+1}{n^2}+\frac{(n+1)^2}{n^3}+\ldots+\frac{(n+1)^n}{n^{n+1}}\right) = e-1.$$

5. α, β and x are real numbers and $x > 1$. Prove that

(i) $x^\alpha > x^\beta$ if and only if $\alpha > \beta$, (ii) $x^{\alpha+\beta} = x^\alpha . x^\beta$.

6. $\lim a_n = l$ where $a_n > 0, l > 0$. Show that for an integer p $\lim a_n^p = l^p$ and $\lim a_n^{1/p} = l^{1/p}$. Deduce that for any real number α $\lim a_n^\alpha = l^\alpha$.

7. α, β and x are real numbers and $x > 0$. Prove that $(x^\alpha)^\beta = x^{\alpha\beta}$.

8. $\{a_n\}$ converges to α. If $x > 0$ prove that $\{x^{a_n}\}$ converges to x^α.

9. Prove that

(i) $\lim 2^{1/n} = 1$. (ii) $\lim (n+n^{\frac{1}{2}})^{1/n} = 1$.

(iii) $\lim (n^2)^{1/n} = 1$. (iv) $\lim n^{n/(n^2+1)} = 1$.

10. Prove that

(i) $\lim (1+n^{-2})^{n^2} = e$, (ii) $\lim (1+n^{-1})^{n^2} = \infty$.

(iii) $\lim (1+n^{-2})^n = 1$, (iv) $\lim (1+a_n)^n = e$

provided that in (iv) $a_n \sim 1/n$.

11. $\{a_n\}$, $\{b_n\}$ are real sequences such that $\lim a_n = \infty$, $\lim b_n = \infty$ and $\{b_n\} \Uparrow$. Show that if $\lim (a_{n+1}-a_n)/(b_{n+1}-b_n)$ exists then so does $\lim (a_n/b_n)$ and the two limits are equal.

12. p is a positive integer. Prove that

$$\lim (1^p + 2^p + \ldots + n^p)/n^{p+1} = 1/(p+1).$$

6

UPPER AND LOWER LIMITS OF REAL SEQUENCES; THE GENERAL PRINCIPLE OF CONVERGENCE

Remark. If a real sequence $\{a_n\}$ is known to be bounded, but it is not known to be convergent, we cannot immediately discuss the behaviour of $\{a_n\}$ in terms of convergence. For this reason two new numbers $\overline{\lim}\, a_n$, $\underline{\lim}\, a_n$ are introduced. They are defined for all real sequences and are either numbers or ∞ or $-\infty$ as the case may be. The sequence converges if and only if they are equal and are not ∞ or $-\infty$. The definitions which follow are applicable to real sequences only and *not to complex sequences.*

Definition 10. (*a*) *If $\{a_n\}$ is bounded above then we define the upper limit of $\{a_n\}$, $\overline{\lim}\, a_n$ or $\overline{\lim}_{n\to\infty} a_n$ by $\overline{\lim}\, a_n = \inf_{p\geqslant 1}\{\sup_{n\geqslant 1} a_{n+p}\}$, and if $\{a_n\}$ is bounded below then we define the lower limit, $\underline{\lim}\, a_n$ or $\underline{\lim}_{n\to\infty} a_n$ by $\underline{\lim}\, a_n = \sup_{p\geqslant 1}\{\inf_{n\geqslant 1} a_{n+p}\}$.*

(*b*) *If $\{a_n\}$ is unbounded above then we say $\overline{\lim}\, a_n = \infty$. If further $a_n \to \infty$ as $n \to \infty$ then we say that $\underline{\lim}\, a_n = \infty$.*

(*c*) *If $\{a_n\}$ is unbounded below then we say that $\underline{\lim}\, a_n = -\infty$. If further $a_n \to -\infty$ as $n \to \infty$ then we say that $\overline{\lim}\, a_n = -\infty$.*

The important definition is (a), the others are variants introduced to deal with unbounded sequences.

In (a) if we write

$$\alpha_n = \sup_{p \geqslant n} a_p, \quad \beta_n = \inf_{p \geqslant n} a_p,$$

then $\{\alpha_n\}\downarrow$ and $\{\beta_n\}\uparrow$. Since $\{a_n\}$ is bounded so are $\{\alpha_n\}$ and $\{\beta_n\}$. By theorem 2, $\lim \alpha_n$ exists and is equal to $\inf \alpha_n$. Thus in (a) we could have written

$$\overline{\lim_{n\to\infty}}\, a_n = \lim \{\sup_{p \geqslant 1} a_{n+p}\}.$$

Similarly $$\underline{\lim_{n\to\infty}}\, a_n = \lim \{\inf_{p \geqslant 1} a_{n+p}\}.$$

Exercises. (1) $\{a_n\}$ is defined by $a_n = (-1)^n$. Show that $\overline{\lim}\, a_n = -\underline{\lim}\, a_n = 1$.

(2) $\{a_n\}$ is defined by $a_n = 1$. Show that $\overline{\lim}\, a_n = \underline{\lim}\, a_n = 1$.

(3) $\{a_n\}$ is defined by $a_{2n} = 1/n$, $a_{2n-1} = 0$ $(n = 1, 2, \ldots)$. Show that $\overline{\lim}\, a_n = \underline{\lim}\, a_n = 0$.

(4) $\{a_n\}$ and $\{b_n\}$ are two bounded sequences of real numbers such that $a_n \geqslant b_n$. Show that $\overline{\lim}\, a_n \geqslant \overline{\lim}\, b_n$ and $\underline{\lim}\, a_n \geqslant \underline{\lim}\, b_n$.

If $\overline{\lim}\, a_n > \underline{\lim}\, a_n$ and both upper and lower limits are numbers then we say that $\{a_n\}$ *oscillates finitely.* If one or other of the upper or lower limits is ∞ or $-\infty$ and the other limit is not equal to it, then we say that $\{a_n\}$ *oscillates infinitely.*

Theorem 3. (a) *If* $\overline{\lim}\, a_n$ *is finite it is characterized by the two properties* (i) $\forall \epsilon > 0\, \exists$ *an integer* N *such that* $a_n < \overline{\lim}\, a_n + \epsilon$ *for all* $n > N$, (ii) $\exists$ *an infinite set of integers* J *such that* $a_n > \overline{\lim}\, a_n - \epsilon$ *for* $n \in J$.

(b) *If* $\underline{\lim}\, a_n$ *is finite it is characterized by the two properties* (i) $\forall \epsilon > 0\, \exists$ *an integer* M *such that* $a_n > \underline{\lim}\, a_n - \epsilon$ *for all* $n > M$, (ii) $\exists$ *an infinite set of integers* K *such that* $a_n < \underline{\lim}\, a_n + \epsilon$ *for* $n \in K$.

(a) First $\overline{\lim}\, a_n$ has the two properties stated. For by definition $\sup_{p \geqslant 1} a_{n+p} \to \overline{\lim}\, a_n$ as $n \to \infty$. Thus $\forall \epsilon > 0\, \exists$ an integer N such that

$$|\sup_{p \geqslant 1} a_{n+p} - \overline{\lim}\, a_n| < \epsilon \quad \text{for all } n > N.$$

Thus $$\sup_{p \geqslant 1} a_{n+p} < \overline{\lim}\, a_n + \epsilon \quad \text{for all } n > N.$$

Now $a_{n+1} \leqslant \sup_{p \geqslant 1} a_{n+p}$ and therefore $a_{n+1} < \overline{\lim}\, a_n + \epsilon$ for all $n > N$, and the first property is established. For the second property we have (as above)

$$\sup_{p\geqslant 1} a_{n+p} > \overline{\lim}\, a_n - \epsilon \quad \text{for all } n > N$$

and thus $\forall\, n > N \,\exists\, m > n$ such that $a_m > \overline{\lim}\, a_n - \epsilon$. Hence $\exists$ an infinite set of integers J such that if $n \in J$ then $a_n > \overline{\lim}\, a_n - \epsilon$, and the second property has been established.

On the other hand if λ is a number with the two stated properties then $\forall\, \epsilon > 0\, \exists$ an integer N such that

$$\sup_{p\geqslant 1} a_{n+p} \leqslant \lambda + \epsilon \quad \text{for } n > N$$

and

$$\sup_{p\geqslant 1} a_{n+p} \geqslant \lambda - \epsilon \quad \text{for all } n.$$

Thus

$$\lambda - \epsilon \leqslant \inf_n (\sup_{p\geqslant 1} a_{n+p}) \leqslant \lambda + \epsilon$$

and since these inequalities are true for all $\epsilon > 0$ it means that

$$\lambda = \inf_n (\sup_{p\geqslant 1} a_{n+p}), \quad \text{i.e.} \quad \lambda = \overline{\lim}\, a_n.$$

(*b*) The proof of (*b*) is similar and is omitted. (*b*) could be deduced from (*a*) by writing $b_n = -a_n$ and applying (*a*) to $\{b_n\}$.

We make the convention that for any number a:

$$\infty > a, \quad a > -\infty, \quad \infty > -\infty.$$

This convention is introduced to avoid the necessity of stating exceptional cases.

Theorem 4. *For any real sequence* $\{a_n\}$

(*a*) $\overline{\lim}\, a_n \geqslant \underline{\lim}\, a_n$,

(*b*) $\{a_n\}$ *converges to a finite limit* l *if and only if* $\overline{\lim}\, a_n = \underline{\lim}\, a_n = l$,

(*c*) $\{a_n\}$ *diverges to* ∞ *if and only if* $\overline{\lim}\, a_n = \underline{\lim}\, a_n = \infty$ *and* $\{a_n\}$ *diverges to* $-\infty$ *if and only if* $\overline{\lim}\, a_n = \underline{\lim}\, a_n = -\infty$.

(*a*) If $\overline{\lim}\, a_n = \infty$ or $\underline{\lim}\, a_n = -\infty$ there is nothing to prove. If $\overline{\lim}\, a_n = \lambda$, $\underline{\lim}\, a_n = \mu$ are both finite then since

$$\sup_{p\geqslant 1} a_{n+p} \geqslant a_{n+p+1} \geqslant \inf_{p\geqslant 1} a_{n+p}$$

it follows that, on taking limits as $n \to \infty$,

$$\overline{\lim}\, a_n = \lim\, (\sup_{p\geqslant 1} a_{n+p}) \geqslant \lim\, (\inf_{p\geqslant 1} a_{n+p}) = \underline{\lim}\, a_n.$$

Finally, if $\overline{\lim}\, a_n = -\infty$ then $\underline{\lim}\, a_n = -\infty$ and if $\underline{\lim}\, a_n = \infty$ then $\overline{\lim}\, a_n = \infty$. Thus (*a*) is true in these cases also.

(*b*) If $\underline{\lim}\, a_n = \overline{\lim}\, a_n = l$, then

$$\lim\,(\sup_{p\geqslant 1} a_{n+p}) = \lim\,(\inf_{p\geqslant 1} a_{n+p}) = l.$$

Thus $\forall\, \epsilon > 0\, \exists\, N$ such that

$$\sup_{p\geqslant 1} a_{n+p} \leqslant l+\epsilon \quad \text{for all } n > N$$

and

$$\inf_{p\geqslant 1} a_{n+p} \geqslant l-\epsilon \quad \text{for all } n > N,$$

$$\therefore \quad l-\epsilon \leqslant a_n \leqslant l+\epsilon \quad \text{for all } n > N+1.$$

Thus $a_n \to l$ as $n \to \infty$.

On the other hand, if $a_n \to l$ as $n \to \infty$, then $\forall\, \epsilon > 0\, \exists\, N$ such that

$$l-\epsilon < a_n < l+\epsilon \quad \text{all } n \geqslant N.$$

$$\therefore \quad l-\epsilon \leqslant \inf_{p\geqslant 1} a_{n+p} \leqslant \sup_{p\geqslant 1} a_{n+p} \leqslant l+\epsilon \quad \text{for all } n \geqslant N.$$

$$\therefore \quad l-\epsilon \leqslant \lim\,(\inf_{p\geqslant 1} a_{n+p}) \leqslant \lim\,(\sup_{p\geqslant 1} a_{n+p}) \leqslant l+\epsilon.$$

But these inequalities are true for every $\epsilon > 0$ and thus

$$\underline{\lim}\, a_n = \overline{\lim}\, a_n = l.$$

(*c*) This part of the theorem follows from definition 10.

Remark. In practice it is often more convenient to use the properties of $\overline{\lim}\, a_n$ and $\underline{\lim}\, a_n$ stated in theorem 3 rather than the definitions of these numbers. These properties apply only when $\overline{\lim}\, a_n$ and $\underline{\lim}\, a_n$ are finite, but there are appropriate modifications if $\overline{\lim}\, a_n = \infty$ or $\overline{\lim}\, a_n = -\infty$, etc. For example, $\overline{\lim}\, a_n = \infty$ is equivalent to

$\forall$ a real number K $\exists$ an infinite set of integers J such that $a_n > K$ for $n \in J$,

and $\overline{\lim}\, a_n = -\infty$ is equivalent to $a_n \to -\infty$ as $n \to \infty$. If $\{a_n\}$ is bounded then $\overline{\lim}\, a_n$ and $\underline{\lim}\, a_n$ are both finite. In fact

$$\inf a_n \leqslant \underline{\lim}\, a_n \leqslant \overline{\lim}\, a_n \leqslant \sup a_n.$$

The general principle of convergence

If $\{a_n\}$ converges, and we do not know and are not able to guess the limit to which it converges, then we are unable to establish the convergence of $\{a_n\}$ directly. It is often essential to know whether or not $\{a_n\}$ converges *even though we have no means of ascertaining its limit*. The general principle of convergence enables us to do just this. It applies both to complex and real sequences.

Theorem 5. *The general principle of convergence. The necessary and sufficient condition that $\{a_n\}$ should converge as $n \to \infty$ is that $\forall \epsilon > 0 \; \exists$ an integer N such that $|a_n - a_m| < \epsilon$ for all integers n, m both greater than N.*

Necessity. If $\{a_n\}$ converges to l as $n \to \infty$ then $\forall \epsilon > 0 \; \exists$ an integer N such that $|a_n - l| < \frac{1}{2}\epsilon$ for all $n > N$. By the modulus inequality if $n > N$ and $m > N$ then

$$|a_n - a_m| \leqslant |a_n - l| + |l - a_m| < \epsilon$$

and the condition has been established.

Sufficiency. If $\{a_n\}$ satisfies the given condition, apply the condition with $\epsilon = 1$. Then $\exists$ an integer N such that

$$|a_n - a_{N+1}| < 1 \quad \text{for all } n > N.$$

(We have taken $m = N + 1$.) Thus

$$|a_n| \leqslant \max\{|a_1|, \ldots, |a_N|, |a_{N+1}| + 1\}$$

that is to say, $\{a_n\}$ is bounded.

Write $b_n = \mathscr{R}(a_n)$, $c_n = \mathscr{I}(a_n)$. Then $\{b_n\}$ and $\{c_n\}$ are bounded real sequences. Write $\beta = \overline{\lim}\, b_n$.

Now $|b_n - b_m| \leqslant |a_n - a_m|$ and thus we have the following property: $\forall \epsilon > 0 \; \exists N$ such that $|b_n - b_m| < \epsilon$ if both $n > N$ and $m > N$ where N, m, n are integers. By theorem 3(*a*) $\exists M > N$ such that $|\beta - b_M| < \epsilon$. Hence by the modulus inequality we obtain for $n > N$

$$|b_n - \beta| \leqslant |b_n - b_M| + |b_M - \beta| < 2\epsilon.$$

Since N can be found corresponding to each $\epsilon > 0$ it follows that $\{b_n\}$ converges to β as $n \to \infty$.

Similarly $\{c_n\}$ converges. Denote its limit by γ and let l be the complex number whose real part is β and whose imaginary part is γ. Then by the modulus inequality

$$0 \leqslant |a_n - l| < |b_n - \beta| + |c_n - \gamma|.$$

As $n \to \infty$, $b_n \to \beta$, $c_n \to \gamma$ and thus $|a_n - l| \to 0$, i.e. $a_n \to l$ as $n \to \infty$.

The theorem is proved.

Worked examples

(i) *$\{a_n\}$ is a bounded real sequence and X is defined by*

$$X = \{x \mid x \leqslant a_n \text{ for an infinite set of integers } n\}.$$

Prove that $\overline{\lim}\, a_n = \sup X$.

Since $\{a_n\}$ is bounded $\sup X$ is finite. $\forall \epsilon > 0 \; \exists x \in X$ such that $x > \sup X - \epsilon$ (from the definition of $\sup X$). Since $x \in X$,

$$a_n \geqslant x > \sup X - \epsilon$$

for an infinite set of integers n. Thus

$$\sup_{p \geqslant 1} a_{n+p} \geqslant \sup X - \epsilon \quad \text{for all } n.$$

Hence $$\overline{\lim}\, a_n = \lim \left(\sup_{p \geqslant 1} a_{n+p}\right) \geqslant \sup X - \epsilon.$$

Again the number $\sup X + \epsilon$ does not belong to X. Thus there are at most a finite set of integers n such that $a_n \geqslant \sup X + \epsilon$. Hence $\exists$ an integer N such that

$$a_n < \sup X + \epsilon \quad \text{for all } n > N.$$

Thus $$\sup_{p \geqslant 1} a_{n+p} \leqslant \sup X + \epsilon \quad \text{for all } n > N$$

and $$\overline{\lim}\, a_n \leqslant \sup X + \epsilon.$$

Combining the two inequalities for $\overline{\lim}\, a_n$ and remembering that ϵ is any positive number, we have $\overline{\lim}\, a_n = \sup X$.

Remark. If a real sequence $\{a_n\}$ is such that for each integer n $a_n < k$, then we can deduce that

$$\inf a_n < k, \quad \sup a_n \leqslant k, \quad \underline{\lim}\, a_n \leqslant k, \quad \overline{\lim}\, a_n \leqslant k,$$

and these relations are the most that can be deduced. In particular it does *not* follow that $\underline{\lim}\, a_n < k$; it is quite possible that $\underline{\lim}\, a_n = k$, for example, if $a_n = -1/n$ and $k = 0$.

(ii) *$\{a_n\}$ is a real bounded sequence. Show that there exists a subsequence $\{a_{n_i}\}$ such that $a_{n_i} \to \underline{\lim}\, a_n$ as $i \to \infty$.*

Let $\{\epsilon_i\}$ be a positive null sequence. By combining the two statements of theorem 3 (*a*) $\exists$ an infinite set of integers J_1 such that

$$|a_n - \underline{\lim}\, a_n| < \epsilon_1 \quad \text{if} \quad n \in J_1.$$

Select one integer of J_1 and denote it by n_1. A sequence of integers is now defined inductively. Suppose that $n_1, n_2, \ldots, n_{k-1}$ have been defined so that $n_1 < n_2 < \ldots < n_{k-1}$ and $|a_{n_i} - \underline{\lim}\, a_n| < \epsilon_1$ for $(i = 1, 2, \ldots, k-1)$. By theorem 3 (*a*) $\exists$ an infinite set of integers J_k such that if $n \in J_k$, then $|a_n - \underline{\lim}\, a_n| < \epsilon_k$. Select n_k an integer belonging to J_k and greater than n_{k-1}. Then the strictly increasing sequence of integers $\{n_k\}$ is defined, inductively, for all k. $\{a_{n_k}\}$ is a subsequence of $\{a_n\}$ and $a_{n_k} \to \underline{\lim}\, a_n$ as $k \to \infty$.

(iii) $\{a_n\}$ *is a real bounded sequence and* $\overline{\lim}\,(a_{n+1}-a_n) = 0$. λ *satisfies* $\underline{\lim}\,a_n < \lambda < \overline{\lim}\,a_n$. *Show that there is a subsequence* $\{a_{n_i}\}$ of $\{a_n\}$ *such that* $a_{n_i} \to \lambda$ *as* $i \to \infty$.

Remark. (ii) contains a similar result with λ replaced by $\underline{\lim}\,a_n$ and without needing the condition $\overline{\lim}\,(a_{n+1}-a_n) = 0$. The corresponding result with λ replaced by $\overline{\lim}\,a_n$ is also true by an argument similar to that used in (ii). The idea here is to use (ii) to find large integers $N, M, N < M$, such that a_N is close to $\underline{\lim}\,a_n$ and a_M is close to $\overline{\lim}\,a_n$, and then to use $\overline{\lim}\,(a_{n+1}-a_n) = 0$ to obtain an integer P such that $N < P < M$ and a_P is close to λ.

Write $c = \underline{\lim}\,|a_n-\lambda|$. Then $c \geqslant 0$ and we shall show that $c = 0$. The result will then follow from (ii), since by (ii) $\exists$ a sequence of integers $\{n_i\}$ such that

$$|a_{n_i}-\lambda| \to \underline{\lim}\,|a_n-\lambda| \quad \text{as} \quad i \to \infty,$$

i.e. such that $a_{n_i}-\lambda \to 0$ as $i \to \infty$. Thus $a_{n_i} \to \lambda$ as $i \to \infty$ and the required result is established.

We show then that $c = 0$. We suppose that $c > 0$ and show that this assumption leads to a contradiction. If $c > 0\,\exists$ an integer N such that $|a_n-\lambda| > \frac{1}{2}c$ for all $n > N$. (This cannot be asserted if $c = 0$ for what we really have is $\forall\,\epsilon > 0\,\exists\,N$ such that

$$|a_n-\lambda| > c-\epsilon$$

for all $n > N$ and we have taken $\epsilon = \frac{1}{2}c$ which is permissible if and only if $c > 0$.) If, further for some $n > N$ we had $a_{n+1} \geqslant \lambda \geqslant a_n$, then

$$a_{n+1}-a_n = a_{n+1}-\lambda+\lambda-a_n = |a_{n+1}-\lambda|+|\lambda-a_n| > c.$$

But $\overline{\lim}\,(a_{n+1}-a_n) = 0$ and thus $\exists$ an integer M such that

$$a_{n+1}-a_n < c \quad \text{for all } n > M.$$

Thus combining the two results above if $n > \max(N, M)$, then either (a) $a_n \geqslant \lambda$, or (b) $\exists\,N_0 > \max(N, M)$ such that $a_n \leqslant \lambda$ for $n > N_0$.

(a) implies that $\underline{\lim}\,a_n \geqslant \lambda$, (b) implies that $\overline{\lim}\,a_n \leqslant \lambda$ and neither of these possibilities can occur since

$$\underline{\lim}\,a_n < \lambda < \overline{\lim}\,a_n.$$

Hence $\underline{\lim}\,|a_n-\lambda| = 0$ and the result follows.

(iv) $\{a_n\}$ *is a bounded real sequence and* $\{a_{n_i}\}$ *is a subsequence. Prove that*

$$\underline{\lim}\,a_n \leqslant \underline{\lim}_{i\to\infty}\,a_{n_i} \leqslant \overline{\lim}_{i\to\infty}\,a_{n_i} \leqslant \overline{\lim}\,a_n.$$

We show that $\underline{\lim}\, a_n \leqslant \underline{\lim}_{i\to\infty}\, a_{n_i}$. The second inequality is contained in theorem 4 (*a*) and the third may be proved by an argument similar to that given below.

$$\begin{aligned}\underline{\lim}\, a_n &= \lim_{n\to\infty} (\inf_{p\geqslant 1} a_{n+p}) \\ &= \lim_{i\to\infty} (\inf_{p\geqslant 1} a_{n_i+p})\end{aligned}$$

by worked example (i), chapter 4. Now

$$\inf_{p\geqslant 1} a_{n_i+p} \leqslant \inf_{p\geqslant 1} a_{n_{i+p}},$$

since the integers of the form n_{i+p}, $p \geqslant 1$, are a subset of the integers of the form n_i+p, $p \geqslant 1$. Since $\underline{\lim}_{i\to\infty}\, a_{n_i} = \lim_{i\to\infty} (\inf_{p\geqslant 1} a_{n_{i+p}})$, we obtain by combining the above

$$\underline{\lim}\, a_n \leqslant \underline{\lim}_{i\to\infty}\, a_{n_i}.$$

(v) *$\{a_n\}$ is a bounded real sequence. Prove that if $a_n \leqslant \underline{\lim}\, a_n$ for infinitely many n and $\{b_n\}$ is the subsequence of $\{a_n\}$ formed by those terms for which $a_n \leqslant \underline{\lim}\, a_n$ then $\{b_n\}$ converges to $\underline{\lim}\, a_n$.*

Since $\{b_n\}$ is a subsequence of $\{a_n\}$, by (iv) $\underline{\lim}\, a_n \leqslant \underline{\lim}\, b_n$. Since $b_n \leqslant \underline{\lim}\, a_n$ for every n, $\overline{\lim}\, b_n \leqslant \underline{\lim}\, a_n$. By theorem 4 (*a*) it follows that $\overline{\lim}\, b_n = \underline{\lim}\, b_n = \underline{\lim}\, a_n$ and this by theorem 4 (*b*) is the required result.

Exercises

1. $\{a_n\}$ and $\{b_n\}$ are real bounded sequences. Show that

$$\underline{\lim}\, a_n + \underline{\lim}\, b_n \leqslant \underline{\lim}\, (a_n+b_n) \leqslant \begin{pmatrix}\underline{\lim}\, a_n + \overline{\lim}\, b_n \\ \overline{\lim}\, a_n + \underline{\lim}\, b_n\end{pmatrix}$$
$$\leqslant \overline{\lim}\, (a_n+b_n) \leqslant \overline{\lim}\, a_n + \overline{\lim}\, b_n.$$

Deduce that if a_n is a real bounded sequence then

$$\underline{\lim}\, (a_n - a_{n+1}) \leqslant 0 \leqslant \overline{\lim}\, (a_n - a_{n+1}).$$

2. Construct $\{a_n\}$ and $\{b_n\}$ such that strict inequality holds in all the inequalities of the first part of exercise 1.

3. $\{a_n\}$ and $\{b_n\}$ are real positive bounded sequences. Prove that

$$\underline{\lim}\, a_n . \underline{\lim}\, b_n \leqslant \underline{\lim}\, (a_n b_n) \leqslant \underline{\lim}\, a_n\, \overline{\lim}\, b_n \leqslant \overline{\lim}\, (a_n b_n) \leqslant \overline{\lim}\, a_n . \overline{\lim}\, b_n.$$

4. Construct $\{a_n\}$, $\{b_n\}$ such that $a_{n+3} = a_n$, $b_{n+3} = b_n$, $a_n \neq 0$, $b_n \neq 0$, $\underline{\lim}\, a_n \neq 0$, $\{a_n b_n\}$ and $\{b_n\}$ do not converge but

$$\underline{\lim}\, a_n b_n = \underline{\lim}\, a_n . \overline{\lim}\, b_n = \overline{\lim}\, a_n . \underline{\lim}\, b_n.$$

5. Construct $\{a_n\}$ such that $\overline{\lim}\, a_n > \underline{\lim}\, a_n > 0$ and either

$$(a)\quad \overline{\lim}\, (a_n a_{n+1}) = \underline{\lim}\, (a_n a_{n+1}),$$

or

$$(b)\quad \overline{\lim}\, (a_n a_{n+1} a_{n+2}) = \underline{\lim}\, (a_n a_{n+1} a_{n+2}).$$

Show that such a sequence cannot satisfy both (a) and (b).

6. The real bounded sequence $\{a_n\}$ is such that $|a_n| \to l$ as $n \to \infty$ where l is a number and $\overline{\lim}\, a_n \neq \underline{\lim}\, a_n$. Show that $l \neq 0$ and $\overline{\lim}\, a_n = -\underline{\lim}\, a_n$.

7. $\{a_n\}$ is a bounded sequence and z, w given numbers (a_n, z, w, may be complex). Show that

$$\overline{\lim}\, |a_n - z| \leqslant |z-w| + \overline{\lim}\, |a_n - w|,$$

$$\underline{\lim}\, |a_n - z| \geqslant |z-w| - \overline{\lim}\, |a_n - w|.$$

8. $\{a_n\}$ is a null sequence and $|z_n| \to M$ as $n \to \infty$. Show that

$$|a_n + z_n| \to M \quad \text{as} \quad n \to \infty.$$

9. $\{a_n\}$ is a bounded positive sequence and $\{b_n\}$ is a convergent positive sequence. Show that

$$\overline{\lim}\, (a_n b_n) = \lim b_n . \overline{\lim}\, a_n,$$

$$\underline{\lim}\, (a_n b_n) = \lim b_n . \underline{\lim}\, a_n.$$

10. $\{a_n\}$ is a real bounded sequence and $\{b_n\}$ is a real convergent sequence. Show that

$$\overline{\lim}\, (a_n + b_n) = \overline{\lim}\, a_n + \lim b_n,$$

$$\underline{\lim}\, (a_n + b_n) = \underline{\lim}\, a_n + \lim b_n.$$

11. $\{b_n\}$ is a real bounded sequence such that for *any* real bounded sequence $\{a_n\}$, $\overline{\lim}\, (a_n + b_n) = \overline{\lim}\, a_n + \overline{\lim}\, b_n$. Show that $\{b_n\}$ converges.

12. $\{b_n\}$ is a positive bounded sequence such that for *any* positive bounded sequence $\{a_n\}$, $\overline{\lim}\, a_n b_n = \overline{\lim}\, a_n\, \overline{\lim}\, b_n$. Show that $\{b_n\}$ converges.

13. Construct positive divergent sequences $\{a_n\}$, $\{b_n\}$, $\{c_n\}$ such that each sequence $\{a_n b_n\}$, $\{b_n c_n\}$, $\{a_n c_n\}$ converges.

14. $\{a_n\}$ is a real bounded sequence. Prove that

$$\underline{\lim}\, a_n \leqslant \underline{\lim}\, ((a_1 + \ldots + a_n)/n) \leqslant \overline{\lim}\, ((a_1 + \ldots + a_n)/n) \leqslant \overline{\lim}\, a_n.$$

If also $a_n > 0$, $\underline{\lim}\, a_n > 0$, show that

$$\underline{\lim}\, (1/a_n) = 1/(\overline{\lim}\, a_n),$$

$$\overline{\lim}\, (1/a_n) = 1/\underline{\lim}\, a_n$$

and for $\alpha > 0$, $\qquad \overline{\lim}\,(a_n)^\alpha = (\overline{\lim}\,a_n)^\alpha,$

$$\underline{\lim}\,(a_n)^\alpha = (\underline{\lim}\,a_n)^\alpha.$$

15. $\{a_n\}$ is such that $a_n \geqslant 0$ and $\underline{\lim}\,a_n = 0$. Show that $\exists$ a decreasing subsequence of $\{a_n\}$.

Deduce that every bounded real sequence contains a convergent subsequence.

16. (Bolzano–Weiestrass theorem, see also chapter 13.) A is a bounded enumerable infinite set of real numbers. Prove that there exists a fixed real number λ such that $\forall \epsilon > 0 \,\exists\, a \in A$ satisfying $\lambda - \epsilon < a < \lambda + \epsilon$, $a \neq \lambda$.

17. $\{a_n\}$ is a real bounded sequence and B the class of all convergent sequences $\{b_n\}$ for which $b_n \geqslant a_n$. Prove that $\overline{\lim}\,a_n = \inf_B(\lim b_n)$.

18. $\{a_n\}$ is a real bounded sequence and $\{a^{(1)}\}, \ldots, \{a^{(k)}\}$ are k subsequences of $\{a_n\}$ such that every a_n occurs as a term in at least one of these subsequences. Show that if

$$\overline{\lim_{n\to\infty}}\, a_n^{(j)} \leqslant a_m \quad \text{for} \quad (j = 1, 2, \ldots, k;\ m = 1, 2, \ldots),$$

then $\{a_n\}$ converges.

19. $\{a_n\}$ is a real sequence such that $a_{m+n} < a_m + a_n$ $(m, n = 1, 2, 3, \ldots)$. Show that

$$\overline{\lim_{q\to\infty}}\, \frac{a_{ql+r}}{ql+r} \leqslant \frac{a_l}{l} \quad (r = 0, 1, \ldots, l-1).$$

If in addition, $\{a_n/n\}$ is a bounded sequence then deduce that it converges.

20. $\{a_n\}$ is a real bounded positive sequence. Show that if for a given strictly increasing sequence of integers $\{p_n\}$, $\{a_n \,.\, a_{n+p_n}\}$ converges to a limit then the value of the limit is $\overline{\lim}\,a_n \,.\, \underline{\lim}\,a_n$.

7

CONVERGENCE OF SERIES; ABSOLUTE CONVERGENCE

If we are given a sequence $\{a_n\}$ then the sequence $\{S_n\}$ where $S_n = a_1 + a_2 + \ldots + a_n$ is called *the series* $\sum_1^\infty a_n$. It is denoted by Σa_n. a_n is the nth *summand* of Σa_n and S_n is the nth *partial sum* of Σa_n. If every summand of a series has a certain property then we refer to the series as having this property. Thus we use the phrases *real series*, *positive series*, etc. If $S_n \to S$ as $n \to \infty$ then the symbol $\sum\limits_1^\infty a_n$ is used instead of S. For example, we shall write equations such as

$$\text{(i)}\ \sum_1^\infty r^n = r/(1-r) \quad (0 < r < 1),$$

$$\text{(ii)}\ \sum_1^\infty r^n = r + \sum_1^\infty r^{n+1} \quad (0 < r < 1).$$

The meanings of these equations are:

(i) the sequence $\{(r+r^2+\ldots+r^n)\}$ converges to $r/(1-r)$ if $0 < r < 1$,

(ii) the sequence $\{(r+r^2+\ldots+r^n)\}$ converges, say to S_1, as $n \to \infty$; the sequence $\{r^2+r^3+\ldots+r^{n+1}\}$ converges, say to S_2, as $n \to \infty$ and $S_1 = r + S_2$ provided that $0 < r < 1$.

The reader *must* distinguish between the *sequence* $\{a_n\}$ and the *series* Σa_n. Just as the sequence $\{a_n\}$ is sometimes written as $a_1, a_2, a_3, \ldots$ so the series Σa_n is sometimes written as $a_1 + a_2 + a_3 + \ldots$.

When $S_n = a_1 + \ldots + a_n$ converges to S we say that the series Σa_n converges and that its *sum* is S. Similarly we say that Σa_n diverges to ∞ or to $-\infty$ according as $\{S_n\}$ diverges to ∞ or to $-\infty$.

To avoid ambiguity we shall whenever possible use 'Σa_n' for the sequence $\{S_n\}$ $S_n = a_1 + \ldots + a_n$, and in those cases where it is essential to indicate the suffix of summation we shall use '*the series* $\sum\limits_1^\infty a_n$'. Neither of these notations presupposes the convergence of the series and neither is used for the sum if the series does converge. The symbol '$\sum\limits_1^\infty a_n$' without the words '*the series*' will be used only when the series is convergent and will then indicate its sum. It is sometimes convenient to start the series at some summand other than the first; we shall use '*the series* $\sum\limits_r^\infty a_n$' to mean 'Σb_n' where $b_n = a_{n+r-1}$.

Similarly by '*the series* $\sum\limits_{-1}^{-\infty} a_n$' we mean '$\Sigma a_{-n}$' and by '*the series* $\sum\limits_{-\infty}^{\infty} a_n$' we mean the two series Σa_n and Σa_{1-n}. The symbol '$\sum\limits_{-\infty}^{\infty} a_n$' is used only if both Σa_n and Σa_{1-n} converge and then stands for the sum of the two sums of these series.

Definition 11. *If Σa_n is such that $\{S_n\}$, $S_n = a_1 + \ldots + a_n$ $(n = 1, 2, \ldots)$ converges to S, then we say that Σa_n converges and that S is its sum. In all other cases we say that Σa_n diverges. If the sum-*

mands a_n are real and $S_n \to \infty$ as $n \to \infty$, then we say that Σa_n diverges to ∞. Similary if the summands are real and $S_n \to -\infty$, then we say that Σa_n diverges to $-\infty$. If $\{S_n\}$ is bounded and does not tend to a limit then we say that Σa_n oscillates finitely.

Theorem 5 can be reformulated to give a general principle of convergence for series as follows.

Theorem 5a. *The necessary and sufficient condition that Σa_n should converge is that*

$$\forall \epsilon > 0 \, \exists \text{ an integer } N \text{ such that}$$

$$|a_n + a_{n+1} + \ldots + a_m| < \epsilon \quad \text{for all } n, m \text{ such that } m \geqslant n > N.$$

Remark. The alteration of a finite number of summands of a series will not alter the fact of convergence or divergence of the series but may alter the sum of the series. An alteration to a finite number of terms of a sequence will not alter its convergence or divergence nor, if the sequence converges, will it alter the limit to which it converges. This difference is because the mth summand of Σa_n occurs in the rth term of $\{(a_1 + \ldots + a_n)\}$ if $r \geqslant m$ and an alteration in value of one summand of the series alters the values of all but a finite number of terms in the sequence of partial sums.

Exercises. (1) Σa_n converges. Show that $\Sigma k a_n$ and Σa_{n+1} converge.

(2) Σa_n converges and each summand is an integer. Show that $\exists$ an integer N such that $a_n = 0$ for $n > N$.

If Σa_n converges then from theorem 5*a* with $m = n$ it follows that $a_n \to 0$ as $n \to \infty$. THE CONVERSE OF THIS STATEMENT IS FALSE. For example, if $a_n = n^{-\frac{1}{2}}$ then $a_n \to 0$ as $n \to \infty$, but

$$a_1 + a_2 + \ldots + a_n \geqslant n^{\frac{1}{2}}$$

and Σa_n diverges to ∞.

Definition 12. *If $\Sigma |a_n|$ converges then we say that Σa_n is absolutely convergent.*

This idea is a great help in discussing the convergence or divergence of given series because (*a*) absolute convergence implies convergence (theorem 6 below) and (*b*) positive series, which include series of the form $\Sigma |a_n|$, are easy to deal with.

Theorem 6. *The absolute convergence of Σa_n implies its convergence.*

$\Sigma |a_n|$ converges and thus by the necessity part of theorem 5*a* $\forall \epsilon > 0 \, \exists$ an integer N such that $||a_n| + \ldots + |a_m|| < \epsilon$ for all n, m integers satisfying $m \geqslant n > N$. By the modulus inequality

$$|a_n + \ldots + a_m| \leqslant ||a_n| + \ldots + |a_m||$$

and thus by the sufficiency part of theorem 5*a* Σa_n converges.

The theorem is proved.

If Σa_n is absolutely convergent and $\{b_n\}$ is bounded then $\Sigma a_n b_n$ is absolutely convergent. This is a simple consequence of theorem 5*a*.

Exercise. Σa_n is defined by $a_{2n} = -a_{2n-1} = n^{-\frac{1}{2}}$. Show that Σa_n is convergent but not absolutely convergent.

Non-negative real series

If $a_n \geqslant 0$ then $\{(a_1 + \ldots + a_n)\}\uparrow$ and by theorem 2 Σa_n converges if and only if $\{(a_1 + \ldots + a_n)\}$ is bounded. Thus

Theorem 7. *The necessary and sufficient condition that the non-negative real series Σa_n converges is that $\{(a_1 + \ldots + a_n)\}$ be bounded.*

There is a similar result for non-positive series.

There are few important series whose convergence properties can be determined directly. One such series is the *geometric* series Σr^n, r real. Since

$$\sum_1^N r^n = r(1-r^N)/(1-r) \quad \text{if} \quad (r \neq 1)$$
$$= N \quad \text{if} \quad (r = 1)$$

it follows that Σr^n converges if $-1 < r < 1$ and diverges if $r \leqslant -1$ or $r \geqslant 1$.

Tests for the convergence of non-negative series

The comparison principle

If $a_n \geqslant 0$, $b_n \geqslant 0$, Σb_n converges and $a_n = O(b_n)$ then Σa_n converges.
If $a_n \geqslant 0$, $b_n \geqslant 0$, Σb_n diverges and $b_n = O(a_n)$ then Σa_n diverges.

The first statement follows directly from theorem 7 and the second statement follows from the first.

The ratio test

If $b_n > 0$, $a_n \geqslant 0$, Σb_n converges and $\{a_n/b_n\}\downarrow$ then Σa_n converges.
If $b_n > 0$, $a_n > 0$, Σb_n diverges and $\{a_n/b_n\}\uparrow$ then Σa_n diverges.

If $\{a_n/b_n\}\downarrow$ then $a_n/b_n \leqslant a_1/b_1$, and i.e. $a_n = O(b_n)$. By the comparison principle the convergence of Σb_n implies that of Σa_n. The second part of the test follows similarly from the second part of the comparison principle.

D'Alembert's test

If $a_n > 0$ and $\overline{\lim}\,(a_{n+1}/a_n) < 1$ then Σa_n converges.
If $a_n > 0$ and $\underline{\lim}\,(a_{n+1}/a_n) > 1$ then Σa_n diverges.

Suppose that $\overline{\lim}\,(a_{n+1}/a_n) = \mu < 1$ and $b_n = (\frac{1}{2}(1+\mu))^n$. Then Σb_n converges as it is of the form Σr^n with $0 \leqslant r < 1$. Also

$$b_{n+1}/b_n = \tfrac{1}{2}(1+\mu) > \overline{\lim}\,(a_{n+1}/a_n).$$

Thus $\exists$ N such that $a_{n+1}/a_n < b_{n+1}/b_n$ for all $n > N$. By the ratio test Σa_{n+N} converges and hence Σa_n converges.

The second part of the test is proved similarly.

An important series

$\Sigma n^{-\alpha}$, α real converges if $\alpha > 1$ and diverges if $\alpha \leqslant 1$.

If $\alpha > 0$ and $N+1 \leqslant n \leqslant 2N$ then $(2N)^{-\alpha} \leqslant n^{-\alpha} \leqslant N^{-\alpha}$. Thus

$$N^{1-\alpha}\,2^{-\alpha} \leqslant \sum_{N+1}^{2N} n^{-\alpha} \leqslant N^{1-\alpha}.$$

If $\alpha > 1$ by taking $N = 2^j, j = 0, 1, \ldots, k-1$ it follows that

$$\sum_{2}^{2^k} n^{-\alpha} \leqslant \sum_{j=0}^{k-1} 2^{(1-\alpha)j} \leqslant \sum_{j=0}^{\infty} 2^{(1-\alpha)j} = \frac{1}{1-2^{(1-\alpha)}},$$

where the series $\Sigma 2^{(1-\alpha)j}$ is a convergent geometric series because $\alpha > 1$ and therefore $2^{(1-\alpha)} < 1$. For any $N \,\exists\, k$ such that $2^k \geqslant N$ and then as each summand is positive

$$\sum_{1}^{N} n^{-\alpha} \leqslant 1 + 1/(1-2^{(1-\alpha)}).$$

From theorem 7 $\Sigma n^{-\alpha}$ converges.

If $\alpha = 1$ it follows from the inequality above that

$$\sum_{N+1}^{2N} n^{-\alpha} \geqslant \tfrac{1}{2}$$

and the general principle of convergence (with any $\epsilon < \frac{1}{2}$, $n = N+1$, $m = 2N$) shows that Σn^{-1} diverges.

If $\alpha < 1$ then $n^{-\alpha} > n^{-1}$ and by the comparison principle $\Sigma n^{-\alpha}$ diverges.

Worked examples

(i) *$\{a_n\}\downarrow$ and Σa_n converges. Show that $na_n \to 0$ as $n \to \infty$. Deduce that if $\alpha \leqslant 1$ then $n^{-\alpha}$ diverges.*

Since Σa_n converges, $a_n \to 0$ and since $\{a_n\}\downarrow$ it follows that $a_n \geqslant 0$ and thus $\underline{\lim}\, na_n \geqslant 0$. If $\overline{\lim}\, na_n = \lambda > 0$ then $na_n > \frac{1}{2}\lambda$

for an infinite set of integers n. If N is one such integer and $N > 4$, then since $\{a_n\}\downarrow$ we have $a_n > \lambda/2N$ for $n \leqslant N$. Thus if

$$\tfrac{1}{4}N \leqslant M \leqslant \tfrac{1}{2}N,$$

M an integer, then

$$\sum_{M}^{N} a_n > (N-M)\lambda/2N \geqslant \tfrac{1}{4}\lambda.$$

By the general principle of convergence, theorem $5a$, it follows that Σa_n diverges. Hence if Σa_n converges then $\overline{\lim}\, na_n \leqslant 0$. Thus $\overline{\lim}\, na_n = \underline{\lim}\, na_n = 0$, i.e. $na_n \to 0$ as $n \to \infty$.

Take $a_n = n^{-\alpha}$, $\alpha \leqslant 1$. Since $na_n \not\to 0$ $\Sigma n^{-\alpha}$ diverges.

(ii) *Construct series with the following properties:*

(*a*) *Σa_n converges and Σa_n^2 diverges,*

(*b*) *Σb_n converges and Σb_n^3 diverges,*

(*c*) *Σc_n converges and both Σc_n^2 and Σc_n^3 diverge.*

Remark. The idea is to consider series Σa_n which converge because of the delicate cancellation properties of consecutive summands or triads of summands. Then cancellation properties will be destroyed when we replace a_n by a_n^2 or a_n^3. Consecutive summands or triads of summands will nearly cancel out but we can arrange that the cumulative effect of these near cancellations is quite large.

(*a*) Take $a_{2n} = -a_{2n-1} = n^{-\frac{1}{2}}$ $(n = 1, 2, \ldots)$.

(*b*) Take $b_{3n} = b_{3n-1} = -\tfrac{1}{2}n^{-\frac{1}{3}}$, $b_{3n-2} = n^{-\frac{1}{3}}$ $(n = 1, 2, \ldots)$.

(*c*) Take $c_n = b_n$ as defined in (*b*).

Exercises

1. Σa_n converges and Σb_n diverges. Show that $\Sigma(a_n + b_n)$ diverges and that Σc_n, where $c_{2n} = a_n$, $c_{2n-1} = b_n$ $(n = 1, 2, \ldots)$, also diverges.

2. Σa_n converges to s. Show that Σa_{n+1} converges to $s - a$, and that Σka_n converges to ks.

3. Σa_n converges to s. Show that $\Sigma(a_n + a_{n+1})$ converges to $2s - a_1$. Construct a divergent series Σb_n such that $\Sigma(b_n + b_{n+1})$ converges.

4. $\Sigma(a_{2n} + \lambda a_{2n-1})$ and $\Sigma(a_{2n} + \mu a_{2n-1})$ where $\lambda \neq \mu$ both converge. Show that Σa_n converges.

5. Σa_n converges absolutely and b_n satisfies $(1 + a_n)(1 - b_n) = 1$. Show that Σb_n converges absolutely.

6. $\Sigma |a_n|$ converges. Show that Σa_n^2 converges and that if $a_n \neq 1$ $(n = 1, 2, \ldots)$, then $\Sigma a_n/(1 - a_n)$ converges.

7. State of each of the following assertions (in which a_n, b_n are real numbers) whether it is true or false. Justify each answer by proving the assertion if it is true or by giving a contrary example if it is false.

(i) If $a_n \sim b_n$ then Σa_n and Σb_n both converge or both diverge.

(ii) If $a_{n+1}/a_n > k > 1$ for infinitely many integers n then Σa_n diverges.

(iii) If $a_{n+1}/a_n < 1$ for all n then Σa_n converges.

8. Prove that the following series converge for all values of x excluding the negative integers.

$$\text{(i) } \Sigma(-1)^n/(n+x), \quad \text{(ii) } \Sigma a_n/(n+x),$$

where $a_{3n} = 1$, $a_{3n-1} = a_{3n-2} = -\frac{1}{2}$ $(n = 1, 2, \ldots)$.

9. The positive series Σa_n is such that $\{a_n\}\downarrow$ and b_n is defined by $b_n = 2^n a_{2^n}$ $(n = 1, 2, \ldots)$. Show that Σa_n and Σb_n both converge or both diverge.

By using the inequality

$$1+\tfrac{1}{2}+\tfrac{1}{3}+\ldots+\frac{1}{2^n-1} = 1+(\tfrac{1}{2}+\tfrac{1}{3})+\ldots+\left(\frac{1}{2^{n-1}}+\ldots+\frac{1}{2^n-1}\right) < n \qquad (n > 1),$$

or otherwise, prove that $\Sigma(1/n)\,(1+\frac{1}{2}+\ldots+(1/n))^{-1}$ diverges.

10. The positive series Σa_n is such that $a_n/a_{n+1} \to \mu$ as $n \to \infty$ and $\mu > 1$. Show that (i) Σa_n converges, and (ii) $\overline{\lim}\, a_n^{1/n} < 1$. By considering the series whose terms are $a_{2n} = (2n)^{-2n}$, $a_{2n-1} = (2n-1)^{3-2n}$ show that the converse statements are both false.

11. The positive series Σa_n and Σb_n both converge. Show that $\Sigma(a_n^{\frac{1}{2}} b_n^{\frac{1}{2}})$ converges.

12. The positive series Σa_n is such that $\overline{\lim}\,(a_{n+r}/a_n) < 1$, where r is a fixed positive integer. Show that Σa_n converges.

Show further that if $\underline{\lim}\,(a_{n+r}/a_n) > 1$ then Σa_n diverges.

13. The positive series Σa_n diverges. Do the following series converge or diverge?

$$\text{(i) } \Sigma a_n/(1+a_n), \qquad \text{(ii) } \Sigma a_n/(1+na_n),$$
$$\text{(iii) } \Sigma a_n/(1+n^2a_n), \qquad \text{(iv) } \Sigma a_n/(1+a_n^2),$$

where in (iv) $\{a_n\}$ is bounded.

14. The positive series Σa_n converges. Show that $\Sigma(a_n a_{n+1})^{\frac{1}{2}}$ and $\Sigma(a_n n^{-1-\delta})^{\frac{1}{2}}$, $\delta > 0$, both converge.

15. $\{a_n\}\downarrow$ and $a_n > 0$. $\Sigma(a_n a_{n+1})^{\frac{1}{2}}$ converges. Show that Σa_n converges. Give an example of a positive divergent series Σb_n such that $\Sigma(b_n b_{n+1})^{\frac{1}{2}}$ converges.

16. The positive series Σa_n converges and the positive increasing sequence $\{p_n\}$ diverges to ∞. Show that $(p_1a_1+p_2a_2+\ldots+p_na_n)/p_n \to 0$ as $n \to \infty$.

17. The positive series Σa_n is such that for every positive increasing sequence $\{p_n\}$ that diverges to ∞, $(p_1 a_1 + p_2 a_2 + \ldots + p_n a_n)/p_n \to 0$. Show that Σa_n converges.

18. $\{a_n\}$ is a positive null sequence and

$$\{(a_1 + a_2 + \ldots + a_n - na_n)\}$$

is bounded above. Prove that Σa_n converges.

19. The positive series Σa_n converges. Prove that there exists a positive convergent series Σb_n such that $a_n = o(b_n)$ as $n \to \infty$.

8

CONDITIONAL CONVERGENCE

Although absolute convergence provides a powerful tool for discussing the convergence of series it is possible for a given series to converge without converging absolutely. There are certain tests for convergence of this type and the simplest are given here.

Definition 13. *A series that converges but does not converge absolutely is said to be non-absolutely or conditionally convergent.*

Theorem 8. (*a*) *If Σa_n converges and $\{b_n\}$ is a bounded real monotonic sequence then $\Sigma a_n b_n$ converges.*

(*b*) *If $\{(a_1 + \ldots + a_n)\}$ is bounded and $\{b_n\}$ is a decreasing real null sequence than $\Sigma a_n b_n$ converges.*

(*c*) *If Σa_n and $\Sigma |b_n - b_{n+1}|$ converge then so does $\Sigma a_n b_n$.*

(*d*) *If $\{(a_1 + \ldots + a_n)\}$ is bounded, $\Sigma |b_n - b_{n+1}|$ converges and $b_n \to 0$ then $\Sigma a_n b_n$ converges.*

(*a*) is a particular case of (*c*) and (*b*) a particular case of (*d*). We shall prove (*c*) and (*d*).

In each case a_n can be complex, in (*c*) and (*d*) b_n can be complex.

Proof of (*c*). Define s_n by $s_0 = 0$,

$$s_n = a_1 + a_2 + \ldots + a_n \quad (n = 1, 2, \ldots).$$

Then

$$(*) \quad \sum_1^N a_n b_n = \sum_1^N (s_n - s_{n-1}) b_n = \sum_1^N s_n (b_n - b_{n+1}) + s_N b_{N+1}.$$

Since $\{s_n\}$ converges it is bounded. $\Sigma |b_n - b_{n+1}|$ converges and it follows from the comparison principle (p. 42) that $\Sigma |s_n| \, |b_n - b_{n+1}|$

converges. By theorem 6 (p. 41) $\Sigma s_n(b_n - b_{n+1})$ converges, i.e. $\sum_1^N s_n(b_n - b_{n+1})$ tends to a finite limit as $N \to \infty$. Further by theorem 6 again and the fact that $\Sigma\, |b_n - b_{n+1}|$ converges we have that $\Sigma(b_n - b_{n+1})$ converges. Thus

$$b_{N+1} = b_1 - \sum_1^N (b_n - b_{n+1}) \to b_1 - \sum_1^\infty (b_n - b_{n+1}) \quad \text{as} \quad N \to \infty.$$

Also
$$s_N \to \sum_1^\infty a_n \quad \text{as} \quad N \to \infty.$$

Thus using all these relations in the identity (*),

$$\sum_1^N a_n b_n \to \sum_1^\infty s_n(b_n - b_{n+1}) + \left(\sum_1^\infty a_n\right)\left(b_1 - \sum_1^\infty (b_n - b_{n+1})\right),$$

i.e. $\Sigma a_n b_n$ converges.

Proof of (d). We start from identity (*) as in (*c*). As in (*c*) $\sum_1^N s_n(b_n - b_{n+1})$ tends to a finite limit as $N \to \infty$. But now $\{s_N\}$ is bounded and $b_n \to 0$ as $n \to \infty$. Thus $s_N b_{N+1} \to 0$ as $N \to \infty$. Thus

$$\sum_1^N a_n b_n \to \sum_1^\infty s_n(b_n - b_{n+1}),$$

i.e. $\Sigma a_n b_n$ converges.

The rearrangement of summation used in theorem 8 can also be used to prove Abel's lemma.

Abel's lemma. *If* $s_N = \sum_1^N a_n$ *and* $\{b_n\}$ *is a positive decreasing sequence then*

$$\left|\sum_1^N a_n b_n\right| \leqslant b_1 \max_{1 \leqslant p \leqslant N} |s_p|.$$

As above (*)
$$\sum_1^N a_n b_n = \sum_1^N s_n(b_n - b_{n+1}) + s_N b_{N+1}.$$

Thus
$$\begin{aligned}\left|\sum_1^N a_n b_n\right| &\leqslant \sum_1^N |s_n|\,|b_n - b_{n+1}| + |s_N|\,|b_{n+1}| \\ &\leqslant \max_{1 \leqslant p \leqslant N} |s_p| \left(\sum_1^N |b_n - b_{n+1}| + |b_{N+1}|\right) \\ &= \max_{1 \leqslant p \leqslant N} |s_p| \,.\, b_1,\end{aligned}$$

since $|b_n - b_{n+1}| = b_n - b_{n+1}$ and $|b_{N+1}| = b_{N+1}$. The lemma is proved.

The alternating series test. If $\{a_n\}$ is a positive decreasing null sequence then $\Sigma(-1)^n a_n$ converges.

This follows from theorem 8 (*b*) with a_n replaced by $(-1)^n$ and b_n by a_n.

Worked examples

(i) *Construct a convergent series Σa_n and a positive null sequence $\{b_n\}$ such that $\Sigma a_n b_n$ diverges.*

Take

$$a_n = (-1)^n n^{-\frac{1}{2}}, \quad b_{2n} = (2n)^{-\frac{1}{2}}, \quad b_{2n-1} = (2n-1)^{-1} \quad (n = 1, 2, \ldots).$$

Then Σa_n converges, $\Sigma a_{2n} b_{2n}$ diverges and $\Sigma a_{2n-1} b_{2n-1}$ converges. Thus $\Sigma a_n b_n$ diverges.

(ii) *$\{a_n\}$ is a decreasing sequence of positive numbers. Show that $\Sigma a_{n-1} z^{n-1}$ converges if $|z| \leqslant 1$, $z \neq 1$.*

We apply theorem 8 (*b*) with z^n in place of a_n and a_n in place of b_n. If $z \neq 1$, $1+z+z^2+\ldots+z^N = (1-z^{N+1})/(1-z)$, and if $|z| \leqslant 1$ also then $|1+z+\ldots+z^N| \leqslant 2/|1-z|$. Thus theorem 8 (*b*) can be applied and the result follows.

Exercises

1. $\{a_n\}$ and $\{b_n\}$ are real sequences. $\Sigma |b_n - b_{n+1}|$ converges and $\Sigma a_n b_n$ diverges to ∞. Show that $\overline{\lim}_{N\to\infty} \sum_1^N a_n = \infty$.

2. r is a fixed positive integer. $\Sigma |b_n - b_{n+r}|$ converges and Σa_{nr+l} converges for each $l = 1, 2, \ldots, r$. Show that $\Sigma a_n b_n$ converges.

3. $\Sigma n^{\frac{1}{2}} a_n$ converges. Show that Σa_n converges.

4. Discuss the convergence and the absolute convergence of

$$\text{(i) } \Sigma(-1)^n ((n^2+1)^{\frac{1}{2}} - n) \quad \text{and} \quad \text{(ii) } \Sigma(-1)^n (2n + (-1)^{n+1})^{-\frac{1}{2}}.$$

5. $\{b_n\}$ is a real positive decreasing sequence, a_n is real and

$$s_n = a_1 + a_2 + \ldots + a_n.$$

Show that $\sum_1^p a_n b_n$ lies between $b_1 \max(s_1, s_2, \ldots, s_p)$ and $b_1 \min(s_1, s_2, \ldots, s_p)$.

6. Sequences $\{a_n\}$, $\{b_n\}$, $\{\alpha_n\}$ are such that (i) $\alpha_n > 0$, (ii) $\Sigma |\alpha_n(b_n - b_{n+1})|$ converges, (iii) $a_1 + a_2 + \ldots + a_n = O(\alpha_n)$, (iv) $\alpha_n b_{n+1} \to 0$ as $n \to \infty$. Show that $\Sigma a_n b_n$ converges.

7. $\Sigma(a_1 + a_2 + \ldots + a_n)$ and $\Sigma |b_n - 2b_{n+1} + b_{n+2}|$ converge, $\{b_n\}$ is bounded. Show that $\Sigma a_n b_n$ converges.

8. (a) The positive series Σb_n is such that for any positive null sequence, $\{a_n\}$, $\Sigma a_n b_n$ converges. Show that Σb_n converges.

(b) $\{b_n\}$ is such that if the sequence $\{s_n\}$ converges then so does $\Sigma s_n(b_n - b_{n+1})$. By considering sequences $\{s_n\}$ for which $s_n \to 0$ and $s_n(b_n - b_{n+1})$ is real and positive show that $\Sigma |b_n - b_{n+1}|$ converges.

(c) $\{b_n\}$ is such that the convergence of Σa_n implies that of $\Sigma a_n b_n$. Show that $b_n = O(1)$ and $\Sigma |b_n - b_{n+1}|$ converges.

9

REARRANGEMENT AND MULTIPLICATION OF ABSOLUTELY CONVERGENT SERIES

A function $\mathfrak{f}$ whose range and domain are each the set of positive integers and for which $\mathfrak{f}(n) = \mathfrak{f}(m)$ if and only if $n = m$ is called a *rearranging function*. Alternatively a rearranging function is a 1-1 correspondence of the set of positive integers with itself. If Σa_n is given and $\mathfrak{f}$ is a rearranging function then Σb_n where $b_n = a_{\mathfrak{f}(n)}$ is said to be a *rearrangement* of Σa_n. Since the function $\mathfrak{f}^{-1}$, the inverse of $\mathfrak{f}$ is also a rearranging function, Σa_n is a rearrangement of Σb_n.

For example, the series $1 + \frac{1}{2} + \frac{1}{3} + \frac{1}{4} + \dots$, i.e. Σa_n where $a_n = 1/n$ and $1 + \frac{1}{3} + \frac{1}{2} + \frac{1}{5} + \frac{1}{7} + \dots$; i.e. Σb_n where $b_n = 1/2k$ if $n = 3k$, $b_n = 1/(4k-3)$ if $n = 3k-2$, $b_n = 1/(4k-1)$ if $n = 3k-1$ $(k = 1, 2, 3, \dots)$ are rearrangements of one another. In this case $b_n = a_{\mathfrak{f}(n)}$ where $\mathfrak{f}(3n-2) = 4n-3$, $\mathfrak{f}(3n-1) = 4n-1$, $\mathfrak{f}(3n) = 2n$ $(n = 1, 2, 3, \dots)$.

If Σa_n and all its rearrangements converge to the same sum, then we say that Σa_n is *unconditionally convergent*. We shall justify this nomenclature by showing that unconditional convergence is equivalent to absolute convergence.

Theorem 9. *Σa_n is absolutely convergent if and only if it is unconditionally convergent.*

Suppose that Σa_n converges absolutely and Σb_n is a rearrangement given by $\mathfrak{f}$ where $a_n = b_{\mathfrak{f}(n)}$. Then $\forall \epsilon > 0 \, \exists$ an integer N such that $\sum_N^\infty |a_n| < \epsilon$. Let $M = \max_{1 \leqslant n \leqslant N} \mathfrak{f}(n)$, then for $m \geqslant M$

$$\left| \sum_1^\infty a_n - \sum_1^m b_n \right| \leqslant \sum_N^\infty |a_n| < \epsilon.$$

It follows that Σb_n converges to $\sum_1^\infty a_n$. Since Σb_n is any rearrangement of Σa_n this last series is unconditionally convergent.

Suppose next that Σa_n is unconditionally convergent. Then if $\alpha_n = \mathscr{R}(a_n)$ and $\beta_n = \mathscr{I}(a_n)$ both $\Sigma \alpha_n$ and $\Sigma \beta_n$ are unconditionally convergent. We shall show that $\Sigma \alpha_n$ converges absolutely. Write $s = \sum_1^\infty \alpha_n$. If $\Sigma \alpha_n$ does not converge absolutely then a rearranging function $\mathfrak{f}$ can be defined as follows.

Let $\mathfrak{f}(1)$ be the least positive integer n for which $\alpha_n \geqslant 0$. Generally when $\mathfrak{f}(1) \dots \mathfrak{f}(N-1)$ have been defined we define $\mathfrak{f}(N)$ in one of two ways. If $\sum_1^{N-1} \alpha_{\mathfrak{f}(n)} > s+1$ then $\mathfrak{f}(N)$ is the least positive integer such that $\mathfrak{f}(N)$ is not equal to any of $\mathfrak{f}(1), \dots, \mathfrak{f}(N-1)$. If

$$\sum_1^{N-1} \alpha_{\mathfrak{f}(n)} \leqslant s+1$$

then $\mathfrak{f}(N)$ is the least positive integer n such that $\mathfrak{f}(N)$ is not equal to any of $\mathfrak{f}(1), \dots, \mathfrak{f}(N-1)$ and $\alpha_n \geqslant 0$.

It may be verified that because $\Sigma \alpha_n$ converges, but does not converge absolutely, the function $\mathfrak{f}$ is a rearranging function. But $\Sigma \alpha_{\mathfrak{f}(n)}$ cannot converge to s since a subsequence of its partial sums has values $> s+1$. Thus $\Sigma \alpha_n$ is not unconditionally convergent.

Hence if $\Sigma \alpha_n$ is unconditionally convergent then it is absolutely convergent. Similarly $\Sigma \beta_n$ is absolutely convergent and finally Σa_n converges absolutely.

Multiplication of absolutely convergent series

Theorem 10. *If Σa_n and Σb_n converge absolutely and their sums are A and B respectively and $\{\alpha_n\}$, $\{\beta_n\}$ are two sequences of integers such that every permutation of two positive integers occurs exactly once as (α_n, β_n) then $\Sigma a_{\alpha_n} b_{\beta_n}$ converges absolutely and its sum is AB.*

Let $$\sum_1^\infty |a_n| = A_1, \quad \sum_1^\infty |b_n| = B_1.$$

$\forall$ an integer N write $M = \max\limits_{1 \leqslant n \leqslant N} (\alpha_n)$ and $L = \max\limits_{1 \leqslant n \leqslant N} (\beta_n)$ then

$$\sum_1^N |a_{\alpha_n} b_{\beta_n}| \leqslant \sum_1^M |a_n| \cdot \sum_1^L |b_n| \leqslant A_1 B_1$$

and thus $\Sigma a_{\alpha_n} b_{\beta_n}$ converges absolutely. The terms of this series are obtained by selecting in some order all the elements of the set

$$\begin{vmatrix} a_1 b_1 & a_1 b_2 & a_1 b_3 & \dots \\ a_2 b_1 & a_2 b_2 & a_2 b_3 & \dots \\ a_3 b_1 & a_3 b_2 & a_3 b_3 & \dots \\ \dots & \dots & \dots & \dots \end{vmatrix}.$$

By theorem 9 the sum of this series is unaltered however we rearrange the summands. We can rearrange them so that the first n^2 summands are $a_i b_j$ $(1 \leqslant i \leqslant n, 1 \leqslant j \leqslant n)$. Then the sum of the first n^2 summands is $\left(\sum_1^n a_i\right)\left(\sum_1^n b_i\right) \to AB$ as $n \to \infty$. Thus since we know that the rearranged series converges to the same sum as the original series, and since a subsequence of its partial sums tend to AB, it follows that

$$\sum_1^\infty a_{\alpha_n} b_{\beta_n} = AB.$$

Theorem 11: Cauchy's rule. *If Σa_n and Σb_n both converge absolutely and their sums are A and B then Σc_n where*

$$c_n = a_1 b_n + a_2 b_{n-1} + \dots + a_n b_1$$

converges absolutely and its sum is AB.

For Σc_n is obtained from a series of the type $\Sigma a_{\alpha_n} b_{\beta_n}$ by bracketing a number of terms together.

Remark. It may happen that Σa_n and Σb_n converge whilst Σc_n where $c_n = a_1 b_n + a_2 b_{n-1} + \dots + a_n b_1$ diverges. For example, if $a_n = b_n = (-1)^n n^{-\frac{1}{2}}$ then

$$c_n = (-1)^{n+1} \sum_1^n r^{-\frac{1}{2}}(n+1-r)^{-\frac{1}{2}}$$

and thus $|c_n| > 1$. But if all three series converge then $\sum_1^\infty c_n = \left(\sum_1^\infty a_n\right)\left(\sum_1^\infty b_n\right)$. See Abel's theorem, p. 185, and exercise (7) below.

If, however, Σa_n converges absolutely and Σb_n converges then Σc_n converges and $\sum_1^\infty c_n = \left(\sum_1^\infty a_n\right)\left(\sum_1^\infty b_n\right)$. See worked example (ii) below. In this case Σc_n need not converge absolutely, for example, if $a_n = (-1)^n n^{-1-\frac{1}{4}}$ and $b_n = (-1)^n n^{-\frac{1}{4}}$, then $|c_n| \geqslant |n . n^{-1-\frac{1}{2}}| \geqslant |n^{-\frac{1}{2}}|$ and $\Sigma |c_n|$ diverges.

An important example. If $|z| < 1$ then

$$\frac{1}{1-z} = 1 + z + z^2 + \dots.$$

For by theorem 10, since $1-z$, $1+z+z^2+\dots$ can both be regarded as absolutely convergent series if $|z|<1$ we have

$$(1-z)(1+z+z^2+\dots) = \sum_0^\infty c_n z^n,$$

where $c_n = 0$ if $n \geqslant 1$, $c_0 = 1$, i.e. $1/(1-z) = 1+z+z^2+\dots$.

This is a particular case of the *Binomial expansion*.

Worked examples

(i) *Σa_n converges absolutely and $b_n \to 0$ as $n \to \infty$. Show that $a_1 b_n + a_2 b_{n-1} + \dots + a_n b_1 \to 0$ as $n \to \infty$.*

Write $A = \sum_1^\infty |a_n|$, $K = \sup |b_n|$. Then $\forall\, \epsilon > 0$

$$\exists\, N \text{ such that } \sum_n^m |a_r| < \epsilon/2(K+1) \quad \text{for} \quad m \geqslant n > N,$$

$$\exists\, M \text{ such that } |b_n| < \epsilon/2(A+1) \quad \text{for} \quad n \geqslant M,$$

(we use $A+1$, $K+1$ instead of A or K in case $A = 0$ or $K = 0$). For $n > N+M$

$$|a_1 b_n + \dots + a_n b_1|$$
$$\leqslant |a_1 b_n + \dots + a_N b_{n-N+1}| + |a_{N+1} b_{n-N} + \dots + a_n b_1|.$$

In the first expression on the right-hand side the terms b_r are such that $r > n-N > M$ and thus $|b_r| < \epsilon/2(A+1)$. Thus

$$|a_1 b_n + \dots + a_N b_{n-N+1}| < \epsilon . \sum_1^N |a_n|/2(A+1) < \tfrac{1}{2}\epsilon.$$

In the second expression we put $|b_r| < K$, $\sum_{N+1}^n |a_r| < \epsilon/2(K+1)$. Then

$$|a_{N+1} b_{n-N} + \dots + a_n b_1| < \tfrac{1}{2}\epsilon.$$

Finally $\quad |a_1 b_n + \dots + a_n b_1| < \epsilon \quad$ for all $n > N+M$.

Thus $\quad a_1 b_n + \dots + a_n b_1 \to 0 \quad$ as $\quad n \to \infty$.

(ii) *Σa_n converges absolutely and Σb_n converges. Show that Σc_n where $c_n = a_n b_1 + \dots + a_1 b_n$ converges and that $\sum_1^\infty c_n = \left(\sum_1^\infty a_n\right)\left(\sum_1^\infty b_n\right)$.*

Write

$$A_N = \sum_1^N a_r, \quad A = \sum_1^\infty a_r,$$

$$B_N = \sum_1^N b_r, \quad B = \sum_1^\infty b_r,$$

$$C_N = \sum_1^N c_r.$$

Then $$C_N = a_1 B_N + \ldots + a_N B_1$$

and $$A_N B - C_N = a_1(B - B_N) + \ldots + a_N(B - B_1).$$

By example (i) above it follows that $A_N B - C_N \to 0$ as $N \to \infty$. But $A_N B \to AB$ and thus Σc_n converges to AB.

(iii) *The real series Σa_n converges conditionally. Given a real number s show that there is a rearrangement of Σa_n whose sum is s.*

Write $b_n = \max(a_n, 0)$, $c_n = \min(a_n, 0)$, then

$$a_n = b_n + c_n, \quad |a_n| = b_n - c_n.$$

Since $\Sigma |a_n|$ diverges Σb_n and Σc_n cannot both converge. Since Σa_n converges it is not possible that only one of Σb_n, Σc_n should converge. Thus Σb_n and Σc_n both diverge and there are infinitely many positive and infinitely many negative terms in $\{a_n\}$.

Let $\{p_n\}$ be the subsequence of $\{a_n\}$ formed by its non-negative terms and let $\{q_n\}$ be the subsequence formed by the negative terms. Since Σp_n may be regarded as being obtained from Σb_n by omitting certain zero summands it follows that Σp_n diverges to infinity. Similarly Σq_n diverges to $-\infty$. Write

$$P_N = p_1 + p_2 + \ldots + p_N,$$
$$Q_N = q_1 + q_2 + \ldots + q_N,$$

then $P_N \to \infty$ as $N \to \infty$ and $Q_N \to -\infty$ as $N \to \infty$.

Let N_1 be the least integer such that $P_{N_1} > s$ and M_1 be the least integer such that $P_{N_1} + Q_{M_1} < s$. Generally when N_i, M_i have been defined for $1 \leqslant i < j$ we define N_j to be the least integer such that $P_{N_j} + Q_{M_{j-1}} > s$ and M_j to be the least integer such that $P_{N_j} + Q_{M_j} < s$. Define a rearrangement of Σa_n say Σv_n as follows

$$v_n = p_n \quad (1 \leqslant n \leqslant N_1), \qquad v_{n+N_1} = q_n \quad (1 \leqslant n \leqslant M_1),$$

and generally

$$v_{n+M_{j-1}} = p_n \quad (N_{j-1} < n \leqslant N_j), \quad v_{n+N_j} = q_n \quad (M_{j-1} < n \leqslant M_j)$$
$$(j = 2, 3, \ldots).$$

Write $V_N = v_1 + v_2 + \ldots + v_N$. Then by the definition of the rearrangement Σv_n we have

$$V_{N_j+M_{j-1}} > s \geqslant V_{N_j+M_{j-1}} - p_{N_j},$$
$$V_{N_j+M_j} < s \leqslant V_{N_j+M_j} - q_{M_j},$$

But Σa_n converges and thus $a_n \to 0$ as $n \to \infty$. Hence

$$p_{N_j} \to 0, \quad q_{M_j} \to 0 \quad \text{as} \quad j \to \infty.$$

Thus $\qquad V_{N_j+M_{j-1}} \to s, \quad V_{N_j+M_j} \to s \quad \text{as} \quad j \to \infty.$

But if
$$N_j + M_{j-1} < n < N_j + M_j$$
then
$$V_{N_j+M_{j-1}} \geqslant V_n > V_{N_j+M_j},$$
and if
$$N_j + M_j < n < N_{j+1} + M_j$$
then
$$V_{N_j+M_j} \leqslant V_n < V_{N_{j+1}+M_j}.$$

Thus finally $V_n \to s$ as $n \to \infty$.

Σv_n converges to s and since this series is a rearrangement of Σa_n we have established the required result.

Exercises

1. Prove that if $|x| < 1$ then

(a) $[1+x+x^2+x^3+\ldots]^2 = 1+2x+3x^2+4x^3+\ldots$.

(b) $[1+x+x^2+x^3+\ldots]^n = 1+nx+\dfrac{n(n+1)}{2}x^2+\dfrac{n(n+1)(n+2)}{1.2.3}x^3+\ldots$

$(n = 3, 4, \ldots)$.

2. Prove that $\Sigma(x^n/n!)$ converges absolutely for all x and
$$\left(\sum_0^\infty \frac{x^n}{n!}\right)\left(\sum_0^\infty \frac{y^n}{n!}\right) = \sum_0^\infty \frac{(x+y)^n}{n!}.$$

3. Prove that Σa_n converges absolutely if and only if $\forall \epsilon > 0 \; \exists$ an integer N such that
$$|a_{n_1}+a_{n_2}+\ldots+a_{n_k}| < \epsilon,$$
where $n_1, n_2, \ldots, n_k$ are any finite set of distinct integers all greater than N.

4. State which of the following statements are true and which false. Give a proof of a correct statement and an example to the contrary of a false statement.

(i) $u_n \geqslant 0$ and $u_n^{1/n} < 1$ then $\Sigma(-1)^n u_n$ is convergent.

(ii) Σu_n converges, $u_n > 0$; $p_1, p_2, \ldots, p_n, \ldots$ is a strictly increasing sequence of positive integers and $v_n = u_{p_n}$. Then Σv_n converges and $\sum_1^\infty v_n \leqslant \sum_1^\infty u_n$.

(iii) Σu_n converges absolutely and u_n is real. Then the conclusions of (ii) still hold.

5. Show that the series
$$1-\tfrac{1}{2}+\tfrac{1}{3}-\tfrac{1}{4}+\tfrac{1}{5}-\tfrac{1}{6}+\ldots \quad \text{and} \quad 1-\tfrac{1}{2}-\tfrac{1}{4}+\tfrac{1}{3}-\tfrac{1}{6}-\tfrac{1}{8}+\tfrac{1}{5}-\tfrac{1}{10}-\tfrac{1}{12}+\ldots$$
both converge and are rearrangements of one another. Show that the sum of the second series is one-half the sum of the first series.

6. Σa_n and Σb_n converge to A and B respectively and $C_N = c_1+c_2+\ldots+c_N$ where $c_n = a_1 b_n + a_2 b_{n-1}+\ldots+a_n b_1$. Show that $(C_1+C_2+\ldots+C_N)/N \to AB$ as $N \to \infty$.

Deduce that if Σc_n converges then its sum is AB.

10

DOUBLE SERIES

Remark. We have defined a sequence as a function defined over the positive integers. A natural extension of this idea is a function defined over the class of ordered pairs of positive integers, usually called a double array. Just as we defined a series in terms of an associated sequence so we define a double series in terms of a double array. The properties of double series are more complicated than those of series, but once they have been established they provide powerful techniques for dealing with series.

A function defined over ordered pairs of positive integers is called a *double array*. If the value of the function at (i, j) is a_{ij} then we write it as $\{\{a_{ij}\}\}$ or

$$\left|\begin{array}{cccc} a_{11} & a_{12} & a_{13} & \cdots \\ a_{21} & a_{22} & a_{23} & \cdots \\ a_{31} & a_{32} & a_{33} & \cdots \\ \cdot & \cdot & \cdot & \cdots \end{array}\right|.$$

By $\Sigma\Sigma a_{ij}$ we mean $\{\{S_{ij}\}\}$ where $S_{m,n} = \sum_{i=1}^{m}\sum_{j=1}^{n} a_{ij}$. We refer to $\Sigma\Sigma a_{ij}$ as a *double series* and say that it *converges to* s if and only if $\forall \epsilon > 0 \,\exists\, N$ such that $|s_{m,n} - s| < \epsilon$ for all integers m, n such that $m > N$, $n > N$. A double series which does not converge is said to *diverge*. If $\Sigma\Sigma a_{ij}$ converges we shall use $\sum_{i,j=1}^{\infty} a_{ij}$ in place of s, and call this number its *sum*.

The reader should observe carefully the difference in meaning between the symbols

$$(a)\ \sum_{i,j=1}^{\infty} a_{ij},\quad (b)\ \sum_{i=1}^{\infty} a_{ij},\quad (c)\ \sum_{j=1}^{\infty} a_{ij},$$

$$(d)\ \sum_{i=1}^{\infty}\left(\sum_{j=1}^{\infty} a_{ij}\right),\quad (e)\ \sum_{j=1}^{\infty}\left(\sum_{i=1}^{\infty} a_{ij}\right).$$

(*a*) is the sum of $\Sigma\Sigma a_{ij}$; (*b*) is the sum for fixed j of Σa_{ij}, i suffix of summation; (*c*) is the sum for fixed i of Σa_{ij}, j suffix of summation; (*d*) is the sum of the series whose nth summand is the sum of the nth row of the double array $\{\{a_{ij}\}\}$; (*e*) is the sum of the series whose nth summand is the sum of the nth column of the double array $\{\{a_{ij}\}\}$.

In (*d*) and (*e*) the 'nth row' and the 'nth column' have their obvious meaning. Thus the nth row is the series whose mth summand is a_{nm} and the nth column is the series whose mth summand is a_{mn}. (*d*) and (*e*) are called the *sum by rows* and the *sum by columns* of the double array $\{\{a_{ij}\}\}$, respectively.

Theorem 5b. *The general principle of convergence.* $\Sigma\Sigma a_{ij}$ *converges if and only if* $\forall \epsilon > 0 \, \exists$ *an integer* N *such that*

$$|s_{m,n} - s_{p,q}| < \epsilon, \quad \textit{where} \quad s_{m,n} = \sum_{i=1}^{m} \sum_{j=1}^{n} a_{ij}$$

for all integers p, q, m, n *which are all greater than* N.

Necessity. If $\Sigma\Sigma a_{ij}$ converges to s then $\forall \epsilon > 0 \, \exists$ an integer N such that $|s_{m,n} - s| < \frac{1}{2}\epsilon$ for $m > N$, $n > N$. Hence by the modulus inequality

$$|s_{m,n} - s_{p,q}| \leqslant |s_{m,n} - s| - |s - s_{p,q}| < \epsilon$$

if p, q, m, n are all greater than N.

Thus the stated condition is necessary.

Sufficiency. If the given condition holds then by the general principle of convergence for sequences (theorem 5, p. 34) applied to $\{s_{n,n}\}$ it follows that $\{s_{n,n}\}$ converges to a limit s as $n \to \infty$. Hence $\forall \epsilon > 0 \, \exists$ an integer M such that $|s_{n,n} - s| < \frac{1}{2}\epsilon$ for $n > M$. But by the given condition $\exists$ an integer L such that $|s_{n,n} - s_{p,q}| < \frac{1}{2}\epsilon$ if p, q, n, are all greater than L. This if $N = \max(M, L)$ we have by the modulus inequality

$$|s_{p,q} - s| < \epsilon \quad \text{if} \quad p > N, \, q > N.$$

That is to say $\Sigma\Sigma a_{ij}$ converges to s.

Exercises. (1) $\Sigma\Sigma a_{ij}$ is such that $a_{ij} \geqslant 0$ and $\sum_{i=1}^{M} \sum_{j=1}^{N} a_{ij} < K$, K fixed, for all positive integers M, N. Show that $\Sigma\Sigma a_{ij}$ converges.

(2) $\Sigma\Sigma a_{ij}$ is defined by $a_{1j} = (-1)^j$, $a_{2j} = (-1)^{j+1}$, $a_{ij} = 0$ $(i > 2)$. Show that $\Sigma\Sigma a_{ij}$ converges but that it is not true that given $\epsilon > 0 \, \exists$ an integer N such that $|a_{ij}| < \epsilon$ if $i + j > N$.

Theorem 12. *If* $\Sigma\Sigma a_{ij}$ *converges and the series* $\sum_{j=1}^{\infty} a_{ij}$ *converges for* $i = 1, 2, 3, \ldots$ *then the sum by rows exists and is equal to the sum of* $\Sigma\Sigma a_{ij}$.

Write

$$\sum_{i,j=1}^{\infty} a_{ij} = s, \quad \sum_{j=1}^{\infty} a_{ij} = R_i, \quad \sum_{i=1}^{m} \sum_{j=1}^{n} a_{ij} = s_{mn}.$$

Then $\forall \epsilon > 0 \exists$ an integer N such that $|s_{mn} - s| < \epsilon$ if $n > N$, $m > N$. Now

$$\sum_{i=1}^{m} R_i = \sum_{i=1}^{m}\left(\sum_{j=1}^{\infty} a_{ij}\right) = \sum_{j=1}^{\infty}\left(\sum_{i=1}^{m} a_{ij}\right) = \lim_{n\to\infty} s_{m,n},$$

$$\therefore \quad \left|\sum_{i=1}^{m} R_i - s\right| \leqslant \epsilon \quad \text{if} \quad m > N.$$

Thus the sum by rows of $\{\{a_{ij}\}\}$ exists and is equal to s.

Remark. There is a similar result with rows replaced by columns.

Absolute convergence

$\Sigma\Sigma a_{ij}$ is *absolutely convergent* if and only if $\Sigma\Sigma\,|a_{ij}|$ converges. By an argument similar to that of theorem 7 it can be deduced from the general principle of convergence, theorem 5*b*, that absolute convergence implies convergence.

Theorem 13. *If $\Sigma\Sigma a_{ij}$ is such that either*

(a) $\Sigma\Sigma\,|a_{ij}|$ converges, or

(b) the sum by rows of $\{\{|a_{ij}|\}\}$ exists, or

(c) the sum by columns of $\{\{|a_{ij}|\}\}$ exists,

then the sum of $\Sigma\Sigma a_{ij}$, the sum by rows and the sum by columns of $\{\{a_{ij}\}\}$, all exist and are all equal.

First, the three alternative conditions (*a*), (*b*), (*c*) are all equivalent. For to see that (*a*) implies (*b*) let T be the sum of the double series $\Sigma\Sigma\,|a_{ij}|$. Then

$$\sum_{j=1}^{m} |a_{nj}| \leqslant \sum_{i=1}^{n}\sum_{j=1}^{m} |a_{ij}| \leqslant T.$$

Thus the series $\sum_{j=1}^{\infty} |a_{nj}|$ converges ($n = 1, 2, 3, \ldots$). By theorem 12 the sum by rows of $\{\{|a_{ij}|\}\}$ exists and equals T.

Next (*b*) implies that $\sum_{i=1}^{M}\sum_{j=1}^{N} |a_{ij}|$ is bounded above for all M, N. For if R is the sum by rows of $\{\{|a_{ij}|\}\}$ then

$$\sum_{i=1}^{M}\sum_{j=1}^{N} |a_{ij}| \leqslant R$$

and it follows that $\Sigma\Sigma\,|a_{ij}|$ converges to $\sup_{M,N} \sum_{i=1}^{M}\sum_{j=1}^{N} |a_{ij}| = R$.

Thus (*a*), (*b*) are equivalent. Similarly (*a*) and (*c*) are equivalent. Thus we can assume that each of (*a*), (*b*), (*c*) holds. Moreover, the three sums mentioned in (*a*), (*b*), (*c*) are all equal.

Next since $\Sigma\Sigma\, a_{ij}$ is absolutely convergent, it is also convergent. Since $\sum_{i=1}^{\infty}\left(\sum_{j=1}^{\infty}|a_{ij}|\right)$ exists, $\sum_{j=1}^{\infty}|a_{ij}|$ exists and the series $\Sigma\, a_{ij}$ converges for $i = 1, 2, \ldots$. By theorem 12 the sum by rows of $\{\{a_{ij}\}\}$ exists and is equal to the sum of $\Sigma\Sigma\, a_{ij}$. Similarly the sum by columns of $\{\{a_{ij}\}\}$ exists and is equal to the sum of $\Sigma\Sigma\, a_{ij}$.

Exercise. By using the identity

$$\frac{1}{1-x^n} = 1+x^n+x^{2n}+x^{3n}+\ldots \quad (0 < x < 1)$$

(see exercise (2), chapter 8), show that if $0 < x < 1$ then

$$\sum_{n=1}^{\infty} \frac{(1-x)^n}{1-x^n} = \sum_{m=0}^{\infty} \frac{(1-x)\,x^m}{1-(1-x)\,x^m}.$$

By a *rearranging function* of a double series we mean a function $\mathfrak{f}(m, n)$ defined over the class of all ordered pairs of positive integers, whose values are ordered pairs of positive integers such that every such ordered pair occurs as a value of $\mathfrak{f}(m, n)$ for exactly one ordered pair (m, n), i.e. the function $\mathfrak{f}$ is a 1-1 correspondence of the class of ordered pairs of positive integers onto itself. For such a function $\Sigma\Sigma a_{\mathfrak{f}(m,n)}$ is a *rearrangement* of $\Sigma\Sigma\, a_{m,n}$.

Theorem 14. *The sum of an absolutely convergent double series is equal to the sum of any series obtained from it by a rearrangement.*

Given $\Sigma\Sigma a_{ij}$ and its rearrangement $\Sigma\Sigma b_{ij}$ where $b_{ij} = a_{\mathfrak{f}(i,j)}$ write

$$s_{m,n} = \sum_{i=1}^{m}\sum_{j=1}^{n} a_{ij}, \quad t_{m,n} = \sum_{i=1}^{m}\sum_{j=1}^{n} b_{ij}, \quad u_{m,n} = \sum_{i=1}^{m}\sum_{j=1}^{n} |a_{ij}|,$$

$$S = \sum_{i,j=1}^{\infty} a_{ij}, \quad U = \sum_{i,j=1}^{\infty} |a_{ij}|.$$

$\forall\, \epsilon > 0\; \exists$ an integer N such that $|u_{m,n} - U| < \epsilon$, if $m > N, n > N$. $\exists$ an integer M such that if $p < N$ and $q < N$ then $\mathfrak{f}(i, j) = (p, q)$ where $i < M, j < M$. Then if $k > M, l > M, m > N, n > N$ we have

$$|s_{m,n} - t_{k,l}| \leqslant \sum_{i=N}^{L}\sum_{j=N}^{P} |a_{ij}| < \epsilon,$$

where if $i \leqslant k, j \leqslant l$, then $\mathfrak{f}(i, j) = (p, q)$ with $p \leqslant L, q \leqslant P$. Thus

$$|S - t_{k,l}| \leqslant \epsilon \quad \text{if} \quad k > M, l > M,$$

i.e. $\Sigma\Sigma b_{ij}$ converges to S.

Worked examples

(i) $|a| < 1$, $|x| < 1$ *and* $\mathfrak{f}(a) = \sum_0^\infty a^n/n!$. *Show that*

$$\sum_0^\infty \frac{a^n}{1+x^2a^{2n}}\frac{1}{n!} = \mathfrak{f}(a) - x^2\mathfrak{f}(a^3) + x^4\mathfrak{f}(a^5) - \ldots.$$

Write $b_{ij} = (-1)^{j-1}a^{i-1}x^{2j-2}a^{2(i-1)(j-1)}/(i-1)!$. Consider the sum by rows of $\{\{b_{ij}\}\}$. Now the series $\sum_{j=1}^\infty |b_{ij}|$ converges and its sum is $|a|^{i-1}/(1-|x|^2|a|^{i-1})(i-1)! = k_{i-1}$ say. As $i \to \infty$ $k_i \sim |a|^i/i!$ and thus Σk_i converges. Hence the sum by rows of $\{\{|b_{ij}|\}\}$ exists. By theorem 13 $\Sigma\Sigma b_{ij}$ converges and its sum by rows is equal to its sum by columns, i.e.

$$\sum_{i=0}^\infty \frac{a^i}{1+x^2a^{2i}}\frac{1}{i!} = \sum_{j=0}^\infty (-1)^j x^{2j}\mathfrak{f}(a^{1+2j}).$$

(ii) Σa_n *is given.* Σb_n *is defined by* $b_n = (a_1 + 2a_2 + \ldots + 2^{n-1}a_n)/2^n$. *Show that* (*a*) *if both series converge then they converge to the same sum,* (*b*) *if either series is absolutely convergent then so is the other.*

Consider $\{\{a_{ij}\}\}$ defined by $a_{ij} = 2^{i-1-j}a_i$ $(i \leqslant j)$, $a_{ij} = 0$ $(i > j)$, i.e. the double array

$$\begin{vmatrix} \frac{1}{2}a_1 & \frac{1}{4}a_1 & \frac{1}{8}a_1 & \ldots \\ 0 & \frac{1}{2}a_2 & \frac{1}{4}a_2 & \ldots \\ 0 & 0 & \frac{1}{2}a_3 & \ldots \\ \cdot & \cdot & \cdot & \ldots \end{vmatrix}.$$

The partial sums $s_{M,N} = \sum_{i=1}^M \sum_{j=1}^N a_{ij}$ are given by $s_{M,N} = s_{N,N} = \sum_1^N b_n$ if $M \geqslant N$,

$$s_{M,N} = s_{M,M} + (a_1 + 2a_2 + \ldots + 2^{M-1}a_M)(2^{-M-1} + 2^{-M-2} + \ldots + 2^{-N}) \qquad (M < N).$$

Thus $$\left|s_{M,N} - \sum_1^N b_n\right| \leqslant |b_M| \text{ for all } M, N.$$

(*a*) If Σb_n converges then $b_n \to 0$ as $n \to \infty$ and it follows that $\Sigma\Sigma a_{ij}$ converges to the sum $\sum_1^\infty b_n$.

If Σa_n converges then the sum by rows of $\Sigma\Sigma a_{ij}$ exists and is equal to $\sum_1^\infty a_n$.

From theorem 12 it follows that $\sum_1^\infty b_n = \sum_1^\infty a_n$.

(*b*) This follows from theorem 13.

Exercises

1. (a) $\Sigma\Sigma b_{ij}$ converges, $b_{ij} \geqslant 0$ and $0 \leqslant a_{ij} \leqslant Kb_{ij}$. Show that $\Sigma\Sigma a_{ij}$ converges.

(b) $\Sigma\Sigma b_{ij}$ diverges, $b_{ij} \geqslant 0$ and $b_{ij} < Ka_{ij}$ for some $K > 0$. Show that $\Sigma\Sigma a_{ij}$ diverges.

2. $\Sigma\Sigma a_{ij}$ converges absolutely. Show that $\sup\limits_{i+j \geqslant N} |a_{ij}| \to 0$ as $N \to \infty$.

3. $\Sigma\Sigma a_{ij}$ converges absolutely and b_{ij} is such that $(1+a_{ij})(1-b_{ij}) = 1$. Show that $\Sigma\Sigma b_{ij}$ converges absolutely.

4. $a_{ij} = 1/(i^2-j^2)$ $(i \neq j)$; $a_{ij} = 0$ $(i = j)$. Show that the sum by rows and the sum by columns of $\Sigma\Sigma a_{ij}$ are not equal.

5. If $0 < b < a$ prove that

$$\sum_{n=1}^{\infty} \frac{nb^n}{a^n - b^n} = \sum_{m=1}^{\infty} \frac{a^m b^m}{(a^m - b^m)^2}.$$

6. If $|q| < 1$ prove that

$$1+\frac{q}{1-q}+\frac{2q^2}{1+q^2}+\frac{3q^3}{1-q^3}+\ldots = 1+\frac{q}{(1-q)^2}+\frac{q^2}{(1+q^2)^2}+\frac{q^3}{(1-q^3)^2}+\ldots.$$

7. If $|x| < 1$ prove that

$$\frac{x}{(1+x)^2}+\frac{x^2}{(1+x^2)^2}+\frac{x^3}{(1+x^3)^2}+\ldots = \frac{x}{1-x}-\frac{2x^2}{1-x^2}+\frac{3x^3}{1-x^3}-\ldots.$$

8. Discuss the convergence of $\Sigma\Sigma a_{ij}$ in the following cases:

(i) $a_{ij} = i^{-\alpha}j^{-\beta}$, (ii) $a_{ij} = (hi+kj)^{-\alpha}$ $(h > 0, k > 0)$,

(iii) $a_{ij} = (i^2+j^2)^{-1}$, (iv) $a_{ij} = (i^3+j^3)^{-1}$,

where α, β are real numbers.

9. If x is not a negative integer prove that

$$\sum_{0}^{\infty}\sum_{r=0}^{n} \frac{(-1)^r}{r!\,(n-r)!\,(x+r)} = \left(\sum_{n=0}^{\infty}\frac{1}{n!}\right)\left(\frac{1}{x}-\frac{1}{1!}\frac{1}{(x+1)}+\frac{1}{2!}\frac{1}{(x+2)}-\ldots\right).$$

10. Show that $\Sigma\Sigma x^{i+j-2}$ converges if $0 < x < 1$ and diverges if $x \geqslant 1$.

11. Σa_n converges absolutely. Prove that

$$\sum_{1}^{\infty} a_n = \frac{a_1}{1.2}+\frac{a_1+2a_2}{2.3}+\frac{a_1+2a_2+3a_3}{3.4}+\ldots.$$

12. Show that if $|x| < 1$ then

$$\frac{x}{1-x}+\frac{x^2}{1-x^2}+\frac{x^3}{1-x^3}+\ldots = x\frac{1+x}{1-x}+x^4\frac{1+x^2}{1-x^2}+x^9\frac{1+x^3}{1-x^3}+\ldots = \sum_{1}^{\infty} a_n x^n,$$

where a_n is the number of positive integers (including 1 and n) which divide n.

13. Show that if $x > 1$, then

$$\frac{1}{x^2} = \frac{1}{x(x+1)} + \frac{1!}{x(x+1)(x+2)} + \frac{2!}{x(x+1)(x+2)(x+3)} + \dots,$$

and deduce that

$$\frac{1}{x^2} + \frac{1}{(x+1)^2} + \frac{1}{(x+2)^2} + \dots = \frac{1}{x} + \frac{1!}{2x(x+1)} + \frac{2!}{3x(x+1)(x+2)} + \dots.$$

11

POWER SERIES

A series $\sum_0^\infty a_n(z-b)^n$ or $\Sigma a_{n-1}(z-b)^{n-1}$ is called a *power series.* Here a_n, b are real or complex constants and we are interested in the convergence and the sum of the series for different values of z. The first summand $a_0(z-b)^0$ is undefined when $z = b$, we make the convention that this symbol is to be interpeted as a_0. If a_n, b and z are all real then we have a *real power series*, otherwise it is a *complex power series.* b is called the *centre of expansion* of the series. We shall consider mainly the case $b = 0$ since the corresponding results for other values of b can be obtained by transforming from z to $z-b$.

Theorem 15. *(a) If* $\overline{\lim}\,|a_n|^{1/n} = 0$, *then* $\Sigma a_{n-1} z^{n-1}$ *converges absolutely for all* z.

(b) If $\overline{\lim}\,|a_n|^{1/n} = \infty$, *then* $\Sigma a_{n-1} z^{n-1}$ *diverges for all* z, *except* $z = 0$.

(c) If $0 < \overline{\lim}\,|a_n|^{1/n} < \infty$, *then* $\Sigma a_{n-1} z^{n-1}$ *converges absolutely for those* z *for which* $|z| < (\overline{\lim}\,|a_n|^{1/n})^{-1}$ *and diverges for those* z *for which* $|z| > (\overline{\lim}\,|a_n|^{1/n})^{-1}$.

(a) Let z be given, $z \neq 0$. Then since $\overline{\lim}\,|a_n|^{1/n} = 0 \;\exists$ an integer N such that $|a_n|^{1/n} < 1/2\,|z|$ $(n > N)$. Thus $|a_n z^n| < 1/2^n$ $(n > N)$ and by the comparison principle $\Sigma a_{n-1} z^{n-1}$ converges absolutely. It is trivial that $\Sigma a_{n-1} z^{n-1}$ converges absolutely when $z = 0$.

(b) Let z be given, $z \neq 0$. Then since $\overline{\lim}\,|a_n|^{1/n} = \infty \;\exists$ infinitely many integers n such that $|a_n|^{1/n} > 2/|z|$ and thus $|a_n z^n| > 2^n$ for infinitely many integers n. Hence $\Sigma a_{n-1} z^{n-1}$ diverges. Again it is trivial that $\Sigma a_{n-1} z^{n-1}$ converges absolutely when $z = 0$.

(c) Write

$$R = (\overline{\lim}\,|a_n^{1/n}|)^{-1}.$$

If $|z| < R$ choose λ so that $|z|/R < \lambda < 1$. Since

$$\overline{\lim}\,(|a_n|^{1/n}.|z|) = |z|/R$$

$\exists$ an integer N such that $|a_n|^{1/n}|z| < \lambda$ for all $n > N$. Hence $|a_n z^n| < \lambda^n$ and since $\lambda < 1$ $\Sigma\lambda^{n-1}$ and (by the comparison principle) $\Sigma a_{n-1} z^{n-1}$ converge absolutely.

If $|z| > R \, \exists$ infinitely many integers n such that $|a_n|^{1/n}|z| > 1$ and thus $|a_n z^n| > 1$ for infinitely many integers. Hence $\Sigma a_{n-1} z^{n-1}$ diverges.

Definition 14. *If* $0 < \overline{\lim}\,|a_n|^{1/n} < \infty$, *then* $(\overline{\lim}\,|a_n|^{1/n})^{-1}$ *is called the radius of convergence of the power series* $\Sigma a_{n-1} z^{n-1}$.

The set of complex numbers z for which $|z| < (\overline{\lim}\,|a_n|^{1/n})^{-1}$ is called the *circle of convergence* of $\Sigma a_{n-1} z^{n-1}$.

Remark. If we know that $\Sigma a_{n-1} z_1^{n-1}$ converges and that $\Sigma a_{n-1} z_2^{n-1}$ diverges then we know that the radius of convergence R satisfies $|z_1| \leqslant R \leqslant |z_2|$.

Corollaries to theorem 15

Corollary 1. *If* $|z| < R_1 < (\overline{\lim}\,|a_n|^{1/n})^{-1}$ *then*

$$\left|\sum_0^\infty a_n z^n - \sum_0^{N-1} a_n z^n\right| \leqslant |z|^N \sum_0^\infty |a_n| R_1^{n-N} = K|z|^N,$$

where K is independent of z.

For

$$\left|\sum_0^\infty a_n z^n - \sum_0^{N-1} a_n z^n\right| = \left|\sum_N^\infty a_n z^n\right| \leqslant \sum_N^\infty |a_n|\,|z^n| \leqslant |z|^N \sum_N^\infty |a_n| R_1^{n-N}$$

since all the series involved converge. This is the required form.

Corollary 2. *Uniqueness of power series. If* $\Sigma a_{n-1} z^{n-1}$, $\Sigma b_{n-1} z^{n-1}$ *converge for* $|z| < R$, $R > 0$ *and if*

$$\sum_0^\infty a_n z_i^n = \sum_0^\infty b_n z_i^n \quad (i = 1, 2, 3, \ldots),$$

where $z_i \to 0$ *as* $i \to \infty$, *then* $a_n = b_n$ $(n = 0, 1, 2, \ldots)$.

By corollary (1) with $N = 1$

$$a_0 = \lim_{i\to\infty} \sum_0^\infty a_n z_i^n = \lim_{i\to\infty} \sum_0^\infty b_n z_i^n = b_0.$$

Thus $a_0 = b_0$ and $\sum_1^\infty a_n z_i^{n-1} = \sum_1^\infty b_n z_i^{n-1}$ $(i = 1, 2, 3, \ldots)$. Now $\Sigma a_n z^{n-1}$ is a power series which converges if $|z| < R$ since the convergence of $\Sigma a_{n-1} z^{n-1}$ for a particular z implies the convergence of $\Sigma a_n z^{n-1}$ for the same z. Similarly $\Sigma b_n z^{n-1}$ converges if $|z| < R$. The argument given above can now be applied again to show that $a_1 = b_1$.

Since the process can be repeated it follows by an induction argument that $a_n = b_n$ $(n = 0, 1, 2, \ldots)$.

This result is also known as the *principle of equating coefficients.*

Corollary 3. *Dominated convergence of power series. Suppose that* $\Sigma a_{n-1} z^{n-1}$ *converges if* $|z| < R$ *and that* $|b_{n,m}| \leqslant |a_n|$ *for* $m = 1, 2, 3, \ldots, n = 0, 1, 2, \ldots$. *If* $b_{n,m} \to b_n$ *as* $m \to \infty$ *then* $\Sigma b_{n-1} z^{n-1}$ *converges for* $|z| < R$ *and* $\sum_0^\infty b_{n,m} z^n \to \sum_0^\infty b_n z^n$ *as* $m \to \infty$.

Since $|b_n| \leqslant |a_n|$, $\overline{\lim}\, |b_n|^{1/n} \leqslant \overline{\lim}\, |a_n|^{1/n} \leqslant 1/R$ and thus the radius of convergence of $\Sigma b_{n-1} z^{n-1}$ is at least R.

If $|z| < R$ and R_1 is chosen so that $|z| < R_1 < R$ then

$$\left|\sum_0^\infty b_{n,m} z^n - \sum_0^\infty b_n z^n\right| \leqslant \sum_0^N |b_{n,m} - b_n|\, |z|^n + \sum_{N+1}^\infty 2\, |a_n|\, |z|^n$$

for any positive integer N. $\forall\, \epsilon > 0$ choose N so that $\sum_{N+1}^\infty |a_n|\, |z|^n \leqslant \frac{1}{4}\epsilon$ and then choose an integer M such that

$$|b_{n,m} - b_n| < \tfrac{1}{2}\epsilon(1 + R_1 + R_1^2 + \ldots + R_1^N)^{-1},$$
$$(n = 0, 1, \ldots, N,\ m > M).$$

Then since $|z| < R_1$, it follows that

$$\left|\sum_0^\infty b_{n,m} z^n - \sum_0^\infty b_n z^n\right| \leqslant \tfrac{1}{2}\epsilon + \tfrac{1}{2}\epsilon \leqslant \epsilon \quad (m > M).$$

Hence $$\sum_0^\infty b_{n,m} z^n \to \sum_0^\infty b_n z^n \quad \text{as} \quad m \to \infty.$$

Remark. When z is such that $|z| = (\overline{\lim}\, |a_n|^{1/n})^{-1}$ then $\Sigma a_{n-1} z^{n-1}$ may either converge or diverge depending on the particular values of a_{n-1}, for example, each of the series Σz^{n-1}, $\Sigma z^{n-1}/n$, $\Sigma z^{n-1}/n^2$ has radius of convergence equal to 1, the first series diverges at $z = \pm 1$, the second converges at $z = -1$ and diverges at $z = 1$, the third converges at $z = \pm 1$.

Exercise. Show that $\Sigma z^{n-1}/(n-1)!$ and $\Sigma 2^{n-1} z^{n-1}/(n-1)!$ converge for all z. Show that $\Sigma (n-1)!\, z^{n-1}$ and $\Sigma n^{n^2} z^{n-1}/(n-1)!$ diverge for all z except $z = 0$.

Properties of power series

In what follows $\Sigma a_{n-1} z^{n-1}$ and $\Sigma b_{n-1} z^{n-1}$ have non-zero radii of convergence R_1, R_2 respectively.

(i) $\Sigma (a_{n-1} + b_{n-1}) z^{n-1}$ *converges if* $|z| < \min(R_1, R_2)$.

For both $\Sigma a_{n-1}z^{n-1}$ and $\Sigma b_{n-1}z^{n-1}$ converge if $|z| < \min(R_1, R_2)$.

(ii) *$\Sigma c_{n-1}z^{n-1}$, where $c_n = a_0 b_n + a_1 b_{n-1} + \ldots + a_n b_0$ converges if*

$$|z| < \min(R_1, R_2) \text{ and then } \sum_0^\infty c_n z^n = \left(\sum_0^\infty a_n z^n\right)\left(\sum_0^\infty b_n z^n\right).$$

This follows from Cauchy's rule theorem 11, p. 51, combined with theorem 15.

We write

$$\left(\sum_0^\infty a_n z^n\right)^k \quad \text{as} \quad \sum_0^\infty a_n^{(k)} z^n \quad (k = 1, 2, \ldots; \; |z| < R_1).$$

(iii) *If $a_0 \neq 0 \; \exists$ a power series $\Sigma d_{n-1}z^{n-1}$ which converges for $|z| < R$, some $R > \infty$ and such that if $|z| < \min(R, R_1)$ then*

$$\sum_0^\infty d_n z^n = \left(\sum_0^\infty a_n z^n\right)^{-1}.$$

Write $c_n = a_n/a_0$. By theorem 15, corollary 1, $\exists \, R_3 > 0$ such that if $|z| < R_3$ then $\sum_1^\infty |c_n z^n| < 1$. Then by the Binomial expansion, chapter 9, p. 51,

$$a_0 \left(\sum_1^\infty a_n z^n\right)^{-1} = \left(1 + \sum_1^\infty c_n z^n\right)^{-1}$$

$$= 1 - \sum_1^\infty c_n z^n + \left(\sum_1^\infty c_n z^n\right)^2 - \ldots \quad (|z| < R_3)$$

$$= 1 - \sum_1^\infty c_n z^n + \sum_1^\infty c_n^{(2)} z^n - \ldots.$$

This sum is the sum by rows of the double array

$$\begin{vmatrix} 1 & 0 & 0 & 0 & \ldots \\ 0 & -c_1 z & -c_2 z^2 & -c_3 z^3 & \ldots \\ 0 & c_1^{(2)} z & c_2^{(2)} z^2 & c_3^{(2)} z^3 & \ldots \\ 0 & -c_1^{(3)} z & -c_2^{(3)} z^2 & -c_3^{(3)} z^3 & \ldots \\ . & . & . & . & \ldots \end{vmatrix}.$$

Now $c_n^{(1)} = c_n$, thus $(|c_n|)^{(1)} = |c_n^{(1)}|$. Further

$$c_n^{(k)} = c_0 c_n^{(k-1)} + \ldots + c_n c_0^{(k-1)}$$

and by a simple induction it follows that $|c_n^{(k)}| \leqslant (|c_n|)^{(k)}$. Now the sum by rows of the double array:

$$\left|\begin{matrix} 1 & 0 & 0 & 0 & \dots \\ 0 & |c_1|\,|z| & |c_2|\,|z|^2 & |c_3|\,|z|^3 & \dots \\ 0 & |c_1|^{(2)}\,|z| & |c_2|^{(2)}\,|z^2|^2 & |c_3|^{(2)}\,|z|^3 & \dots \\ \cdot & \cdot & \cdot & \cdot & \dots \end{matrix}\right|$$

exists and is equal to $1\bigg/\left(1-\sum\limits_1^\infty |c_n|\,|z|^n\right)$. It follows from theorem 13, p. 57, that this second array is absolutely convergent. Hence the first array is also absolutely convergent (since $|(-1)^j c_i^{(j)} z^i| \leqslant |c_i|^{(j)}\,|z|^i$). By theorem 13 its sums by rows and by columns are equal. Hence

$$\left(\sum_0^\infty a_n z^n\right)^{-1} = a_0^{-1}\left(1+\left(\sum_{j=1}^\infty (-1)^j c_1^{(j)}\right) z \right.$$
$$\left. +\left(\sum_{j=1}^\infty (-1)^j c_2^{(j)}\right) z^2 + \dots\right) = \sum_0^\infty d_n z^n,$$

where $d_0 = a_0^{-1}$, $\quad d_n = a_0^{-1} \sum\limits_{j=1}^\infty (-1)^j c_n^{(j)} \quad n \geqslant 1$, $\quad$ and $\quad |z| < R_3$.

(iv) *If* $\Sigma d_{n-1} z^{n-1}$ *converges for* $|z| < r$ *and* $d_0 = 0$, *then* $\exists R_4 > 0$ *and a series* $\Sigma l_{n-1} z^{n-1}$ *such that* $\sum\limits_0^\infty a_n (d_1 z + d_2 z^2 + \dots)^n = \sum\limits_0^\infty l_n z^n$ *if* $|z| < R_4$.

$\exists r_1 > 0$ such that $\sum\limits_1^\infty |d_{n-1} z^{n-1}| < R_1$ if $|z| < r_1$. For $|z| < \min(r_1, R_1)$

$$\sum_0^\infty a_n (d_1 z + d_2 z^2 + \dots)^n = \sum_0^\infty a_n \left(\sum_1^\infty d_i^{(n)} z^i\right),$$

and this is the sum by rows of the double array

$$\left|\begin{matrix} a_0 & 0 & 0 & 0 & \dots \\ 0 & a_1 d_1^{(1)} z & a_1 d_2^{(1)} z^2 & a_1 d_3^{(1)} z^3 & \dots \\ 0 & a_2 d_1^{(2)} z & a_2 d_2^{(2)} z^2 & a_2 d_3^{(2)} z^3 & \dots \\ \cdot & \cdot & \cdot & \cdot & \dots \end{matrix}\right|.$$

Now $|a_n d_i^{(n)} z^i| \leqslant |a_n|\,|z^i|\,(|d_i|)^{(n)}$ as in (iii). Since

$$|d_1 z| + |d_2 z^2| + \dots < R_1$$

the sum by rows of the double array

$$\left|\begin{matrix} |a_0| & 0 & 0 & 0 & \dots \\ 0 & |a_1|\,(|d_1|)^{(1)}\,|z| & |a_1|\,(|d_2|)^{(1)}\,|z|^2 & |a_1|\,(|d_3|)^{(1)}\,|z|^3 & \dots \\ 0 & |a_2|\,(|d_1|)^{(2)}\,|z| & |a_2|\,(|d_2|)^{(2)}\,|z|^2 & |a_2|\,(|d_3|)^{(2)}\,|z|^3 & \dots \\ . & . & . & . & \dots \end{matrix}\right|$$

exists and is equal to $\sum_0^\infty |a_n|(|d_1 z| + |d_2 z^2| + \dots)^n$. Thus in this case this second double array is absolutely convergent. Hence the first double array is also absolutely convergent and its sum by rows is equal to its sum by columns (theorem 13). Choose

$$R_4 = \min(r, R_1, r_1)$$

then for $|z| < R_4$ equating the sum by rows and the sum by columns of the first array gives

$$\sum_0^\infty a_n(d_1 z + d_2 z^2 + \dots)^n = a_0 + \left(\sum_1^\infty a_j d_1^{(j)}\right) z + \left(\sum_1^\infty a_j d_2^{(j)}\right) z^2 + \dots$$
$$= \sum_0^\infty l_n z^n.$$

This is the required result.

(v) *If* $|z_1| < R_1 \exists$ *a power series* $\Sigma d_{n-1}(z-z_1)^{n-1}$ *such that* $\sum_0^\infty a_n z^n = \sum_0^\infty d_n(z-z_1)^n$ *for* z *satisfying* $|z-z_1| < R_1 - |z_1|$.

If $|z-z_1| < R_1 - |z_1|$, then by the modulus inequality $|z| < R_1$ and

$$\sum_0^\infty a_n z^n = \sum_0^\infty a_n(z_1 + z - z_1)^n$$
$$= \sum_0^\infty a_n \left(z_1^n + \binom{n}{1} z_1^{n-1}(z-z_1) + \binom{n}{2} z_1^{n-2}(z-z_1)^2 + \dots + \binom{n}{n}(z-z_1)^n\right).$$

This series is the sum by rows of the double array

$$\left|\begin{matrix} a_0 & 0 & 0 & 0 & \dots \\ a_1 z_1 & a_1(z-z_1) & 0 & 0 & \dots \\ a_2 z_1^2 & \binom{2}{1} a_2 z_1(z-z_1) & a_2(z-z_1)^2 & 0 & \dots \\ . & . & . & . & \dots \end{matrix}\right|.$$

Now $|z_1|+|z-z_1|<R$ and thus $\Sigma\,|a_{n-1}|\,(|z_1|+|z-z_1|)^{n-1}$ converges. But on expansion this series is seen to be the sum by rows of the double array

$$\begin{Vmatrix} |a_0| & 0 & 0 & 0 & \dots \\ |a_1 z_1| & |a_1(z-z_1)| & 0 & 0 & \dots \\ |a_2 z_1^2| & \binom{2}{1}|a_2 z_1(z-z_1)| & |a_2(z-z_1)^2| & 0 & \dots \\ \cdot & \cdot & \cdot & \cdot & \dots \end{Vmatrix}.$$

All the terms in this array are positive and its sum by rows exists. Thus it is absolutely convergent. Hence the first array is absolutely convergent and its sum by rows equals its sum by columns, i.e.

$$\sum_0^\infty a_n z^n = \sum_0^\infty a_j z_1^j + \left(\sum_0^\infty \binom{j+1}{1} a_{j+1} z_1^j\right)(z-z_1)$$
$$+\left(\sum_0^\infty \binom{j+2}{2} a_{j+2} z_1^j\right)(z-z_1)^2 + \dots = \sum_0^\infty d_n (z-z_1)^n,$$

where
$$d_n = \sum_{k=0}^\infty \binom{n+k}{n} a_{n+k} z_1^k.$$

Exercises

1. $|a_n/a_{n+1}| \to \lambda$ as $n\to\infty$. Show that if $\lambda>0$ the radius of convergence of $\Sigma a_{n-1}z^{n-1}$ is λ.

2. $\Sigma a_{n-1}z^{n-1}$ and $\Sigma b_{n-1}z^{n-1}$ have radii of convergence R_1 and R_2 respectively. Show that:
(i) the radius of convergence of $\Sigma c_{n-1}z^{n-1}$, where $c_{2n}=a_n$, $c_{2n-1}=b_n$ is $\min(R_1,R_2)$;
(ii) the radius of convergence of $\Sigma(a_{n-1}+b_{n-1})z^{n-1}$ is R_2 if $R_1>R_2$ and when $R_1=R_2$ is at least R_2;
(iii) the radius of convergence of $\Sigma a_{n-1}b_{n-1}z^{n-1}$ is at least R_1R_2.

3. Find the radius of convergence of each of the following series.

(a) $\Sigma(2(n-1))!\,z^{n-1}/((n-1)!)^2$. (b) $\Sigma r^{(n-1)^2} z^{n-1}$ $(r>0)$.

(c) $\Sigma(n-1)!\,z^{n-1}/n^n$. (d) $\Sigma(3n)!\,z^{n-1}/2^3.4^3.\dots.(2n)^3$.

4. Show that

$$(1-z)^{-1}-(2-z)^{-1} = \sum_1^\infty (1-2^{-n})z^{n-1} \quad \text{if} \quad |z|<1,$$
$$= -4\sum_0^\infty 2^{2n}(z-\tfrac{3}{2})^{2n} \quad \text{if} \quad |z-\tfrac{3}{2}|<\tfrac{1}{2}.$$

5. The rational fractions between 0 and 1 are arranged in order

$$\tfrac{1}{2},\ \tfrac{1}{3},\ \tfrac{2}{3},\ \tfrac{1}{4},\ \tfrac{2}{4},\ \tfrac{3}{4},\ \tfrac{1}{5},\ \dots$$

and p_n is the nth fraction in this sequence. Show that $\Sigma(p_n)^{n^2} z^n$ converges for all values of z.

6. $\{a_n\}$ is a decreasing sequence of positive numbers. Show that $\Sigma a_{n-1}z^{n-1}$ converges for $|z| < 1$ and that if its sum is $\mathfrak{f}(z)$ then

$$|(1-z)\,\mathfrak{f}(z)-a_0| < a_0 \quad (|z| < 1).$$

Deduce that $\mathfrak{f}(z)$ does not vanish for $|z| < 1$.

7. Each of the series $\Sigma a_{n-1}z^{n-1}$, $\Sigma b_{n-1}z^{n-1}$, $\Sigma a_{n-1}b_{n-1}z^{n-1}$ has radius of convergence equal to unity. Show that $\Sigma a_{n-1}b_{n-1}^2 z^{n-1}$ also has radius of convergence equal to unity.

12

POINT SET THEORY

In this chapter we consider those properties of sets of numbers which depend only on the function $|x-y|$ defined for every pair of real numbers or for every pair of complex numbers and since the properties developed are equally valid for real or for complex numbers we shall not specify with which we are dealing unless it is necessary to do so. But in each case we deal entirely with real numbers or entirely with complex numbers.

$|x-y|$ is called the *distance* between x and y. If x, y are real then $|x-y| = \max(x-y, y-x)$ and if x, y are complex then

$$|x-y| = [(\mathscr{R}(x)-\mathscr{R}(y))^2+(\mathscr{I}(x)-\mathscr{I}(y))^2]^{\frac{1}{2}}.$$

We refer to x, y, as *points* and by a *sequence of points* mean either a sequence of real numbers or a sequence of complex numbers. A set of real numbers is called a *linear* set; a set of complex numbers is called a *planar* set. We shall use $\mathbf{S}$ to denote the set of all points; either $\mathbf{S} = \mathbf{Z}$ or $\mathbf{S} = \mathbf{R}$ and we are not committed to either alternative.

A sequence $\{x_n\}$ converges to y if and only if $|x_n - y| \to 0$ as $n \to \infty$. We also denote this situation by '$x_n \to y$ as $n \to \infty$' and say that 'x_n tends to y as $n \to \infty$'.

As examples we shall use the following sets.

Subsets of $\mathbf{R}$.

$K_1 = \{x|x > 0\}$, $\qquad K_2 = \{x|x < 0\}$,

$K_3 = \{x|x \geqslant 0\}$, $\qquad K_4 = \{x|x \leqslant 0\}$,

$K_5 = \{x|x = 1/n,\ n = 1, 2, \ldots\}$, $\qquad K_6 = \{x|-1 < x < 1\}$,

$K_7 = \{x|x = 0\}$, $\qquad K_8 = \{x|x = \pm n,\ n = 1, 2, \ldots\}$,

$T_1 = \{x|x \text{ rational}\}$, $\qquad T_2 = \{x|x \text{ irrational}\}$.

Subsets of **Z**.

$$Y_1 = \{x \mid |x| = 1\}, \quad Y_2 = \{x \mid |x| < 1\}, \quad Y_3 = \{x \mid |x| > 1\}.$$

Definition 15. *If X is a subset of $\mathbf{S}$ and $y \in \mathbf{S}$ is such that $\exists \{x_n\}$ with $x_n \in X$, $|x_n - y| \to 0$ as $n \to \infty$, $x_n \neq y$, then y is a limit-point or cluster-point of X.*

Definition 16. *The set of all limit points of X is called the derived set of X and denoted by X'.*

Examples:

$K_1' = K_3' = K_3$, $K_2' = K_4' = K_4$, $K_5' = K_7$,

$K_6' = \{x \mid -1 \leqslant x \leqslant 1\}$, $K_7' = K_8' = \phi$, $T_1' = T_2' = \mathbf{R}$,

$Y_1' = Y_1$, $Y_2' = Y_2 \cup Y_1$, $Y_3' = Y_3 \cup Y_1$.

Definition 17. *$X \cup X'$ is called the closure of X and is denoted by $\overline{X}$. We define $\overline{\phi}$ to be ϕ.*

Examples:

$\overline{K}_1 = \overline{K}_3 = K_3$, $\overline{K}_2 = \overline{K}_4 = K_4$, $\overline{K}_5 = K_5 \cup K_7$, $\overline{K}_6 = K_6'$,

$\overline{K}_7 = K_7$, $\overline{K}_8 = K_8$, $\overline{T}_1 = \overline{T}_2 = \mathbf{R}$, $\overline{Y}_1 = Y_1$,

$\overline{Y}_2 = Y_2'$, $\overline{Y}_3 = Y_3'$.

Definition 18. *The set of points of $\mathbf{S}$ that do not belong to X is called the complement of X and denoted by $\mathscr{C}(X)$.*

The complement of a set of real numbers is again a set of real numbers. Clearly $\mathscr{C}(\mathscr{C}(X)) = X$.

Examples:

$\mathscr{C}(K_1) = K_4$, $\mathscr{C}(K_2) = K_3$, $\mathscr{C}(K_3) = K_2$,

$\mathscr{C}(K_4) = K_1$, $\mathscr{C}(K_5) = \{x \mid x \in \mathbf{R} \text{ and } x \neq 1/n,\ n = 1, 2, \ldots\}$,

$\mathscr{C}(K_6) = \{x \mid x \leqslant -1 \text{ or } x \geqslant 1\}$,

$\mathscr{C}(K_7) = \{x \mid x \in \mathbf{R},\ x \neq 0\}$,

$\mathscr{C}(K_8) = \{x \mid x \in \mathbf{R},\ x \neq \pm n,\ n = 1, 2, \ldots\}$,

$\mathscr{C}(T_1) = T_2$, $\mathscr{C}(T_2) = T_1$, $\mathscr{C}(Y_1) = Y_2 \cup Y_3$,

$\mathscr{C}(Y_2) = Y_1 \cup Y_3$, $\mathscr{C}(Y_3) = Y_1 \cup Y_2$.

Definition 19. *The frontier of X, $\mathscr{F}r(X)$, is defined by*

$$\mathscr{F}r(X) = \overline{X} \cap \overline{\mathscr{C}(X)}.$$

Examples:

$$\mathscr{F}r(K_1) = \mathscr{F}r(K_2) = \mathscr{F}r(K_3) = \mathscr{F}r(K_4) = \mathscr{F}r(K_7) = K_7,$$
$$\mathscr{F}r(K_5) = \bar{K}_5, \qquad \mathscr{F}r(K_6) = \{x \mid x = 1 \text{ or } x = -1\},$$
$$\mathscr{F}r(K_8) = K_8, \qquad \mathscr{F}r(T_1) = \mathscr{F}r(T_2) = \mathbf{R},$$
$$\mathscr{F}r(Y_1) = \mathscr{F}r(Y_2) = \mathscr{F}r(Y_3) = Y_1.$$

Definition 20. *If $X = \bar{X}$ then we say that X is closed.*

Examples: K_3, K_4, K_7, K_8, Y_1 are closed. The other sets are not closed.

Definition 21. *If $x_0 \in \mathbf{S}$ and $\delta > 0$ then $\{x \mid |x - x_0| < \delta\}$ is called a neighbourhood or spherical neighbourhood of x_0 and is denoted by $\mathscr{U}(x_0, \delta)$.*

Definition 22. *If $x_0 \in \mathbf{S}$* and *$\exists\, \delta > 0$ such that $\mathscr{U}(x_0, \delta) \subset X$ then x_0 is said to be an interior point of X. The interior points of X form the interior of X denoted by X^0.*

Examples:

$$K_1^0 = K_3^0 = K_1, \qquad K_2^0 = K_4^0 = K_2, \qquad K_5^0 = K_7^0 = K_8^0 = \phi,$$
$$K_6^0 = K_6, \qquad T_1^0 = T_2^0 = \phi, \qquad Y_1^0 = \phi,$$
$$Y_2^0 = Y_2, \qquad Y_3^0 = Y_3.$$

Definition 23. *X is open if and only if $X = X^0$.*

Examples: K_1, K_2, K_6, Y_2, Y_3 are open. The other sets are not open.

Definition 24. *If $X = X_1 \cup X_2$ where*

$$X_1 \neq \phi, \quad X_2 \neq \phi, \quad \bar{X}_1 \cap X_2 = \phi, \quad X_1 \cap \bar{X}_2 = \phi,$$

then X is not connected. If X cannot be expressed in this form then X is connected.

Examples: K_5, K_8, T_1, T_2 are not connected. The other sets are connected and for some of them proofs will be given later.

Definition 25. *If $X \subset \mathbf{S}$ is such that $\bar{X} = \mathbf{S}$ then X is dense in $\mathbf{S}$. If $X \subset Y \subset \mathbf{S}$ and $\bar{X} \supset Y$ then X is dense in Y.*

Examples: K_i $(i = 1, 2, \ldots, 8)$ are not dense in $\mathbf{R}$. T_1 and T_2 are dense in $\mathbf{R}$. Y_1, Y_2, Y_3 are not dense in $\mathbf{Z}$, but $Y_2 \cup Y_3$ is dense in $\mathbf{Z}$.

We regard $\mathbf{S}$ as a subset of itself; as such it is both open and closed. The void set is also regarded as being both open and closed.

If X_i is a class of sets with $i \in I$ then

$$\mathscr{C}\left(\bigcup_{i \in I} X_i\right) = \bigcap_{i \in I} \mathscr{C}(X_i), \quad \mathscr{C}\left(\bigcap_{i \in I} X_i\right) = \bigcup_{i \in I} \mathscr{C}(X_i).$$

The spherical neighbourhoods are themselves open sets. For if $y \in \mathscr{U}(x, \delta)$ then $|x-y| < \delta$ and the neighbourhood $\mathscr{U}(y, \delta - |x-y|)$ exists. If z belongs to this second neighbourhood then

$$|z-y| < \delta - |x-y|$$

and by the modulus inequality this implies $|z-x| < \delta$, i.e. $z \in \mathscr{U}(x, \delta)$. Hence $\mathscr{U}(y, \delta - |x-y|) \subset \mathscr{U}(x, \delta)$ and thus $y \in (\mathscr{U}(x, \delta))^0$. This is true for every point of $\mathscr{U}(x, \delta)$ which is therefore an open set.

Theorem 16. (i) *The complement of a closed set is an open set.* (ii) *The complement of an open set is a closed set.*

(i) Let X be a given closed set. If $\mathscr{C}(X) = \phi$ there is nothing to prove. Otherwise let $y \in \mathscr{C}(X)$. Suppose that $y \notin (\mathscr{C}(X))^0$ then $X \cap \mathscr{U}(y, 1/n) \neq \phi$ for $n = 1, 2, \ldots$. Choose x_n to be any point of X in $\mathscr{U}(y, 1/n)$ then

$$x_n \in X, \quad |x_n - y| < 1/n \to 0 \quad \text{as} \quad n \to \infty, \quad x_n \neq y$$

(where the last relation is because $y \in \mathscr{C}(X)$). Hence $y \in X'$. But X is closed, therefore $X = X \cup X'$, therefore $y \in X$. This is impossible as $y \in \mathscr{C}(X)$.

This contradiction shows that $y \in (\mathscr{C}(X))^0$. Since this is true of all points of $\mathscr{C}(X)$ it follows that $\mathscr{C}(X)$ is open.

(ii) Let X be a given open set. If $\mathscr{C}(X) = \phi$ there is nothing to prove. Otherwise it is sufficient to show that $(\mathscr{C}(X))' \subset \mathscr{C}(X)$, i.e. that if $\{y_n\}$, y are such that

$$y_n \in \mathscr{C}(X), \quad |y_n - y| \to 0 \quad \text{as} \quad n \to \infty, \quad y_n \neq y,$$

then $y \in \mathscr{C}(X)$. Assume the contrary, i.e. that $y \in X$. Now X is given to be open, therefore $\exists\, \delta > 0$ such that $\mathscr{U}(y, \delta) \subset X$. If $|y_n - y| < \delta$ then $y_n \in \mathscr{U}(y, \delta)$ and therefore $y_n \in X$. Since $|y_n - y| \to 0$ as $n \to \infty$ there are values of n such that $|y_n - y| < \delta$, i.e. for some n, $y_n \in X$. But this is impossible as $y_n \in \mathscr{C}(X)$ for all n.

This contradiction shows that $y \in \mathscr{C}(X)$ and it follows that $\mathscr{C}(X)$ is closed.

Remark. A set need be neither open nor closed. 'Closed' is quite distinct from 'not open' and 'open' from 'not closed'.

Theorem 17. (i) *The union of a collection of open sets is an open set.*

(ii) *The intersection of a collection of closed sets is a closed set.*

(i) Let G_i, $i \in I$ be the collection of open sets. If $x \in \bigcup_{i \in I} G_i$ then $\exists\, i$ such that $x \in G_i$. Since G_i is open $\exists\, \delta > 0$ such that $\mathscr{U}(x, \delta) \subset G_i$. But then $\mathscr{U}(x, \delta) \subset \bigcup G_i$ and $x \in (\bigcup G_i)^0$. Since this is true for all x of $\bigcup G_i$ it follows that $\bigcup G_i$ is an open set.

(ii) Let F_i, $i \in I$ be the collection of closed sets. Then $\mathscr{C}(F_i)$ is open by theorem 16 and since $\bigcap_{i \in I} F_i = \mathscr{C}(\bigcup_{i \in I} \mathscr{C}(F_i))$ it follows that $\bigcap_{i \in I} F_i$ is closed (by (i) and theorem 16).

Remark. An argument such as that in (ii) above is referred to as an argument by complements.

Theorem 18. (i) *The intersection of two open sets is an open set.*

(ii) *The union of two closed sets is a closed set.*

(i) Let the open sets be G_1 and G_2. If $G_1 \cap G_2 = \phi$ then the intersection is open (since ϕ is open). Otherwise let $y \in G_1 \cap G_2$. Since $y \in G_1$ and G_1 is open $\exists\, \delta_1 > 0$ such that $\mathscr{U}(y, \delta_1) \subset G_1$. Similarly $\exists\, \delta_2 > 0$ such that $\mathscr{U}(y, \delta_2) \subset G_2$. Hence if

$$\delta = \min(\delta_1, \delta_2), \quad \mathscr{U}(y, \delta) \subset G_1 \cap G_2.$$

Thus $y \in (G_1 \cap G_2)^0$ and since this is true for every point of $G_1 \cap G_2$ it follows that $G_1 \cap G_2$ is an open set.

(ii) Let the closed sets be F_1, F_2 then

$$F_1 \cup F_2 = \mathscr{C}(\mathscr{C}(F_1) \cap \mathscr{C}(F_2))$$

is closed by (i) and theorem 16.

Theorem 19. *For any set* X, *the sets* (*a*) X', (*b*) $\bar{X}$ *and* (*c*) $\mathscr{F}r X$ *are all closed.*

(*a*) Let G denote the union of all those neighbourhoods $\mathscr{U}(y, \delta)$ such that $\mathscr{U}(y, \delta) \cap X$ is either ϕ or the single point y. As a union of open sets G is an open set. We shall show that $G = \mathscr{C}(X')$. First, $G \supset \mathscr{C}(X')$ for if $p \in \mathscr{C}(X')$ then there is no sequence $\{x_n\}$, with

$$x_n \in X, \quad |x_n - p| \to 0 \quad \text{as} \quad n \to \infty, \quad x_n \neq p$$

and thus for some $\delta > 0$, $\mathscr{U}(p, \delta) \cap X$ is either ϕ or p (for otherwise we could construct such a sequence $\{x_n\}$), therefore $p \in \mathscr{U}(p, \delta) \subset G$ and since this is true for every $p \in \mathscr{C}(X')$ we have $\mathscr{C}(X') \subset G$.

Next $G \subset \mathscr{C}(X')$ for if $p \in G$ then for some y and some $\delta > 0$, $p \in \mathscr{U}(y, \delta)$ where $\mathscr{U}(y, \delta) \cap X$ is y or ϕ. Thus if $\eta < \delta - |y - p|$, $\mathscr{U}(p, \eta) \cap X$ contains at most one point. Hence $p \notin X'$ and $p \in \mathscr{C}(X')$.

By theorem 16, $X' = \mathscr{C}(G)$ is a closed set.

(b) Next let H denote the union of all those neighbourhoods $\mathscr{U}(y, \delta)$ such that $\mathscr{U}(y, \delta) \cap X = \phi$. H is an open set and we show that $H = \mathscr{C}(\bar{X})$. $H \supset \mathscr{C}(\bar{X})$ for if $p \in \mathscr{C}(\bar{X}) \exists \delta > 0$ such that

$$\mathscr{U}(p, \delta) \cap X = \phi$$

for otherwise either $p \in X$ or $\exists \{x_n\}$ such that $x_n \in X$, $|x_n - p| \to 0$ as $n \to \infty$ and $x_n \neq p$, i.e. either $p \in X$ or $p \in X'$. This is not so. By the definition of H it follows that $p \in H$ and since this is true for every $p \in \mathscr{C}(\bar{X})$ we have $H \supset \mathscr{C}(\bar{X})$. Next $\mathscr{C}(\bar{X}) \supset H$ for if $p \in H$ then $p \in \mathscr{U}(y, \delta)$ for some y, $\delta > 0$ such that $\mathscr{U}(y, \delta) \cap X = \phi$, therefore $\mathscr{U}(p, \eta) \cap X = \phi$ where $0 < \eta < \delta - |p-y|$. Hence $p \notin X$ and $p \notin X'$ therefore $p \in \mathscr{C}(\bar{X})$.

By theorem 16 $\bar{X} = \mathscr{C}(H)$ is closed.

(c) $\mathscr{F}r X = \bar{X} \cap \overline{\mathscr{C}(X)}$ is closed by (b) and theorem 17 (ii).

Worked examples

(i) *Show that* (a) $\overline{X_1 \cap X_2} \subset \bar{X}_1 \cap \bar{X}_2$, (b) $(X_1 \cap X_2)^0 = X_1^0 \cap X_2^0$.

(a) If $\overline{X_1 \cap X_2} = \phi$ the result is trivial. Otherwise let $y \in \overline{X_1 \cap X_2}$ then either (i) $y \in X_1 \cap X_2$, or (ii) $y \in (X_1 \cap X_2)'$. In case (i) $y \in X_1$ and $y \in X_2$. But $\bar{X}_1 = X_1 \cup X_1'$ thus $y \in \bar{X}_1$ and similarly $y \in \bar{X}_2$. Hence $y \in \bar{X}_1 \cap \bar{X}_2$. In case (ii) $\exists$ a sequence of points $\{y_n\}$ such that

$$y_n \in X_1 \cap X_2, \quad |y_n - y| \to 0 \quad \text{as} \quad n \to \infty, \quad y_n \neq y.$$

Since $y_n \in X_1$ the existence of the above sequence shows that $y \in X_1'$ and since $X_1' \subset \bar{X}_1$ we have $y \in \bar{X}_1$. Similarly $y \in \bar{X}_2$. Thus $y \in \bar{X}_1 \cap \bar{X}_2$. The required result follows.

(b) Since $X_1 \cap X_2 \subset X_1$ we have $(X_1 \cap X_2)^0 \subset X_1^0$. Similarly $(X_1 \cap X_2)^0 \subset X_2^0$ and thus $(X_1 \cap X_2)^0 \subset X_1^0 \cap X_2^0$. If $X_1^0 \cap X_2^0 = \phi$ it follows that $(X_1 \cap X_2)^0 = \phi$ which gives the result in this case. Otherwise let y be any point of $X_1^0 \cap X_2^0$. Then $\exists \delta_1 > 0$, $\delta_2 > 0$ such that $\mathscr{U}(y, \delta_1) \subset X_1$, $\mathscr{U}(y, \delta_2) \subset X_2$. Write $\delta = \min(\delta_1, \delta_2)$ then $\mathscr{U}(y, \delta) \subset X_1 \cap X_2$. Thus $y \in (X_1 \cap X_2)^0$. Hence $X_1^0 \cap X_2^0 \subset (X_1 \cap X_2)^0$ and finally $(X_1 \cap X_2)^0 = X_1^0 \cap X_2^0$.

(ii) *X is a connected subset and Y a subset of* **S**. *Show that if $X \cap Y \neq \phi$, and $X \cap \mathscr{C}(Y) \neq \phi$ then $X \cap \mathscr{F}r(Y) \neq \phi$.*

Write $X_1 = X \cap Y$, $X_2 = X \cap \mathscr{C}(Y)$. Then $X = X_1 \cup X_2$, $X_1 \neq \phi$ and $X_2 \neq \phi$. Now

$$\bar{X}_1 \cap X_2 = (\overline{X \cap Y} \cap X \cap \mathscr{C}(Y)) \subset (\bar{Y} \cap X \cap \mathscr{C}(Y)) \subset (X \cap \mathscr{F}r(Y))$$

since $\overline{X \cap Y} \subset \bar{X} \cap \bar{Y}$. Similarly $X_1 \cap \bar{X}_2 \subset X \cap \mathscr{F}r(Y)$. Thus if $X \cap \mathscr{F}r(Y) = \phi$ we should have $\bar{X}_1 \cap X_2 = X_1 \cap \bar{X}_2 = \phi$ and X would not be connected. Hence $X \cap \mathscr{F}r(Y) \neq \phi$.

(iii) *Construct a subset X of R such that each set of the sequence $\{X_n\}$ defined by $X_1 = X$, $X_n = (X_{n-1})'$ is distinct.*

Write

$$A_1 = \left\{x \,|\, x = \frac{1}{2^l},\ l = 1, 2, \ldots\right\},$$

$$A_2 = \left\{x \,|\, x = 1 + \frac{1}{2^l} + \frac{1}{2^{l+m}},\ l, m = 1, 2, \ldots\right\}$$

and generally

$$A_r = \left\{x \,|\, x = r - 1 + \frac{1}{2^l} + \frac{1}{2^{l+m}} + \ldots + \frac{1}{2^{l+m+\ldots+p+q}},\quad l, m, \ldots, p, q = 1, 2, \ldots\right\}.$$

Then

$$A_1' = \{x \,|\, x = 0\},$$

$$A_2' = \left\{x \,|\, x = 1 \text{ or } x = 1 + \frac{1}{2^l},\ l = 1, 2, \ldots\right\}$$

and generally

$$A_r = \left\{x \,|\, x = r - 1 \text{ or } x = r - 1 + \frac{1}{2^l} \text{ or } x = r - 1 + \frac{1}{2^l} + \frac{1}{2^{l+m}} \text{ or } x = r - 1 + \frac{1}{2^l} + \frac{1}{2^{l+m}} + \ldots + \frac{1}{2^{l+m+\ldots+p}},\ l, m, \ldots, p = 1, 2, 3, \ldots\right\}.$$

The set $X = \bigcup_{r=1}^{\infty} A_r$ has the required properties.

(iv) *Show that X is a bounded connected subset of* **R** *only if it is a set of one of the forms $\{x|a \leqslant x \leqslant b\}$, $\{x|a < x \leqslant b\}$, $\{x|a \leqslant x < b\}$, $\{x|a < x < b\}$, $\{x|x = a\}$, ϕ.*

$\{x|x = a\}$ is certainly bounded and connected as is also ϕ. Let $\sup X = \lambda$, $\inf X = \mu$. Then if y satisfies $\mu < y < \lambda$ it follows that $y \in X$. For if $y \notin X$ write

$$X_1 = \{x|x \in X, x < y\}, \quad X_2 = \{x|x \in X, x > y\}.$$

Then $X = X_1 \cup X_2$, and as $\mu < y < \lambda$, $X_1 \neq \phi$, $X_2 \neq \phi$. Moreover, $\bar{X}_1 \subset \{x|x \leqslant y\}$ thus $\bar{X}_1 \cap X_2 = \phi$. Similarly $X_1 \cap \bar{X}_2 = \phi$. But then X is not connected. Thus we must have $y \in X$. Hence

$$\{x|\mu < x < \lambda\} \subset X \subset \{x|\mu \leqslant x \leqslant \lambda\}$$

and the various possibilities with a instead of μ and b instead of λ are enumerated above.

Exercises

1. X_1, X_2 are given subsets of S. Show that

(a) if $X_1 \subset X_2$ then $\bar{X}_1 \subset \bar{X}_2$, $X_1^0 \subset X_2^0$, (b) $\overline{X_1 \cup X_2} = \bar{X}_1 \cup \bar{X}_2$,

(c) $(X_1 \cup X_2)^0 \supset X_1^0 \cup X_2^0$, (d) $(X_1^0)^0 = X_1^0$,

(e) $\bar{X}_1 = \mathscr{F}r X_1 \cup (X_1)^0$,

(f) X_1 is closed if and only if $X_1' \subset X_1$,

(g) $(X_1')' \subset X_1'$, (h) $\mathscr{F}r(\mathscr{F}r(X_1)) \subset \mathscr{F}r(X_1)$.

2. Show that $\bar{X}$ is the intersection of all closed sets which contain X and that X^0 is the union of all open sets contained in X.

3. X is a set of real numbers bounded above. Show that the set of upper bounds of X form a closed set.

4. X is a closed set and $y \notin X$. Show that $\inf\limits_{x \in X} |x-y| > 0$.

5. Show that (a) X is open if and only if $X \cap \mathscr{F}r(X) = \phi$ and (b) X is closed if and only if $X \cap \mathscr{F}r(X) = \mathscr{F}r(X)$.

6. $\{X_i\}$ is a sequence of subsets of $\mathbf{S}$. Show that

$$\left(\bigcup_{i=1}^{\infty} X_i\right)' \supset \bigcup_{i=1}^{\infty} X_i', \quad \left(\bigcap_{i=1}^{\infty} X_i\right)' \subset \bigcap_{i=1}^{\infty} X_i'.$$

7. Show that $\mathbf{R}$ is connected.

8. Determine all the connected subsets of $\mathbf{R}$.

9. X_1, X_2 are connected subsets of $\mathbf{S}$ such that $X_1 \cap X_2 \neq \phi$. Show that $X_1 \cup X_2$ is connected.

10. X_1 is a set dense in X_2 and X_2 is dense in X_3. Show that X_1 is dense in X_3.

11. Show that the following sets are dense in $\mathbf{R}$:
(a) all rational numbers;
(b) all irrational numbers;
(c) all rational numbers of the form $p/2^q$, p, q integers (positive or negative).

13

THE BOLZANO–WEIERSTRASS, CANTOR AND HEINE–BOREL THEOREMS

These three theorems are of basic importance and very general forms of them have been established. We shall only consider the simplest cases.

We consider first subsets of **R**. A set of one of the forms

$$\{x|a \leqslant x \leqslant b\}, \quad \{x|a < x < b\}, \quad \{x|a \leqslant x < b\}, \quad \{x|a < x \leqslant b\}$$

is called an *interval*, the first is a *closed interval*, the second an *open interval* and the third and fourth are *half-open intervals*. In all cases a is called the *left-hand end-point* and b is the *right-hand end-point*. We shall use the notation $[a, b]$ to indicate a closed interval and (a, b) to indicate an open interval. Given a closed interval $[a, b]$ both closed intervals $[a, \frac{1}{2}(a+b)]$ and $[\frac{1}{2}(a+b), b]$ are called *half-intervals* of $[a, b]$.

Theorem 20. *If $\{I_n\}$ is a sequence of closed intervals such that I_{n+1} is a half-interval of I_n then $\bigcap_{n=1}^{\infty} I_n$ is a single point.*

Suppose that I_n is $[a_n, b_n]$. Then $a_n < b_n$, $\{a_n\}\uparrow$ and $\{b_n\}\downarrow$. Thus $a_n < b_n \leqslant b_1$. Hence the sequence $\{a_n\}$ is increasing and bounded above. Thus $\{a_n\}$ converges to a limit l (theorem 2, p. 24). Similarly $b_n \to m$ as $n \to \infty$ and

$$l - m = \lim_{n\to\infty} (a_n - b_n) = \lim_{n\to\infty} (a_1 - b_1)/2^n = 0.$$

Thus $l = m$. Moreover $\{a_n\}\uparrow$ and $\{b_n\}\downarrow$. Thus $a_n \leqslant l \leqslant b_n$, i.e. $l \in I_n$. Since this is true for every n, $\bigcap_{n=1}^{\infty} I_n$ contains the point l. Also given $\epsilon > 0\, \exists\, N$ such that for $n > N$ $b_n - a_n < \epsilon$. Thus

$$I_n \subset (l-\epsilon, l+\epsilon)$$

for $n > N$. Since $\bigcap_{n=1}^{\infty} I_n \subset I_n$ it follows that $\bigcap_{n=1}^{\infty} I_n \subset (l-\epsilon, l+\epsilon)$ and this is true for every $\epsilon > 0$. Thus $\bigcap_{n=1}^{\infty} I_n$ is the single point l.

Theorem 21. *The Bolzano–Weierstrass theorem. If X is a subset of* **R** *that is both bounded and contains infinitely many points then $X' \neq \phi$.*

Since X is bounded it is contained in a certain closed interval I_1. Let I_2 be a half-interval of I_1 such that $I_2 \cap X$ is an infinite set. Generally when I_n has been defined let I_{n+1} be a half-interval of I_n such that $I_{n+1} \cap X$ is an infinite set. It is clear that it is always possible to select one such half-interval of I_n and if both are possible selections we take that one which has the same left-hand end-point as I_n. By theorem 20 $\bigcap_{n=1}^{\infty} I_n$ is a single point say l. Since

$I_n \cap X$ is an infinite set we may select one point of it say x_n such that $x_n \neq l$. Then

$$x_n \in X, \quad |x_n - l| \leqslant |b_1 - a_1|/2^n \to 0 \quad \text{as} \quad n \to \infty \quad \text{and} \quad x_n \neq l.$$

Thus $l \in X'$ and the theorem is proved.

Remark. This argument is known as the subdivision argument. For any closed interval I for which $I \cap X$ is an infinite set we say that 'I has property P'. Then the argument depends upon the fact that if a closed interval has property P so does at least one of its half-intervals. We shall use this type of argument again with different properties P.

Definition 26. *A set is compact if and only if it is both bounded and closed.*

Definition 27. *A collection of open sets is called a covering of the set X if every point of X belongs to at least one member of the collection.*

Both the above definitions apply both to subsets of $\mathbf{R}$ and to subsets of $\mathbf{Z}$.

Theorem 22. *Cantor's theorem. If $\{F_n\}$ is a sequence of non-empty compact subsets of $\mathbf{R}$ and $F_n \supset F_{n+1}$ $(n = 1, 2, \ldots)$, then*

$$\bigcap_{n=1}^{\infty} F_n \neq \phi.$$

Write $a_n = \inf F_n$. Since F_n is closed $a_n \in F_n$ and since $F_n \supset F_{n+1}$, $a_n \leqslant a_{n+1}$. But $F_n \subset F_1$ and F_1 is bounded; thus $\{a_n\}$ is bounded above. Hence $\{a_n\}$ is an increasing sequence bounded above and as such converges to a limit l.

For $m \geqslant n$ $a_m \in F_m$ and $F_m \subset F_n$ thus $a_m \in F_n$. Consider the sequence $\{a_{n+N}\}$. We have $a_{n+N} \in F_N$ $(n = 1, 2, \ldots)$, $a_{n+N} \to l$ as $n \to \infty$ and F_N is closed. Thus $l \in F_N$. This is true for every positive integer N. Hence $l \in \bigcap_{n=1}^{\infty} F_n$ and the theorem is proved.

Theorem 23. *The Heine–Borel theorem. If X is a compact subset of $\mathbf{R}$ and $\mathscr{G}$ is a covering of X then we can select a finite number of members of $\mathscr{G}$ which form a covering of X.*

Denote by K the class of all closed intervals I such that $I \cap X$ can not be covered by any finite subcollection of $\mathscr{G}$. Let I_1 be a closed interval that contains X. The theorem will be proved if we can show that $I_1 \notin K$. We assume the contrary and obtain a contradiction.

A sequence of intervals $\{I_n\}$ is defined inductively. When I_n has been defined, $I_n \in K$, we select a half-interval of I_n which also belongs to K and denote it by I_{n+1}. It is clear that if $I_n \in K$ then at least one half-interval of I_n also belongs to K. If both half intervals of I_n belong to K we select that one with the same left-hand end-point as I_n.

By theorem 20 $\bigcap\limits_{n=1}^{\infty} I_n$ is a single point say l. Now $I_n \cap X \neq \phi$ for otherwise we should not have $I_n \in K$. X is closed and thus $l \in X$. Hence l is contained in at least one member of $\mathscr{G}$. Suppose that $G \in \mathscr{G}$ and $\mathscr{G} \ni l$. G is open and therefore $\exists\, \delta > 0$ such that $\mathscr{U}(l, \delta) \subset G$. If $I_n = [a_n, b_n]$ then $a_n \to l$, $b_n \to l$ as $n \to \infty$. Hence $\exists\, N$ such that $|a_n - l| < \delta$, $|b_n - l| < \delta$ for $n > N$. Thus, for example, $I_{N+1} \subset \mathscr{U}(l, \delta)$. But this gives the required contradiction. On the one hand by its definition $I_{N+1} \in K$. On the other hand

$$I_{N+1} \subset \mathscr{U}(l, \delta) \subset G.$$

Thus $I_{N+1} \cap X$ is covered by a finite subcollection of $\mathscr{G}$, namely, the subcollection consisting of the single member G. Hence $I_{N+1} \notin K$.

This contradiction shows that the original assumption that $I_1 \in K$ is false. Hence $I_1 \notin K$ and the theorem is proved.

We next consider subsets of $\mathbf{Z}$ and prove results analogous to theorems 20–23.

A set of the form $I = \{z \mid a \leqslant \mathscr{R}(z) \leqslant b,\ c \leqslant \mathscr{I}(z) \leqslant d\}$ is called a *closed planar interval* and is denoted by $[a, b; c, d]$. Any one of the four closed planar intervals

$$[a, \tfrac{1}{2}(a+b); c, \tfrac{1}{2}(c+d)], \quad [a, \tfrac{1}{2}(a+b); \tfrac{1}{2}(c+d), d],$$

$$[\tfrac{1}{2}(a+b), b; c, \tfrac{1}{2}(c+d)], \quad [\tfrac{1}{2}(a+b), b; \tfrac{1}{2}(c+d), d]$$

is called a *half-interval* of I. We shall also use the words *rectangle* and *half-rectangle* instead of *closed planar interval* and *half-interval.*

For any two real sets, X, Y we denote by $X \times Y$ the set of complex numbers whose real parts belong to X and whose imaginary parts belong to Y. $X \times Y$ is called the *Cartesian product* of X and Y. If either X or Y is empty then $X \times Y$ is taken to be empty.

Lemma. *For any two sequences of real sets* $\{X_n\}$, $\{Y_n\}$

$$\bigcap_{n=1}^{\infty} (X_n \times Y_n) = \left(\bigcap_{n=1}^{\infty} X_n\right) \times \left(\bigcap_{n=1}^{\infty} Y_n\right).$$

If z belongs to the left-hand side of this relation and $\mathscr{R}(z) = x$,

$\mathscr{I}(z) = y$, then $z \in X_n \times Y_n$ for $n = 1, 2, \ldots$, i.e. $x \in X_n$ $(n = 1, 2, \ldots)$, $y \in Y_n$ $(n = 1, 2, \ldots)$. Thus

$$x \in \bigcap_{n=1}^{\infty} X_n, \quad y \in \bigcap_{n=1}^{\infty} Y_n \quad \text{and} \quad z \in \left(\bigcap_{n=1}^{\infty} X_n\right) \times \left(\bigcap_{n=1}^{\infty} Y_n\right).$$

If $z \in \left(\bigcap_{n=1}^{\infty} X_n\right) \times \left(\bigcap_{n=1}^{\infty} Y_n\right)$, then $x \in X_n$, $y \in Y_n$ $(n = 1, 2, \ldots)$. Thus $z \in X_n \times Y_n$ $(n = 1, 2, \ldots)$ and finally $z \in \bigcap_{n=1}^{\infty} (X_n \times Y_n)$.

For the rectangle $[a, b; c, d]$ we have

$$[a, b; c, d] = [a, b] \times [c, d].$$

Theorem 20a. *If $\{I_n\}$ is a sequence of rectangles such that I_{n+1} is a half-rectangle of I_n then $\bigcap_{n=1}^{\infty} I_n$ is a single point.*

Write $I_n = J_n \times K_n$ where J_n, K_n are real intervals. By theorem 20 $\bigcap_{n=1}^{\infty} J_n$, $\bigcap_{n=1}^{\infty} K_n$ are single points, say l, m. By the lemma $\bigcap_{n=1}^{\infty} I_n$ is the single complex number whose real part is l and imaginary part m.

Theorem 21a. *The Bolzano–Weierstrass theorem. If X is a bounded infinite set of complex numbers then $X' \neq \phi$.*

The proof is the same as that of theorem 21 with intervals and half-intervals replaced by rectangles and half-rectangles. There are now four half-rectangles to choose from in defining I_{n+1}, but we can both choose at least one half-rectangle with the required property and if more than one half-rectangle has the required property we can lay down a rule as to which one we shall choose.

Theorem 22a. *Cantor's theorem. If $\{F_n\}$ is a sequence of non-empty compact sets of complex numbers and $F_n \supset F_{n+1}$ then $\bigcap_{n=1}^{\infty} F_n \neq \phi$.*

Choose a point $x_n \in F_n$ for $n = 1, 2, 3, \ldots$. If the same point say a happens to be chosen for infinitely many n then $a \in F_n$ for all n. (Since $F_n \supset F_m$ for $n < m$ if we had $a \notin F_N$ say then we should have $a \notin F_n$, $n > N$ and thus a would belong to at most a finite number of the F_n.) Thus $a \in \bigcap_{n=1}^{\infty} F_n$ and the theorem is proved. Otherwise there are infinitely many distinct points amongst the $x_1, x_2, \ldots$. Write $X = \bigcup_{n=1}^{\infty} x_n$. By theorem 21a $X' \neq \phi$. Let y be a point of X'

and $\{y_n\}$ a subsequence of $\{x_n\}$ such that $y_n \to l$ as $n \to \infty$. Now $y_n = x_m$ for some m and given $M \exists N$ such that if $n \geqslant N$ then $m \geqslant M$. Thus if $n \geqslant N$, $y_n = x_m \in F_m \subset F_M$ and it follows that F_M contains all of $\{y_n\}$ except at most a finite number of terms. Thus $l \in F'_M$ and as F_M is closed $l \in F_M$. This is true for every integer M. Hence $l \in \bigcap_{n=1}^{\infty} F_n$ and the theorem is proved.

Theorem 23a. *The Heine–Borel theorem. If X is a compact set of complex numbers and $\mathscr{G}$ is a covering of X then we can select a finite number of members of $\mathscr{G}$ which form a covering of X.*

The proof is the same as that of theorem 23 with intervals and half-intervals replaced by rectangles and half-rectangles respectively.

Worked example

Deduce Cantor's theorem from the Heine–Borel theorem.

We are given a sequence of non-empty compact sets $\{F_n\}$ such that $F_n \supset F_{n+1}$ and we have to show that $\bigcap_{n=1}^{\infty} F_n \neq \phi$. We shall assume that $\bigcap_{n=1}^{\infty} F_n = \phi$ and show that this assumption leads to a contradiction.

By theorem 16, p. 71, the sets $\mathscr{C}(F_n)$ $(n = 1, 2, \ldots)$ are open. Also

$$\bigcup_{n=1}^{\infty} \mathscr{C}(F_n) = \mathscr{C} \bigcap_{n=1}^{\infty} F_n = \mathscr{C}(\phi) = \mathbf{S},$$

where $\mathbf{S} = \mathbf{R}$ or $\mathbf{Z}$ as the case may be. Thus the sets $\mathscr{C}(F_n)$ $(n = 1, 2, \ldots)$ form a covering of $\mathbf{S}$ and therefore also of F_1 a subset of $\mathbf{S}$. By the Heine–Borel theorem $\exists$ a finite collection of the sets $\mathscr{C}(F_n)$ which covers F_1. Suppose that the sets $\mathscr{C}(F_{i_1})$, $\mathscr{C}(F_{i_2})$, ..., $\mathscr{C}(F_{i_N})$, where $i_1 < i_2 < \ldots < i_N$ cover F_1. Now $F_n \supset F_{n+1}$ thus

$$\mathscr{C}(F_n) \subset \mathscr{C}(F_{n+1}) \quad \text{and thus} \quad \mathscr{C}(F_{i_j}) \subset \mathscr{C}(F_{i_N}) \quad (j = 1, 2, \ldots, N).$$

Hence $\bigcup_{j=1}^{N} \mathscr{C}(F_{i_j}) = \mathscr{C}(F_{i_N})$ and since $\mathscr{C}(F_{i_j})$ $(j = 1, 2, \ldots N)$, cover F_1 we must have

$$\mathscr{C}(F_{i_N}) \supset F_1.$$

But $F_1 \supset F_{i_N}$ and thus $\mathscr{C}(F_{i_N}) \supset F_{i_N}$. Thus $\phi = \mathscr{C}(F_{i_N}) \cap F_{i_N} = F_{i_N}$. But we are given that $F_{i_N} \neq \phi$. This contradiction shows that the assumption $\bigcap_{n=1}^{\infty} F_n = \phi$ is false and thus $\bigcap_{n=1}^{\infty} F_n \neq \phi$.

Exercises

1. Deduce from the Bolzano–Weierstrass theorem that a real bounded monotonic sequence converges.

2. Deduce from Cantor's theorem that an enumerably infinite collection of open sets that covers a compact set X contains a finite subcollection also covering X.

3. Deduce the Bolzano–Weierstrass theorem from the Heine–Borel theorem.

4. F is a closed set of complex numbers and $x \notin F$. Show that

$$\inf_{y \in F} |x-y| > 0$$

and that $\exists\, z \in F$ such that $|x-z| = \inf_{y \in F} |x-y|$.

5. Give examples of sequences of real sets $\{X_n\}$ with the following properties:

(i) X_n open, $X_n \supset X_{n+1}$, $\bigcap_{n=1}^{\infty} X_n = \phi$.

(ii) X_n open, $\bigcup_{n=1}^{\infty} X_n \supset (0,1)$, no finite subcollection of X_n $(n = 1, 2, \ldots)$, covers $(0,1)$.

(iii) X_n closed $\bigcup_{n=1}^{\infty} X_n \supset [0,1]$, no finite subcollection of X_n $(n = 1, 2, \ldots)$, covers $[0,1]$.

(iv) X_n open, $\bigcup_{n=1}^{\infty} X_n \supset \{x | x \geqslant 0\}$, no finite subcollection of X_n $(n = 1, 2, \ldots)$, covers $\{x | x \geqslant 0\}$.

6. X is a compact set of complex numbers and $\mathfrak{f}$ is a real function defined over X such that for each $z \in X$ a number K and a number $\delta > 0$ can be found such that $|\mathfrak{f}(x)| < K$ if $x \in \mathscr{U}(z, \delta)$. Show that $\mathfrak{f}$ is bounded over X.

14

FUNCTIONS DEFINED OVER REAL OR COMPLEX NUMBERS

In discussing the behaviour of a function near a point, it is convenient to have available the idea of a neighbourhood $\mathscr{U}(y, \delta)$ of y with the point y itself excluded, i.e. a set such as $\mathscr{U}(y, \delta) \cap \mathscr{C}(y)$. Such a set is called a *punctured neighbourhood*.

We shall use these punctured neighbourhoods, first, to extend the idea of a limit already defined for certain types of functions, i.e. sequences.

Definition 28. *$\mathfrak{f}$ is defined in a punctured neighbourhood of $y \in \mathbf{S}$. If $\forall \epsilon > 0 \, \exists \delta > 0$ and l such that $|\mathfrak{f}(x) - l| < \epsilon$ for all x satisfying $0 < |x-y| < \delta$ then we say that $\mathfrak{f}(x) \to l$ as $x \to y$.*

We shall also write $l = \lim_{x \to y} \mathfrak{f}(x)$ or even $l = \lim \mathfrak{f}(x)$ if it is clear from the context which point y is involved, cf. definition 7*a*, chapter 4.

We shall also use $\mathscr{V}(y, \delta)$ for the punctured neighbourhood $\{x | 0 < |x-y| < \delta\}$.

Exercise. $\mathfrak{f}(x) \to l$ and $\mathfrak{g}(x) \to m$ as $x \to y$. Show that $\mathfrak{f}(x) + \mathfrak{g}(x) \to l + m$, $\mathfrak{f}(x) . \mathfrak{g}(x) \to l . m$ as $x \to y$. If further $l \neq 0$ show that $1/\mathfrak{f}(x) \to 1/l$ as $x \to y$.

Definition 29. *$\mathfrak{f}(x)$ is defined and real for $x \in \mathscr{V}(y, \delta)$.*

If $\forall X$, a real number, $\exists \delta > 0$ such that $\mathfrak{f}(x) > X$ for $x \in \mathscr{V}(y, \delta)$, then $\mathfrak{f}(x) \to \infty$ as $x \to y$ (or $\lim_{x \to y} \mathfrak{f}(x) = \infty$ or $\lim \mathfrak{f}(x) = \infty$).

If $\forall X$, a real number, $\exists \delta > 0$ such that $\mathfrak{f}(x) < X$ for $x \in \mathscr{V}(y, \delta)$, then $\mathfrak{f}(x) \to -\infty$ as $x \to y$ (or $\lim_{x \to y} \mathfrak{f}(x) = -\infty$ or $\lim \mathfrak{f}(x) = -\infty$).

Exercises. (1) $\mathfrak{f}$ and $\mathfrak{g}$ are real functions defined in a neighbourhood y; $\mathfrak{f}(x) \to \infty$ as $x \to y$, $\mathfrak{g}(x) \to l$ as $x \to y$ when $l \neq -\infty$. Show that $\mathfrak{f}(x) + \mathfrak{g}(x) \to \infty$ as $x \to y$. If $l > 0$ show that $\mathfrak{f}(x) . \mathfrak{g}(x) \to \infty$ as $x \to y$.

(2) $\mathfrak{f}$ is real and defined for $x \in \mathscr{V}(y, \delta)$ and $\mathfrak{f}(x) \to \infty$ as $x \to y$. Show that $1/\mathfrak{f}(x) \to 0$ as $x \to y$.

Definition 30. *$\mathfrak{f}(x)$ is defined and real for $x \in \mathscr{V}(y, \delta)$. If $\mathfrak{f}$ is bounded then the upper and lower limits as $x \to y$ are defined by*

$$\overline{\lim_{x \to y}} \mathfrak{f}(x) = \overline{\lim} \mathfrak{f}(x) = \inf_{\delta > 0} \left(\sup_{0 < |x-y| < \delta} \mathfrak{f}(x) \right),$$

$$\underline{\lim_{x \to y}} \mathfrak{f}(x) = \underline{\lim} \mathfrak{f}(x) = \sup_{\delta > 0} \left(\inf_{0 < |x-y| < \delta} \mathfrak{f}(x) \right).$$

If $\mathfrak{f}$ is unbounded above in every $\mathscr{V}(y, \delta)$, $0 < \delta < \delta_0$ then $\overline{\lim} \mathfrak{f}(x) = \infty$. If $\mathfrak{f}$ is unbounded below in every $\mathscr{V}(y, \delta)$, $0 < \delta < \delta_0$ then

$$\underline{\lim} \mathfrak{f}(x) = -\infty.$$

If $\lim \mathfrak{f}(x) = \infty$ then we write $\underline{\lim} \mathfrak{f}(x) = \infty$ and if $\lim \mathfrak{f}(x) = -\infty$ then we write $\underline{\lim} \mathfrak{f}(x) = -\infty$.

For example, if $\mathbf{S} = \mathbf{R}$ and $\mathfrak{f}(x)$ is defined by

$$\mathfrak{f}(x) = 1 \quad (x \text{ rational}), \qquad \mathfrak{f}(x) = 0 \quad (x \text{ irrational}),$$

then $\overline{\lim_{x \to y}} \mathfrak{f}(x) = 1$, $\underline{\lim_{x \to y}} \mathfrak{f}(x) = 0$ for every $y \in \mathbf{R}$. If $\mathfrak{f}(x) = 1/x$ for $x \neq 0$, $\mathfrak{f}(0) = 0$, then $\overline{\lim_{x \to 0}} \mathfrak{f}(x) = \infty$, $\underline{\lim_{x \to 0}} \mathfrak{f}(x) = -\infty$ and for $y \neq 0$

$$\underline{\lim_{x \to y}} \mathfrak{f}(x) = \overline{\lim_{x \to y}} \mathfrak{f}(x) = 1/y.$$

We shall also write sometimes, $\lim \mathfrak{f}$, $\underline{\lim}\, \mathfrak{f}$, $\overline{\lim}\, \mathfrak{f}$.

Exercises. In the following exercises $\mathfrak{f}$, $\mathfrak{g}$ are real bounded functions defined in a neighbourhood of $y \in \mathbf{S}$.

(1) Show that $\overline{\lim_{x \to y}} \mathfrak{f}(x) \geqslant \underline{\lim_{x \to y}} \mathfrak{f}(x)$.

(2) Show that as $x \to y$:

$$\overline{\lim}\, \mathfrak{f}(x) + \overline{\lim}\, \mathfrak{g}(x) \geqslant \overline{\lim}\, (\mathfrak{f}(x) + \mathfrak{g}(x)) \geqslant \overline{\lim}\, \mathfrak{f}(x) + \underline{\lim}\, \mathfrak{g}(x)$$
$$\geqslant \underline{\lim}\, (\mathfrak{f}(x) + \mathfrak{g}(x)).$$

(3) If $\mathfrak{f}(x) \geqslant \mathfrak{g}(x)$ all x concerned show that

$$\overline{\lim_{x \to y}} \mathfrak{f}(x) \geqslant \overline{\lim_{x \to y}} \mathfrak{g}(x), \quad \underline{\lim_{x \to y}} \mathfrak{f}(x) \geqslant \underline{\lim_{x \to y}} \mathfrak{g}(x).$$

(4) $\{x_n\}$ is a sequence of points such that $x_n \to y$ as $n \to \infty$. Show that

$$\overline{\lim_{x \to y}} \mathfrak{f}(x) \geqslant \overline{\lim_{n \to \infty}} \mathfrak{f}(x_n) \geqslant \underline{\lim_{n \to \infty}} \mathfrak{f}(x_n) \geqslant \underline{\lim_{x \to y}} \mathfrak{f}(x).$$

Definition 31. *$\mathfrak{f}$ is defined in a neighbourhood of the point $y \in \mathbf{S}$. We say that $\mathfrak{f}$ is continuous at y if $\forall \epsilon > 0 \, \exists \, \delta > 0$ such that*

$$|\mathfrak{f}(x) - \mathfrak{f}(y)| < \epsilon \quad \textit{for all} \quad x \in \mathscr{U}(y, \delta).$$

Equivalently $\mathfrak{f}$ is continuous at y if $\lim_{x \to y} \mathfrak{f}(x)$ exists and is equal to $\mathfrak{f}(y)$.

We have already met one class of continuous functions, namely, the sums of convergent power series inside their circles of convergence. If $\Sigma a_{n-1} z^{n-1}$ converges for $|z| < R$ and if y is such that $|y| < R$, then by (v) of chapter 11, p. 66

$$\sum_0^\infty a_n z^n = \sum_0^\infty d_n (z-y)^n \quad \text{if} \quad |z| + |z-y| < R$$

and by theorem 15, corollary 1, p. 62,

$$d_0 = \lim_{z \to y} \sum_0^\infty d_n (z-y)^n = \lim_{z \to y} \sum_0^\infty a_n z^n.$$

Since $d_0 = \sum_0^\infty a_n y^n$ (chapter 11, (v), p. 66) it follows that $\sum_0^\infty a_n z^n$ is continuous at $z = y$.

Exercises. (1) $\mathfrak{f}$, $\mathfrak{g}$ are continuous at $y \in \mathbf{S}$. Show that $\mathfrak{f}+\mathfrak{g}$ and $\mathfrak{f}.\mathfrak{g}$ are also continuous at y.

(2) $\mathfrak{f}$ is continuous at $y \in \mathbf{S}$ and $\mathfrak{f}(y) \neq 0$. Show that $1/\mathfrak{f}$ is continuous at y.

Definition 32. *$\mathfrak{f}$ is defined on a subset E of $\mathbf{S}$ and $y \in E$. We say that $\mathfrak{f}$ is continuous in E at y if $\forall \epsilon > 0 \exists \delta > 0$ such that $|\mathfrak{f}(x) - \mathfrak{f}(y)| < \epsilon$ for all $x \in E \cap \mathscr{U}(y, \delta)$.*

Exercise. $\mathfrak{f}$ is continuous in E_1 at y and in E_2 at y. Show that $\mathfrak{f}$ is continuous in $E_1 \cup E_2$ at y.

We say that $\mathfrak{f}$ is *continuous over the set E* if and only if it is continuous in E at each point of E.

In what follows it is assumed that the functions concerned are defined over appropriate neighbourhoods of the points at which their limiting behaviour is being considered.

Theorem 24. *$\mathfrak{f}(x) \to l$ as $x \to y$ if and only if for every sequence $\{x_n\}$ such that $x_n \neq y$, $x_n \to y$ as $n \to \infty$ we have $\mathfrak{f}(x_n) \to l$ as $n \to \infty$.*

We shall consider the case when $l \neq \infty$, $l \neq -\infty$. The proofs in the other cases are modifications of the one given here and are left to the reader.

If $\mathfrak{f}(x) \to l$ as $x \to y$ then $\forall \epsilon > 0 \exists \delta > 0$ such that $|\mathfrak{f}(x) - l| < \epsilon$ if $0 < |x - y| < \delta$. Since $x_n \to y$ as $n \to \infty \exists N$ such that $|x_n - y| < \delta$ for all $n > N$. Since $x_n \neq y$, $0 < |x_n - y|$ for all n. Combining these statements we have $|\mathfrak{f}(x_n) - l| < \epsilon$ for all $n > N$, i.e. $\mathfrak{f}(x_n) \to l$ as $n \to \infty$.

If '$\mathfrak{f}(x) \to l$ as $x \to y$' is false, then for some $\epsilon > 0$ and any $\delta > 0 \exists$ a particular x (depending on δ) such that $|\mathfrak{f}(x) - l| \geqslant \epsilon$ and $0 < |x - y| < \delta$. Let δ take the values $1/n$ $(n = 1, 2, \ldots)$ and x_n be the corresponding values of x. Then $|\mathfrak{f}(x_n) - l| \geqslant \epsilon$ and thus '$\mathfrak{f}(x_n) \to l$ as $n \to \infty$' is false, but $x_n \neq y$ and $x_n \to y$ as $n \to \infty$. Thus if $\mathfrak{f}(x_n) \to l$ as $n \to \infty$ for every sequence $\{x_n\}$ such that $x_n \neq y$, $x_n \to y$ as $n \to \infty$, then $\mathfrak{f}(x) \to l$ as $x \to y$.

Corollary 1. *$\mathfrak{f}(x)$ is continuous at y if and only if $\mathfrak{f}(x_n) \to \mathfrak{f}(y)$ for every sequence $\{x_n\}$ such that $x_n \to y$ as $n \to \infty$.*

This is an immediate consequence of the above argument.

Corollary 2. *It is sufficient to know that $\{\mathfrak{f}(x_n)\}$ converges for every $\{x_n\}$ such that $x_n \neq y$, $x_n \to y$ as $n \to \infty$ in order to conclude that $\mathfrak{f}(x) \to l$ as $x \to y$.*

For $\lim_{n\to\infty} \mathfrak{f}(x_n)$ must be independent of the particular sequence $\{x_n\}$. In fact if we had two different values of this limit for sequences $\{x_n^{(1)}\}$ and $\{x_n^{(2)}\}$, then for sequence $\{x_n\}$ defined by

$$x_{2n} = x_n^{(1)}, \quad x_{2n-1} = x_n^{(2)} \quad (n = 1, 2, \ldots)$$

$\{\mathfrak{f}(x_n)\}$ would not converge.

Exercise. Show that the necessary and sufficient condition that $\mathfrak{f}$ be continuous in E at y is that for every sequence $\{x_n\}$ such that $x_n \in E$, $x_n \neq y$, $x_n \to y$ as $n \to \infty$ we have $\mathfrak{f}(x_n) \to \mathfrak{f}(y)$ as $n \to \infty$.

Theorem 25. *$\mathfrak{f}$ is defined and continuous over a closed subset E of $\mathbf{S}$ and $\mathbf{r}(\mathfrak{f}) \subset \mathbf{S}_1$. If X is a closed subset of $\mathbf{S}_1$ then $\mathfrak{f}^{-1}(X)$ is closed.*

If $(\mathfrak{f}^{-1}(X))' \neq \phi$ let $y \in (\mathfrak{f}^{-1}(X))'$. Then $\exists \{x_n\}$ such that

$$x_n \in \mathfrak{f}^{-1}(X), \quad |x_n - y| \to 0 \quad \text{as} \quad n \to \infty, \quad x_n \neq y.$$

Since $x_n \in \mathfrak{f}^{-1}(X)$, $\mathfrak{f}(x_n) \in X$. Also $y \in E' \subset E$ since E is closed and thus $\mathfrak{f}(y)$ is defined. By continuity and theorem 24 $\mathfrak{f}(x_n) \to \mathfrak{f}(y)$ as $n \to \infty$. But $\mathfrak{f}(x_n) \in X$, X is closed and thus $\mathfrak{f}(y) \in X$. Hence $y \in \mathfrak{f}^{-1}(X)$.

Thus either $(\mathfrak{f}^{-1}(X))' = \phi$ or every point of $(\mathfrak{f}^{-1}(X))'$ belongs to $\mathfrak{f}^{-1}(X)$ and in either case $\mathfrak{f}^{-1}(X)$ is closed.

Corollary 1. *With the hypothesis of the theorem, if E_1 is a connected subset of E then $\mathfrak{f}(E_1)$ is connected.*

For if $\mathfrak{f}(E_1) = X_1 \cup X_2$ where $\bar{X}_1 \cap X_2 = X_1 \cap \bar{X}_2 = \phi$, then

$$\mathfrak{f}^{-1}(\bar{X}_1) \cap \mathfrak{f}^{-1}(X_2) = \mathfrak{f}^{-1}(X_1) \cap \mathfrak{f}^{-1}(\bar{X}_2) = \phi.$$

Now $\mathfrak{f}^{-1}(\bar{X}_1)$ and $\mathfrak{f}^{-1}(\bar{X}_2)$ are closed sets containing $\mathfrak{f}^{-1}(X_1)$ and $\mathfrak{f}^{-1}(X_2)$, respectively, and therefore containing also $\overline{\mathfrak{f}^{-1}(X_1)}$ and $\overline{\mathfrak{f}^{-1}(X_2)}$, respectively. Thus

$$E_1 = \mathfrak{f}^{-1}(X_1) \cup \mathfrak{f}^{-1}(X_2),$$

where $$\overline{\mathfrak{f}^{-1}(X_1)} \cap \mathfrak{f}^{-1}(X_2) = \mathfrak{f}^{-1}(X_1) \cap \overline{\mathfrak{f}^{-1}(X_2)} = \phi.$$

Now E_1 is connected; thus either $\mathfrak{f}^{-1}(X_1) = \phi$ or $\mathfrak{f}^{-1}(X_2) = \phi$. Hence either $X_1 = \phi$ or $X_2 = \phi$ and this implies that $\mathfrak{f}(E_1)$ is connected.

Corollary 2. *If $\mathfrak{f}$ is continuous over $\mathbf{S}$ and G is an open subset of the range of $\mathfrak{f}$ then $\mathfrak{f}^{-1}(G)$ is an open subset of $\mathbf{S}$.*

For $\mathfrak{f}^{-1}(G) = \mathscr{C}[\mathfrak{f}^{-1}\{\mathscr{C}(G)\}]$. By theorem 16 $\mathscr{C}(G)$ is closed: by theorem 25 $\mathfrak{f}^{-1}(\mathscr{C}(G))$ is a closed subset of S, and by theorem 16 again $\mathfrak{f}^{-1}(G)$ is open.

Remark. $\mathfrak{f}$, $\mathfrak{g}$ are both real or both complex and in either case continuous at $y \in \mathbf{S}$. Then $\mathfrak{f}+\mathfrak{g}$, $\mathfrak{f}.\mathfrak{g}$ and $1/\mathfrak{f}$ (if $\mathfrak{f}(y) \neq 0$) are continuous at y. This may be proved directly or by combining theorem 24 with the properties of limits of sequences contained in chapter 4.

Composite functions

Suppose that $\mathfrak{f}$ is a function defined over a neighbourhood of $x_0 \in \mathbf{S}$ and $\mathfrak{f}(x_0) = y_0$. Suppose further that $\mathfrak{g}(y)$ is a function defined over a neighbourhood of y_0 in $\mathbf{R}$ or $\mathbf{Z}$ as the case may be. If $\mathfrak{f}$ is continuous at x_0 and $\mathfrak{g}$ is continuous at y_0 then the composite function $\mathfrak{g}(\mathfrak{f}(x))$ is continuous at x_0.

Since $\mathfrak{g}$ is continuous at $y_0 \forall \epsilon > 0 \exists \delta > 0$ such that

$$|\mathfrak{g}(y)-\mathfrak{g}(y_0)| < \epsilon \quad \text{if} \quad |y-y_0| < \delta.$$

Since $\mathfrak{f}$ is continuous at $x_0 \forall \delta > 0$, $\exists \delta_1 > 0$ such that

$$|\mathfrak{f}(x)-\mathfrak{f}(x_0)| < \delta \quad \text{if} \quad |x-x_0| < \delta_1.$$

Thus combining these statements, if $|x-x_0| < \delta_1$, then

$$|\mathfrak{g}(\mathfrak{f}(x))-\mathfrak{g}(\mathfrak{f}(x_0))| < \epsilon$$

and this means that $\mathfrak{g}(\mathfrak{f}(x))$ is continuous at x_0.

Worked example

$\mathfrak{f}$ and $\mathfrak{g}$ are real and continuous over $E \subset \mathbf{S}$. Show that $\mathfrak{h} = \max(\mathfrak{f}, \mathfrak{g})$ and $\mathfrak{k} = \min(\mathfrak{f}, \mathfrak{g})$ are continuous over E.

Take $y \in E$. Since $\mathfrak{g}$ is continuous at $y \forall \epsilon > 0 \exists \delta_1 > 0$ such that $|\mathfrak{g}(x)-\mathfrak{g}(y)| < \epsilon$ if $x \in E$ and $|x-y| < \delta_1$, i.e.

$$\mathfrak{g}(y)-\epsilon < \mathfrak{g}(x) < \mathfrak{g}(y)+\epsilon \quad \text{if} \quad x \in E, \quad |x-y| < \delta_1.$$

Similarly $\exists \delta_2 > 0$ such that

$$\mathfrak{f}(y)-\epsilon < \mathfrak{f}(x) < \mathfrak{f}(y)+\epsilon \quad \text{if} \quad x \in E, \quad |x-y| < \delta_2.$$

Then if $x \in E$ and $|x-y| < \min(\delta_1, \delta_2)$ we have

$$\mathfrak{g}(x) < \mathfrak{h}(y)+\epsilon, \quad \mathfrak{f}(x) < \mathfrak{h}(y)+\epsilon.$$

Thus $\quad \mathfrak{h}(x) < \mathfrak{h}(y)+\epsilon \quad \text{if} \quad x \in E, \quad |x-y| < \min(\delta_1, \delta_2).$

Similarly $\quad \mathfrak{g}(y)-\epsilon < \mathfrak{h}(x), \quad \mathfrak{f}(y)-\epsilon < \mathfrak{h}(x)$

and thus

$$\mathfrak{h}(y)-\epsilon < \mathfrak{h}(x) \quad \text{if} \quad x \in E, \quad |x-y| < \min(\delta_1, \delta_2).$$

Hence $\quad |\mathfrak{h}(x)-\mathfrak{h}(y)| < \epsilon \quad \text{if} \quad |x-y| < \min(\delta_1, \delta_2).$

It follows that $\mathfrak{h}(x)$ is continuous in E at y. This is true for every $y \in E$ and thus $\mathfrak{h}(x)$ is continuous over E.

That $\mathfrak{k}(x)$ is continuous over E follows from the fact that

$$\mathfrak{k}(x) = -\max(-\mathfrak{f}(x), -\mathfrak{g}(x))$$

combined with the result proved above.

Exercises

1. $Y \subset \mathbf{S}$ and for $x \in \mathbf{S}$ $\mathfrak{f}(x) = \inf_{y \in Y} |x-y|$. Prove that $\mathfrak{f}$ is continuous.

2. $Y \subset \mathscr{U}(x_0, \delta) \subset \mathbf{S}$ and $\mathfrak{f}(x) = \sup_{y \in Y} |x-y|$. Prove that $\mathfrak{f}$ is continuous.

3. $\mathfrak{f}$ is real and defined over $\mathbf{S}$ and such that if G is an open real set then $\mathfrak{f}^{-1}(G)$ is open. Show that $\mathfrak{f}$ is continuous over $\mathbf{S}$.

4. $\mathfrak{f}$ is real and bounded in $\mathscr{U}(y, \eta)$ and functions $\mathfrak{g}$, $\mathfrak{h}$ are defined by

$$\mathfrak{g}(\delta) = \sup_{x \in \mathscr{U}(y, \delta)} \mathfrak{f}(x), \quad \mathfrak{h}(\delta) = \inf_{x \in \mathscr{U}(y, \delta)} \mathfrak{f}(x), \quad 0 < \delta < \eta.$$

Show that as $\delta \to 0$, $\delta > 0$

$$\mathfrak{g}(\delta) \to \mathfrak{f}(y) \quad \text{or} \quad \overline{\lim_{x \to y}}\, \mathfrak{f}(x), \qquad \mathfrak{h}(\delta) \to \mathfrak{f}(y) \quad \text{or} \quad \underline{\lim_{x \to y}}\, \mathfrak{f}(x).$$

5. $\mathfrak{f}$ is real and bounded in $\mathscr{U}(y, \eta)$ and $\mathfrak{f}(y) > \overline{\lim_{x \to y}}\, \mathfrak{f}(x)$. Show that $\exists\, \delta > 0$ such that if $x \in \mathscr{V}(y, \delta)$ then $\mathfrak{f}(x) < \mathfrak{f}(y)$.

Deduce that if X is the set of points y for which $\mathfrak{f}(y) > \overline{\lim_{x \to y}}\, \mathfrak{f}(x)$ then we can write $X = \bigcup_{n=1}^{\infty} X_n$ where $X'_n = \phi$ $(n = 1, 2, \ldots)$.

6. $\mathfrak{f}$, $\mathfrak{g}$ are real positive and bounded in $\mathscr{U}(y, \eta)$. Prove that

$$\underline{\lim}\, \mathfrak{f} . \underline{\lim}\, \mathfrak{g} \leqslant \underline{\lim}\, \mathfrak{f} . \mathfrak{g} \leqslant \underline{\lim}\, \mathfrak{f} . \overline{\lim}\, \mathfrak{g} \leqslant \overline{\lim}\, \mathfrak{f} . \mathfrak{g} \leqslant \overline{\lim}\, \mathfrak{f} . \overline{\lim}\, \mathfrak{g},$$

where the limits are taken as $x \to y$.

7. $\mathfrak{f}$ is bounded for $x \in \mathscr{U}(y, \delta)$. Show that $\exists\, E \subset \mathbf{S}$ such that $y \in E'$ and

$$\lim_{\substack{x \to y \\ x \in E}} \mathfrak{f}(x) = \underline{\lim_{x \to y}}\, \mathfrak{f}(x).$$

8. $\mathfrak{f}$, $\mathfrak{g}$ are real and bounded in $\mathscr{U}(y, \delta)$. $\mathfrak{g}$ is continuous at y. Show that

$$\overline{\lim_{x \to y}}\, (\mathfrak{f}(x) + \mathfrak{g}(x)) = \overline{\lim_{x \to y}}\, \mathfrak{f}(x) + \mathfrak{g}(y), \quad \underline{\lim_{x \to y}}\, (\mathfrak{f}(x) + \mathfrak{g}(x)) = \underline{\lim_{x \to y}}\, \mathfrak{f}(x) + \mathfrak{g}(y).$$

If further $\mathfrak{f}(x) > 0$, $\mathfrak{g}(y) > 0$ show that

$$\overline{\lim_{x \to y}}\,(\mathfrak{f}(x) \cdot \mathfrak{g}(x)) = (\overline{\lim_{x \to y}}\,\mathfrak{f}(x)) \cdot \mathfrak{g}(y), \quad \underline{\lim_{x \to y}}\,(\mathfrak{f}(x) \cdot \mathfrak{g}(x)) = (\underline{\lim_{x \to y}}\,\mathfrak{f}(x)) \cdot \mathfrak{g}(y).$$

9. $E \subset \mathbf{S}$ is such that for any continuous function $\mathfrak{f}$ defined over $\mathbf{S}$, the set $\mathfrak{f}(E)$ is closed. Show that E is closed.

15

FUNCTIONS OF A SINGLE REAL VARIABLE; LIMITS AND CONTINUITY

We shall examine the behaviour of functions defined over sets of real numbers in much more detail than those defined over complex sets. In what follows we consider such functions only unless the contrary is expressly stated. The functions may be often real- or complex-valued, where it is essential that they should be real-valued we shall say so.

Definition 33. *$\mathfrak{f}$ is defined over $(y, y+\eta)$ for some $\eta > 0$. If $\forall \epsilon > 0 \, \exists \delta > 0$ such that $|\mathfrak{f}(x) - l| < \epsilon$ for $x \in (y, y+\delta)$ then we say that $\mathfrak{f}(x)$ has the limit l on the right at y and write*

$$\mathfrak{f}(x) \to l \quad \textit{as} \quad x \to y+ \quad \textit{or} \quad \lim_{x \to y+} \mathfrak{f}(x) = l.$$

Similarly when $\mathfrak{f}$ is defined over $(y-\eta, y)$ for some $\eta > 0$ and $\forall \epsilon > 0 \, \exists \delta > 0$ such that $|\mathfrak{f}(x) - l| < \epsilon$ for $x \in (y-\delta, y)$, then we say that $\mathfrak{f}(x)$ has the limit l on the left at y and write

$$\mathfrak{f}(x) \to l \quad \textit{as} \quad x \to y- \quad \textit{or} \quad \lim_{x \to y-} \mathfrak{f}(x) = l.$$

Such limits are called one-sided limits.

Definition 34. *If $\forall \epsilon > 0 \, \exists X$ such that $|\mathfrak{f}(x) - l| < \epsilon$ for all x satisfying $x > X$, then we write $\mathfrak{f}(x) \to l$ as $x \to \infty$ or $\lim_{x \to \infty} \mathfrak{f}(x) = l$.*

Similarly we may define $\lim_{x \to -\infty} \mathfrak{f}(x) = l$.

$$\mathfrak{f}(x) \to l \quad \text{as} \quad x \to \infty \quad \text{if and only if} \quad \mathfrak{f}(-x) \to l \quad \text{as} \quad x \to -\infty.$$

Finally, if $\mathfrak{f}$ is real-valued we can define

$$\lim_{x \to y+} \mathfrak{f}(x) = \infty, \quad \lim_{x \to y-} \mathfrak{f}(x) = \infty, \quad \lim_{x \to y+} \mathfrak{f}(x) = -\infty, \quad \lim_{x \to y-} \mathfrak{f}(x) = -\infty,$$

$$\lim_{x \to \infty} \mathfrak{f}(x) = \infty, \quad \lim_{x \to \infty} \mathfrak{f}(x) = -\infty, \quad \lim_{x \to -\infty} \mathfrak{f}(x) = \infty \quad \lim_{x \to -\infty} \mathfrak{f}(x) = -\infty.$$

These definitions are left to the reader.

Exercises. It may be assumed that the functions concerned are defined over appropriate subsets of **R**.

(1) $\mathfrak{f}(x) \to l$ and $\mathfrak{g}(x) \to m$ as $x \to y+$. Show that

$$\mathfrak{f}(x)+\mathfrak{g}(x) \to l+m, \quad \mathfrak{f}(x).\mathfrak{g}(x) \to l.m \quad \text{as} \quad x \to y+$$

and if $l \neq 0$
$$1/\mathfrak{f}(x) \to 1/l \quad \text{as} \quad x \to y+.$$

(2) $\mathfrak{f}(x) \to l$ as $x \to y+$. Show that $\mathfrak{f}(-x) \to l$ as $x \to (-y)-$.

For each of the limiting operations defined above there is a result analogous to theorem 24. For example

Theorem 24*a*. *$\mathfrak{f}$ is defined over $\{x|x > a\}$. The necessary and sufficient condition that $\mathfrak{f}(x) \to l$ as $x \to \infty$ is that for any sequence $\{x_n\}$ for which $x_n \to \infty$ as $n \to \infty$ we have $\mathfrak{f}(x_n) \to l$ as $n \to \infty$.*

The proof is an appropriate modification of theorem 24. For each of the forms of convergence given above there is a criterion analogous to the general principle of convergence theorem 5, p. 34, which enables us to assert the existence of a finite limit even when we do not know what that limit is. The two most important cases are as follows.

Theorem 5*c*. *The necessary and sufficient condition that $\mathfrak{f}(x)$ tends to a finite limit as $x \to \infty$ is that $\forall \epsilon > 0 \,\exists X$ such that*

$$|\mathfrak{f}(x_1)-\mathfrak{f}(x_2)| < \epsilon \quad \textit{if} \quad x_1 > X \quad \textit{and} \quad x_2 > X.$$

Necessity. If $\mathfrak{f}(x) \to l$ as $x \to \infty$ then $\forall \epsilon > 0 \,\exists X$ such that $|\mathfrak{f}(x)-l| < \frac{1}{2}\epsilon$ for $x > X$. Thus $|\mathfrak{f}(x_1)-\mathfrak{f}(x_2)| < \epsilon$ if $x_1 > X$, $x_2 > X$ by the modulus inequality and this is the required condition.

Sufficiency. If the condition is applied to a particular sequence $\{x_n\}$, where $x_n \to \infty$ as $n \to \infty$, then we conclude from theorem 5 that $\{\mathfrak{f}(x_n)\}$ converges. By theorem 24*a*, together with the analogue of corollary 2, theorem 24, it follows that $\mathfrak{f}(x)$ tends to a finite limit as $x \to \infty$.

Theorem 5*d*. *The necessary and sufficient condition that $\mathfrak{f}(x)$ tends to a finite limit as $x \to a$ is that $\forall \epsilon > 0 \,\exists \delta > 0$ such that $|\mathfrak{f}(x_1)-\mathfrak{f}(x_2)| < \epsilon$ if $0 < |x_1-a| < \delta$ and $0 < |x_2-a| < \delta$.*

The proof is similar to theorem 5*c*, based on an appropriate analogue of theorem 24. It is left to the reader.

Continuity is defined as in chapter 14. An important special case is continuity over an interval $[a, b]$.

Definition 35. *If $\lim_{x \to y} \mathfrak{f}(x) = \mathfrak{f}(y)$ for $a < y < b$, $\lim_{x \to a+} \mathfrak{f}(x) = \mathfrak{f}(a)$ and $\lim_{x \to b-} \mathfrak{f}(x) = \mathfrak{f}(b)$, then we say that $\mathfrak{f}$ is continuous over $[a, b]$.*

If $\lim_{x \to a+} \mathfrak{f}(x) = \mathfrak{f}(a)$ we say that $\mathfrak{f}$ is *continuous on the right at* a. Similarly we can define *continuous on the left at* b.

When $\mathfrak{f}$ is real we can extend the use of $\overline{\lim}$ and $\underline{\lim}$ as follows.

Definition 36.

$$\overline{\lim_{x\to\infty}}\, \mathfrak{f}(x) = \inf_X (\sup_{x>X} \mathfrak{f}(x)), \qquad \underline{\lim_{x\to\infty}}\, \mathfrak{f}(x) = \sup_X (\inf_{x>X} \mathfrak{f}(x));$$

$$\overline{\lim_{x\to-\infty}}\, \mathfrak{f}(x) = \inf_X (\sup_{x<X} \mathfrak{f}(x)), \qquad \underline{\lim_{x\to-\infty}}\, \mathfrak{f}(x) = \sup_X (\inf_{x<X} \mathfrak{f}(x));$$

$$\overline{\lim_{x\to y+}}\, \mathfrak{f}(x) = \inf_{\delta>0} (\sup_{y<x<y+\delta} \mathfrak{f}(x)), \qquad \underline{\lim_{x\to y+}}\, \mathfrak{f}(x) = \sup_{\delta>0} (\inf_{y<x<y+\delta} \mathfrak{f}(x));$$

etc.

In this case we also define semi-continuity.

Definition 37. $\mathfrak{f}$ *defined in a neighbourhood of* y *is upper-semi-continuous at* y *if* $\overline{\lim_{x\to y}}\, \mathfrak{f}(x) \leqslant \mathfrak{f}(y)$ *and is lower semi-continuous at* y *if* $\underline{\lim_{x\to y}}\, \mathfrak{f}(x) \geqslant \mathfrak{f}(y)$.

The fundamental properties of a real continuous function of a real variable

Theorem 26 *If* $\mathfrak{f}$ *is real and continuous over* $[a, b]$ *then*

(i) $\mathfrak{f}$ *is bounded over* $[a, b]$;

(ii) $\mathfrak{f}$ *attains its bounds*;

(iii) $\mathfrak{f}$ *is uniformly continuous over* $[a, b]$, *i.e.* $\forall \epsilon > 0 \, \exists \delta > 0$ *such that* $|\mathfrak{f}(x) - \mathfrak{f}(y)| < \epsilon$ *if* $|x-y| < \delta$ *and* $x, y \in [a, b]$.

(i) Write $I_1 = [a, b]$. We assume that $\mathfrak{f}$ is unbounded over I_1 and show that this leads to a contradiction.

When I_n has been defined so that $\mathfrak{f}$ is unbounded over I_n, let I_{n+1} be a half-interval of I_n over which $\mathfrak{f}$ is unbounded. (There must be at least one such half-interval or $\mathfrak{f}$ would be bounded over I_n.) If both half-intervals have this property, choose that with the same left-hand end-point as I_n. By theorem 20 $\bigcap_{n=1}^{\infty} I_n$ is a single point, say l. Now $l \in [a, b]$ and $\mathfrak{f}$ is continuous at l (on the right or on the left of l if l happens to be a or b). Thus $\forall \epsilon > 0 \, \exists \delta > 0$ such that $|\mathfrak{f}(x) - \mathfrak{f}(l)| < \epsilon$ for x satisfying $|x-l| < \delta$ and $a \leqslant x \leqslant b$. Hence $|\mathfrak{f}(x)| < |\mathfrak{f}(l)| + \epsilon$ if $|x-l| < \delta$ and $a \leqslant x \leqslant b$. But for n sufficiently large $I_n \subset (l-\delta, l+\delta)$ and since $I_n \subset [a, b]$ for all n, it follows that $|\mathfrak{f}(x)| < |\mathfrak{f}(l)| + \epsilon$ for $x \in I_n$ and n sufficiently large. This contradicts the selection of I_n as an interval over which $\mathfrak{f}$ is unbounded. Thus the assumption that $\mathfrak{f}$ is unbounded over I_1 leads to a contradiction. This assumption is therefore false and (i) has been proved.

(ii) Write $M = \sup \mathfrak{f}(x)$, $a \leqslant x \leqslant b$. If $\mathfrak{f}(x) < M$ for all $x \in [a,b]$, then $\mathfrak{g}(x) = 1/(M - \mathfrak{f}(x))$ is continuous over $[a,b]$ and therefore by (i) is bounded over $[a,b]$. Hence $\exists\, N$ such that

$$1/(M - \mathfrak{f}(x)) < N \quad (a \leqslant x \leqslant b),$$

i.e.
$$1 < NM - N\mathfrak{f}(x) \quad (a \leqslant x \leqslant b),$$

i.e.
$$\mathfrak{f}(x) < M - 1/N \quad (a \leqslant x \leqslant b)$$

(since $M - \mathfrak{f}(x) > 0$ and therefore $N > 0$). But this is impossible as M is the *least* upper bound of $\mathfrak{f}$ and $M - 1/N$, less than M, cannot be an upper bound of $\mathfrak{f}$. The assumption that $\mathfrak{f}(x) < M$ for all $x \in [a,b]$ is thus false and hence for some $x \in [a,b]$, $\mathfrak{f}(x) = M$.

Similarly $\mathfrak{f}(x)$ attains its greatest lower bound in $[a,b]$.

(iii) If $\mathfrak{f}$ is not uniformly continuous over $[a,b]$, then for some $\epsilon > 0\,\exists$ sequences $\{x_n\}$, $\{y_n\}$ such that $|x_n - y_n| \to 0$,

$$|\mathfrak{f}(x_n) - \mathfrak{f}(y_n)| > \epsilon, \quad \text{where} \quad x_n, y_n \in [a,b].$$

By the Bolzano–Weierstrass theorem, p. 76, $\exists$ a convergent subsequence of $\{x_n\}$, say $\{x_{n_i}\}$. If $x_{n_i} \to l$ as $i \to \infty$ then $y_{n_i} \to l$ as $i \to \infty$. Also $a \leqslant l \leqslant b$ and by the continuity of $\mathfrak{f}$ at l $\mathfrak{f}(y_{n_i}) \to \mathfrak{f}(l)$, $\mathfrak{f}(x_{n_i}) \to \mathfrak{f}(l)$ as $i \to \infty$. In particular for sufficiently large i $|\mathfrak{f}(y_{n_i}) - \mathfrak{f}(x_{n_i})| < \epsilon$. This contradicts the definition of $\{x_n\}$, $\{y_n\}$. Thus the assumption that $\mathfrak{f}$ is not uniformly continuous leads to a contradiction. Hence $\mathfrak{f}$ is uniformly continuous over $[a,b]$.

Corollary. *A continuous function of a real or complex variable defined over a closed bounded set E is uniformly continuous over E.*

The argument is exactly as in (iii). The sequences $\{x_n\}$, $\{y_n\}$, and the limit l all belong to E. We use theorem 21, p. 76 or 21a, p. 79.

Remark. The result (iii) above is more important than it seems to be. If we can prove that a certain function is continuous in $[a,b]$ at each point of $[a,b]$, then in later applications we can use the fact of uniform continuity. We could of course attempt to establish uniform continuity directly. But this is much more difficult because we have to consider $\mathfrak{f}(x) - \mathfrak{f}(y)$, a function of two independent variables x and y, instead of $\mathfrak{f}(x) - \mathfrak{f}(y)$, where y is kept fixed and only x varies, i.e. a function of one variable. It is the case that functions of two variables are much more difficult to handle than functions of one variable.

The class of continuous functions has appropriate algebraic properties, that is to say if it contains $\mathfrak{f}$ and $\mathfrak{g}$, then it contains $-\mathfrak{f}$, $\mathfrak{f} + \mathfrak{g}$, $\mathfrak{f} \cdot \mathfrak{g}$ and $1/\mathfrak{f}$ if

$\mathfrak{f}(x) \neq 0$ any x. A more natural and intuitive idea than continuity is that of Darboux continuity defined below. However, Darboux continuity is little more than a curiosity largely because the class of Darboux continuous functions does not have suitable algebraic properties.

Exercises. (1) Show that $1/x$ is continuous but not uniformly continuous over $0 < x \leqslant 1$.

(2) Show that $1/x$ is uniformly continuous over $\{x|x > 1\}$.

(3) Show that if $\mathfrak{f}$ is uniformly continuous over the bounded set E then $\mathfrak{f}$ is bounded over E.

Darboux continuity

If $\mathfrak{f}$ is real and defined over $[a, b]$ and for each subinterval $I = [c, d]$ of $[a, b]$, $\mathfrak{f}(I) \supset \{x|\mathfrak{f}(c) \leqslant x \leqslant \mathfrak{f}(d)$ or $\mathfrak{f}(d) \leqslant x \leqslant \mathfrak{f}(c)\}$, then $\mathfrak{f}$ is said to be *Darboux continuous* over $[a, b]$. Thus a Darboux continuous function takes every value between any two which it takes. It follows from the corollary 1 to theorem 25, p. 85, that a continuous function is Darboux continuous. An alternative proof is given later in worked example (i).

The O, o, $\sim$ notation

Definition of O. If $\mathfrak{f}$, $\mathfrak{g}$ are two functions of x such that $\exists\, x_0$, K and

$$|\mathfrak{f}(x)| < K\,|\mathfrak{g}(x)| \quad \text{for all} \quad x \geqslant x_0,$$

then we write

$$\mathfrak{f} = O(|\mathfrak{g}|) \quad \text{as} \quad x \to \infty \quad \text{or simply} \quad \mathfrak{f} = O(|\mathfrak{g}|),$$

if the fact that $x \to \infty$ is clear from the context. Similarly $\mathfrak{f} = O\,|\mathfrak{g}|$ as $x \to -\infty$ if $\exists\, x_0$, K and $|\mathfrak{f}(x)| < K\,|\mathfrak{g}(x)|$ all $x \leqslant x_0$; $\mathfrak{f} = O\,|\mathfrak{g}|$ as $x \to a$ if $\exists\, \delta > 0$, K and $|\mathfrak{f}(x)| < K\,|\mathfrak{g}(x)|$ all x satisfying $0 < |x-a| < \delta$.

Definition of o. If $\mathfrak{f}$, $\mathfrak{g}$ are such that $\mathfrak{f}(x)/\mathfrak{g}(x) \to 0$ as $x \to \infty$, then we write $\mathfrak{f} = o(\mathfrak{g})$ as $x \to \infty$ or perhaps simply $\mathfrak{f} = o(\mathfrak{g})$. $\mathfrak{f} = o(\mathfrak{g})$ as $x \to -\infty$ or as $x \to a$ are also defined in the obvious manner.

Definition of $\sim$. If $\mathfrak{f}$, $\mathfrak{g}$ are such that $\mathfrak{f}(x)/\mathfrak{g}(x) \to 1$ as $x \to \infty$, then we write $\mathfrak{f} \sim \mathfrak{g}$ as $x \to \infty$ or perhaps $\mathfrak{f} \sim \mathfrak{g}$. $\mathfrak{f} \sim \mathfrak{g}$ as $x \to -\infty$ or as $x \to a$ are also defined similarly.

For example

$$7x^2 + 10x = O(x^2) \quad \text{as} \quad x \to \infty;$$

$$7x^2 + 10x = o(x^3) \quad \text{as} \quad x \to \infty;$$

$$7x^2 + 10x \sim 7x^2 \quad \text{as} \quad x \to \infty;$$

$$7x^2 + 10x \sim 10x \quad \text{as} \quad x \to 0, \quad 7x^2 + 10x = O(x) \quad \text{as} \quad x \to 0.$$

Note that $\mathfrak{f}-\mathfrak{g}\to 0$ as $x\to\infty$ does not imply nor is it implied by $\mathfrak{f}\sim\mathfrak{g}$ as $x\to\infty$ (cf. exercises 15, 16, p. 22). For example $1/x-1/x^2\to 0$ as $x\to\infty$, but it is not true that $1/x\sim 1/x^2$. Again $x+x^2\sim x^2$ as $x\to\infty$, but it is not true that $(x+x^2)-x^2\to 0$ as $x\to\infty$.

Exercise. Verify the following relations between O, o, $\sim$ as $x\to\infty$ where $\mathfrak{g}(x)\geqslant 0$.

(*a*) If $\mathfrak{f}=o(\mathfrak{g})$ then $\mathfrak{f}=O(\mathfrak{g})$.
(*b*) If $\mathfrak{f}\sim\mathfrak{g}$ then $\mathfrak{f}=O(\mathfrak{g})$.
(*c*) If $\mathfrak{f}=O(\mathfrak{g})$ and $\mathfrak{g}=O(\mathfrak{h})$ where $\mathfrak{h}(x)\geqslant 0$ then $\mathfrak{f}=O(\mathfrak{h})$.
(*d*) If $\mathfrak{f}=O(\mathfrak{g})$ and $\mathfrak{g}=o(\mathfrak{h})$ then $\mathfrak{f}=o(\mathfrak{h})$.
(*e*) If $\mathfrak{f}\sim\mathfrak{g}$ and $\mathfrak{g}\sim\mathfrak{h}$ then $\mathfrak{f}\sim\mathfrak{h}$.

Remark. We often use $\mathfrak{f}=O(1)$ to mean that for some x_0 and all $x\geqslant x_0$ $\mathfrak{f}(x)$ is bounded. Similarly $\mathfrak{f}=o(1)$ means that $\mathfrak{f}(x)\to 0$ as $x\to\infty$. The '1' in these symbols stands for the constant function whose value is 1. One sometimes sees such phrases as '$\mathfrak{f}(x)=O(1)$ in $a\leqslant x\leqslant b$' meaning simply that $\mathfrak{f}$ is bounded in $[a,b]$, but it is better to avoid this indiscriminate use of the symbols and to retain them exclusively for the case when x is tending to some limit.

Worked examples

(i) *If* $\mathfrak{f}$ *is continuous over* $[a,b]$ *then it is Darboux continuous over* $[a,b]$.

Suppose that $a\leqslant x_0<x_1\leqslant b$ and $\mathfrak{f}(x_0)=\alpha$, $\mathfrak{f}(x_1)=\beta$. Let γ be a number between α and β. For definiteness assume that $\alpha<\beta$. The argument in the case $\alpha>\beta$ is similar and will be omitted. Let X be the subset of $[x_0,x_1]$ defined by $X=\{x\,|\,x_0\leqslant x\leqslant x_1,\ \mathfrak{f}(x)\leqslant\gamma\}$ and write $x_2=\sup X$. There exists a sequence of points $\{y_n\}$ such that $y_n\in X$, $y_n\to x_2$ as $n\to\infty$. By the continuity of $\mathfrak{f}$ at x_2 $\mathfrak{f}(y_n)\to\mathfrak{f}(x_2)$. Since $\mathfrak{f}(y_n)\leqslant\gamma$ it follows that $\mathfrak{f}(x_2)\leqslant\gamma$. Next $\mathfrak{f}(x_1)=\beta>\gamma$. Thus $x_1\neq x_2$ and we must have $x_1>x_2$. Then $\exists$ a sequence $\{z_n\}$ such that $x_2<z_n<x_1$ and $z_n\to x_2$ as $n\to\infty$. Then since $z_n>\sup X$ and $x_0\leqslant z_n\leqslant x_1$ we must have $\mathfrak{f}(z_n)>\gamma$. But $z_n\to x_2$ and by continuity of $\mathfrak{f}$ at x_2 $\mathfrak{f}(x_2)\geqslant\gamma$. Hence $\mathfrak{f}(x_2)=\gamma$.

It follows that $\mathfrak{f}$ is Darboux continuous over $[a,b]$.

(ii) $\mathfrak{f}$ *is a complex-valued function of the real variable* x *defined and continuous over* $[a,b]$. *If the set of values of* $\mathfrak{f}$ *is at most enumerable show that* $\mathfrak{f}$ *is a constant.*

Suppose that $\mathfrak{f}$ is not constant. Write

$$\mathfrak{u}(x)=\mathscr{R}(\mathfrak{f}(x)),\quad \mathfrak{v}(x)=\mathscr{I}(\mathfrak{f}(x)),$$

then at least one of $\mathfrak{u}$, $\mathfrak{v}$ is not constant. Suppose that it is $\mathfrak{u}$ and that $\mathfrak{u}(x_0)\neq\mathfrak{u}(x_1)$. Now $\mathfrak{u}$ is continuous and therefore by (i) is Darboux continuous. Thus the range of $\mathfrak{u}$ contains the interval with end-points $\mathfrak{u}(x_0)$, $\mathfrak{u}(x_1)$. Corresponding to each point of this interval say y, select a point x of $[x_0, x_1]$ such that $\mathfrak{u}(x) = y$. Then the corresponding values of $\mathfrak{f}(x)$ are all distinct and since there is one corresponding to each point of the interval $[x_0, x_1]$ they form a non-enumerable set (cf. chapter 3, p. 14).

(iii) $\mathfrak{f}$ *is real bounded and defined over* $[a, b]$. *Show that the set of points* $\{x \mid \mathfrak{f}(x) > \overline{\lim_{y \to x}} \mathfrak{f}(y)\}$ *is enumerable.*

Let $X_n = \{x \mid \mathfrak{f}(x) > \mathfrak{f}(y)$ for all y satisfying $|y - x| < 1/n\}$. If $x_1 \in X_n$ and $x_2 \in X_n$, then $|x_1 - x_2| \geqslant 1/n$, thus the number of points in X_n is not more than $n(b-a)+1$. Hence X_n is a finite set and $\bigcup_{n=1}^{\infty} X_n$ is an enumerable set. This last set contains all the points at which $\mathfrak{f}(x) > \overline{\lim_{y \to x}} \mathfrak{f}(y)$. Thus the required property has been established.

(iv) *If* $\mathfrak{f}$, $\mathfrak{g}$ *are real bounded functions defined in a neighbourhood of* a, *then*

$$\underline{\lim_{x \to a}} (\mathfrak{f}(x) + \mathfrak{g}(x)) \leqslant \underline{\lim_{x \to a}} \mathfrak{f}(x) + \overline{\lim_{x \to a}} \mathfrak{g}(x)$$

$\forall \epsilon > 0 \, \exists \, \delta > 0$ such that

$$\mathfrak{f}(x) + \mathfrak{g}(x) > \underline{\lim_{x \to a}} (\mathfrak{f}(x) + \mathfrak{g}(x)) - \epsilon \quad \text{if } x \text{ satisfies } 0 < |x - a| < \delta.$$

Further $\exists \, \delta_1 > 0$ such that

$$\mathfrak{g}(x) < \overline{\lim_{x \to a}} \mathfrak{g}(x) + \epsilon \quad \text{if } x \text{ satisfies } 0 < |x - a| < \delta_1.$$

Moreover, $\exists \, x_0$ with $0 < |x_0 - a| < \min(\delta, \delta_1)$ and such that

$$\mathfrak{f}(x_0) < \underline{\lim_{x \to a}} \mathfrak{f}(x) + \epsilon.$$

Combining these inequalities we have

$$\underline{\lim_{x \to a}} (\mathfrak{f}(x) + \mathfrak{g}(x)) - \epsilon < \mathfrak{f}(x_0) + \mathfrak{g}(x_0) < \overline{\lim_{x \to a}} \mathfrak{g}(x) + \epsilon + \underline{\lim_{x \to a}} \mathfrak{f}(x) + \epsilon.$$

But ϵ is any positive number. Hence

$$\underline{\lim_{x \to a}} (\mathfrak{f}(x) + \mathfrak{g}(x)) \leqslant \underline{\lim_{x \to a}} \mathfrak{f}(x) + \overline{\lim_{x \to a}} \mathfrak{g}(x).$$

Exercises

1. For each real number x let $\mathfrak{N}(x)$ be the integer nearest to x (the least such integer if there are two). $\mathfrak{f}$ is defined by

$$\mathfrak{f}(0) = 0, \quad \mathfrak{f}(x) = |1/x - \mathfrak{N}(1/x)| \quad (x > 0).$$

Prove that

(a) $\mathfrak{f}$ is discontinuous on the right at $x = 0$;

(b) if $\mathfrak{h}(x)$ is continuous at $x = 0$, then $\mathfrak{h}(x).\mathfrak{f}(x) \to 0$ as $x \to 0+$ if and only if $\mathfrak{h}(0) = 0$.

2. (a) $\mathfrak{f}$ if defined by $\mathfrak{f}(0) = 0$,

$$\mathfrak{f}(x) = \frac{(|x| - x - x^2)}{2x} \qquad (x \neq 0, x \text{ real}).$$

For what values of x are $\mathfrak{f}$, $x.\mathfrak{f}(x)$, $\mathfrak{f}(x^2)$ continuous?

(b) $\mathfrak{g}$ is defined by

$$\mathfrak{g}(x) = \lim_{n\to\infty} \frac{x^{2n+1} + x^2}{x^{2n} + 1} \qquad (x \text{ real}).$$

Is $\mathfrak{g}$ continuous at $x = +1$ or at $x = -1$?

3. $\mathfrak{f}$ is real and bounded in a neighbourhood of $x = a$. Show that $\exists$ real sequences $\{x_n\}$, $\{y_n\}$ with $x_n \to a$, $y_n \to a$ as $n \to \infty$ such that

$$\lim_{n\to\infty} \mathfrak{f}(x_n) = \overline{\lim_{x\to a}}\, \mathfrak{f}(x), \quad \lim_{n\to\infty} \mathfrak{f}(y_n) = \underline{\lim_{x\to a}}\, \mathfrak{f}(x).$$

4. A sequence of real functions $\{\mathfrak{f}_n(x)\}$ satisfies

(a) $\exists$ a number A such that $|\mathfrak{f}_n(x)| \leqslant A$ $(n = 1, 2, \ldots,$ all $x)$;

(b) $\forall \epsilon > 0 \, \exists \delta$ such that $|\mathfrak{f}_n(x) - \mathfrak{f}_n(y)| < \epsilon$ $(n = 1, 2, \ldots)$ provided that $|x - y| < \delta$ and δ is independent of n. Show that $\sup_n \mathfrak{f}_n(x)$ is continuous.

5. $\mathfrak{f}$ is real, continuous for all x and such that $\mathfrak{f}(x + y) = \mathfrak{f}(x) + \mathfrak{f}(y)$. Show that $\mathfrak{f}(x) = kx$ for some constant k.

6. $\mathfrak{f}$ is upper semi-continuous in $[a, b]$. Show that $\mathfrak{f}$ is bounded above in $[a, b]$ and that it attains its least upper bound in $[a, b]$. (Hint: apply a subdivision method to each part of the question. For the second part select half-intervals with the same upper bound as the interval.)

7. A sequence $\{\mathfrak{f}_n(x)\}$ of real upper semi-continuous functions is such that $\mathfrak{f}_n(x) > \mathfrak{f}_m(x) \geqslant 0$ if $n < m$, $a \leqslant x \leqslant b$. Show that $\mathfrak{f}_n(x) \to \mathfrak{f}(x)$ as $n \to \infty$ and that $\mathfrak{f}(x)$ is upper semi-continuous for $x \in (a, b)$.

8. (a) $\mathfrak{f} = O(1)$, $\mathfrak{f}(x) \neq 0$ and $\mathfrak{f} \sim \mathfrak{g}$ as $x \to \infty$. Show that $\mathfrak{f} - \mathfrak{g} = o(1)$.

(b) $1/\mathfrak{f} = O(1)$ and $\mathfrak{f} - \mathfrak{g} = o(1)$ as $x \to \infty$. Show that $\mathfrak{f} \sim \mathfrak{g}$ as $x \to \infty$.

9. $\mathfrak{f}$, $\mathfrak{g}$ are positive functions such that

$$\mathfrak{g} = o(\mathfrak{f} + \mathfrak{g}) \quad \text{as} \quad x \to \infty.$$

Show that

$$\mathfrak{g} = o(\mathfrak{f}).$$

10. $\mathfrak{f}$ is a non-constant function defined over $[a, b]$. Show that $\exists\, x \in [a, b]$ such that any neighbourhood of x contains a point y in $[a, b]$ for which $\mathfrak{f}(y) \neq \mathfrak{f}(x)$.

11. $\mathfrak{N}(x)$ is the integer nearest to x (the least such integer if there are two) $\mathfrak{f}$, $\mathfrak{g}$ are defined by

$$\mathfrak{f}(0) = 0, \quad \mathfrak{f}(x) = |1/x - \mathfrak{N}(1/x)| \quad (x \neq 0);$$

$$\mathfrak{g}(0) = \tfrac{1}{2}, \quad \mathfrak{g}(x) = \mathfrak{f}(x) \quad (x \neq 0).$$

Show that $\mathfrak{f}$, $\mathfrak{g}$ are both Darboux continuous but that $\mathfrak{f} - \mathfrak{g}$ is not Darboux continuous over $[-1, 1]$.

16

MONOTONIC FUNCTIONS; FUNCTIONS OF BOUNDED VARIATION

Throughout this chapter all functions are real-valued functions of a real variable. In considering the properties of functions we consider particular classes of functions both for their own interest and for the information that can be obtained from using these functions. Two such classes of functions are monotonic functions and functions of bounded variation.

Definition 38. *A function $\mathfrak{f}$ is said to be increasing if $\mathfrak{f}(x_1) \geqslant \mathfrak{f}(x_2)$ whenever $x_1 \geqslant x_2$ and x_1, x_2 both belong to the domain of definition of $\mathfrak{f}$. We shall write in this case $\mathfrak{f}\uparrow$.*

$\mathfrak{f}$ is strictly increasing if $\mathfrak{f}(x_1) > \mathfrak{f}(x_2)$ when $x_1 > x_2$ and x_1, x_2 belong to the domain of definition of $\mathfrak{f}$. In this case we write $\mathfrak{f}\Uparrow$.

Definition 39. *$\mathfrak{f}$ is decreasing or strictly decreasing if $\mathfrak{f}(x_1) \leqslant \mathfrak{f}(x_2)$ when $x_1 \geqslant x_2$ or $\mathfrak{f}(x_1) < \mathfrak{f}(x_2)$ when $x_1 > x_2$ respectively and x_1, x_2 belong to the domain of definition of $\mathfrak{f}$. We write $\mathfrak{f}\downarrow$ for decreasing and $\mathfrak{f}\Downarrow$ for strictly decreasing.*

Definition 40. *If either $\mathfrak{f}\uparrow$ or $\mathfrak{f}\downarrow$ then $\mathfrak{f}$ is monotonic. If either $\mathfrak{f}\Uparrow$ or $\mathfrak{f}\Downarrow$ then $\mathfrak{f}$ is strictly monotonic.*

Exercises. (1) $\mathfrak{f}\uparrow$ and $\mathfrak{g}\uparrow$ for $a \leqslant x \leqslant b$. Show that $\mathfrak{f} + \mathfrak{g}\uparrow$ and $\max(\mathfrak{f}(x), \mathfrak{g}(x))\uparrow$.

(2) For each n $\mathfrak{f}_n(x)\uparrow$ for $a \leqslant x \leqslant b$ and $|\mathfrak{f}_n(x)| \leqslant M$ $(n = 1, 2, \ldots)$, $a \leqslant x \leqslant b$. Show that $\sup\limits_n \mathfrak{f}_n(x)\uparrow$.

Theorem 27. *The one-sided limits of a monotonic bounded function defined over* $[a, b]$ *always exist.*

Suppose that $\mathfrak{f}\uparrow$ for $x \in [y-\delta, y]$, $\delta > 0$. Then write $l = \sup\limits_{y-\delta \leqslant x < y} \mathfrak{f}(x)$. $\forall \epsilon > 0 \; \exists x_0$ such that $\mathfrak{f}(x_0) > l - \epsilon$ and $y - \delta \leqslant x_0 < y$. But $\mathfrak{f}\uparrow$ therefore $\mathfrak{f}(x) > l - \epsilon$ for $x_0 \leqslant x < y$. By the definition of l, $\mathfrak{f}(x) \leqslant l$ for $x_0 \leqslant x < y$. Thus $|\mathfrak{f}(x) - l| \leqslant \epsilon$ for $x_0 \leqslant x < y$, that is to say $\lim\limits_{x \to y-} \mathfrak{f}(x) = l$.

The other cases when $\mathfrak{f}\downarrow$ or limits on the right are considered can be dealt with similarly.

Theorem 28. *The continuous function* $\mathfrak{f}\Uparrow$ *over* $[a, b]$. *Then* $\exists$ *a continuous function* $\mathfrak{g}\Uparrow$ *over* $[\mathfrak{f}(a), \mathfrak{f}(b)]$ *such that*

$$\mathfrak{g}(\mathfrak{f}(x)) = x \quad \textit{if} \quad a \leqslant x \leqslant b \quad \textit{and} \quad \mathfrak{f}(\mathfrak{g}(y)) = y \quad \textit{if} \quad \mathfrak{f}(a) \leqslant y \leqslant \mathfrak{f}(b).$$

$\mathfrak{f}$ is continuous and therefore Darboux continuous (see p. 92) thus $\forall y$ satisfying $\mathfrak{f}(a) \leqslant y \leqslant \mathfrak{f}(b) \; \exists x$ satisfying $a \leqslant x \leqslant b$ and $y = \mathfrak{f}(x)$. But $\mathfrak{f}\Uparrow$ and thus $\exists$ at most one such x. Hence x is a single-valued function of y which we denote by $\mathfrak{g}(y)$ defined over $[\mathfrak{f}(a), \mathfrak{f}(b)]$.

By the definition of $\mathfrak{g}$ we have $y = \mathfrak{f}(\mathfrak{g}(y))$, if $y \in [\mathfrak{f}(a), \mathfrak{f}(b)]$. Also $\mathfrak{g}(\mathfrak{f}(x))$ is that number say x' such that $\mathfrak{f}(x') = \mathfrak{f}(x)$. But $\mathfrak{f}\Uparrow$ and thus $x' = x$. Hence $\mathfrak{g}(\mathfrak{f}(x)) = x$ if $x \in [a, b]$.

If $y_1 > y_2$ and $\mathfrak{g}(y_1) = x_1$, $\mathfrak{g}(y_2) = x_2$, then $y_1 = \mathfrak{f}(x_1)$, $y_2 = \mathfrak{f}(x_2)$ and $\mathfrak{f}(x_1) > \mathfrak{f}(x_2)$. Since $\mathfrak{f}\Uparrow$ this means that $x_1 > x_2$, i.e. $\mathfrak{g}(y_1) > \mathfrak{g}(y_2)$. Thus $\mathfrak{g}\Uparrow$.

Finally, to see that $\mathfrak{g}$ is continuous we observe that since it is monotonic its one-sided limits exist. Suppose $\mathfrak{f}(a) \leqslant p < \mathfrak{f}(b)$ and $l = \lim\limits_{y \to p+} \mathfrak{g}(y)$. $\mathfrak{f}$ is continuous thus $\lim\limits_{y \to p+} \mathfrak{f}(\mathfrak{g}(y)) = \mathfrak{f}(l)$, i.e. $\lim\limits_{y \to p+} y = \mathfrak{f}(l)$, i.e. $p = \mathfrak{f}(l)$. Thus $l = \mathfrak{g}(p)$. Similarly if $\mathfrak{f}(a) < p \leqslant \mathfrak{f}(b)$,

$$\lim_{y \to p-} \mathfrak{g}(y) = \mathfrak{g}(p).$$

Thus $\mathfrak{g}$ is continuous over $[\mathfrak{f}(a), \mathfrak{f}(b)]$.

$\mathfrak{f}(x)$ and $\mathfrak{g}(y)$ are called *inverse functions*. It will be clear from the context when we use the term *inverse function* whether this usage or one defined in the preliminaries, pp. 3, 7, is intended.

Remark. A result similar to theorem 28 holds if $\mathfrak{f}\Downarrow$. In this case $\mathfrak{g}\Downarrow$.

Functions of bounded variation

Let $\mathscr{D}$ be any finite set of points contained in $[a, b]$ and containing both a and b, then we say that $\mathscr{D}$ is a *dissection* of the interval $[a, b]$. If $\mathfrak{f}$ is defined over $[a, b]$ then $V^{\mathfrak{f}}_{\mathscr{D}}$, $P^{\mathfrak{f}}_{\mathscr{D}}$, $N^{\mathfrak{f}}_{\mathscr{D}}$ are functions defined by

$$V^{\mathfrak{f}}_{\mathscr{D}} = \sum_{i=0}^{k-1} |\mathfrak{f}(x_{i+1}) - \mathfrak{f}(x_i)|,$$

$$P^{\mathfrak{f}}_{\mathscr{D}} = \sum_{i=0}^{k-1} \max(\mathfrak{f}(x_{i+1}) - \mathfrak{f}(x_i), 0),$$

$$N^{\mathfrak{f}}_{\mathscr{D}} = \sum_{i=0}^{k-1} \min(\mathfrak{f}(x_{i+1}) - \mathfrak{f}(x_i), 0),$$

where $a = x_0 < x_1 < \ldots < x_{k-1} < x_k = b$ are the points of $\mathscr{D}$.

Definition 41. *The least upper bounds of $V^{\mathfrak{f}}_{\mathscr{D}}$, $P^{\mathfrak{f}}_{\mathscr{D}}$, $-N^{\mathfrak{f}}_{\mathscr{D}}$ taken over all possible dissections $\mathscr{D}$ are denoted by $V^{\mathfrak{f}}$, $P^{\mathfrak{f}}$, $-N^{\mathfrak{f}}$ and are called the total variation, the positive variation and the negative variation of $\mathfrak{f}$ respectively.*

Exercise. Show that if $\mathscr{D}_1 \supset \mathscr{D}_2$ then

$$V^{\mathfrak{f}}_{\mathscr{D}_1} \geqslant V^{\mathfrak{f}}_{\mathscr{D}_2}, \quad P^{\mathfrak{f}}_{\mathscr{D}_1} \geqslant P^{\mathfrak{f}}_{\mathscr{D}_2}, \quad -N^{\mathfrak{f}}_{\mathscr{D}_1} \geqslant -N^{\mathfrak{f}}_{\mathscr{D}_2}.$$

Definition 42. *A function for which $V^{\mathfrak{f}}_{\mathscr{D}}$ is bounded above for all $\mathscr{D}$ is said to be of bounded variation over $[a, b]$.*

To denote that $\mathfrak{f}$ is of bounded variation over $[a, b]$ we write $\mathfrak{f} \in \mathsf{B}(a, b)$.

Exercises. (1) $\mathfrak{f} \in \mathsf{B}(a, b)$ and $a \leqslant c < d \leqslant b$. Show that $\mathfrak{f} \in \mathsf{B}(c, d)$.

(2) $\mathfrak{f} \in \mathsf{B}(a, b)$ and $I_1, I_2, \ldots, I_j$ are disjoint closed intervals contained in $[a, b]$. Show that the sum of the total variations over the intervals I_k, $1 \leqslant k \leqslant j$ is not greater than the total variation over $[a, b]$.

(3) $a < c < b$ and $\mathfrak{f} \in \mathsf{B}(a, c)$, $\mathfrak{f} \in \mathsf{B}(c, b)$. Show that $\mathfrak{f} \in \mathsf{B}(a, b)$.

(4) $\mathfrak{f}$, $\mathfrak{g} \in \mathsf{B}(a, b)$. Show that $\mathfrak{f} + \mathfrak{g}$, $\mathfrak{f} \cdot \mathfrak{g}$ also belong to $\mathsf{B}(a, b)$.

Theorem 29. *$\mathfrak{f} \in \mathsf{B}(a, b)$ if and only if $\mathfrak{f}(x) = \mathfrak{g}(x) - \mathfrak{h}(x)$ where $\mathfrak{g}\uparrow$, $\mathfrak{h}\uparrow$ and $\mathfrak{g}$, $\mathfrak{h}$ are both bounded over $[a, b]$.*

If $\mathfrak{f}(x) = \mathfrak{g}(x) - \mathfrak{h}(x)$ where $\mathfrak{g}$, $\mathfrak{h}$ are bounded and $\mathfrak{g}\uparrow$, $\mathfrak{h}\uparrow$, then for any dissection $\mathscr{D}$

$$V^{\mathfrak{f}}_{\mathscr{D}} \leqslant V^{\mathfrak{g}}_{\mathscr{D}} + V^{\mathfrak{h}}_{\mathscr{D}} = \mathfrak{g}(b) - \mathfrak{g}(a) + \mathfrak{h}(b) - \mathfrak{h}(a)$$

and thus $\mathfrak{f} \in \mathsf{B}(a, b)$. Conversely by the definitions

$$P^{\mathfrak{f}}_{\mathscr{D}} = \mathfrak{f}(b) - \mathfrak{f}(a) - N^{\mathfrak{f}}_{\mathscr{D}}$$

and if $V^{\mathfrak{f}}_{\mathscr{D}} < K$ for all $\mathscr{D}$, then

$$\sup_{\mathscr{D}} P^{\mathfrak{f}}_{\mathscr{D}} = \mathfrak{f}(b) - \mathfrak{f}(a) + \sup_{\mathscr{D}} (-N^{\mathfrak{f}}_{\mathscr{D}}),$$

i.e. $$(1) \quad P^{\mathfrak{f}} = \mathfrak{f}(b) - \mathfrak{f}(a) - N^{\mathfrak{f}}.$$

But $\mathfrak{f} \in \mathsf{B}(a, x)$ if $a < x \leqslant b$ and thus there is a relation corresponding to (1) for each $a < x \leqslant b$, i.e.

$$(2) \quad \mathfrak{g}(x) = \mathfrak{f}(x) - \mathfrak{f}(a) + \mathfrak{h}(x),$$

where $\mathfrak{g}(x)$, $\mathfrak{h}(x)$ are respectively the positive and negative variations of $\mathfrak{f}$ over $[a, x]$. Again from the definitions $\mathfrak{g}\uparrow$, $\mathfrak{h}\uparrow$ and $0 \leqslant \mathfrak{g}(x) \leqslant V^{\mathfrak{f}}$, $0 \leqslant \mathfrak{h}(x) \leqslant V^{\mathfrak{f}}$. Since (2) can be rewritten

$$\mathfrak{f}(x) = (\mathfrak{g}(x) + \mathfrak{f}(a)) - \mathfrak{h}(x)$$

we have proved the required result.

Exercise. $\mathfrak{f} \in \mathsf{B}(a, b)$ and $a < c < b$. Prove the existence of $\lim_{x \to c+} \mathfrak{f}(x)$, $\lim_{x \to c-} \mathfrak{f}(x)$.

Remark. A function of bounded variation need not be continuous, for example, $\mathfrak{f}(x) = 0$, $x < \frac{1}{2}$, $\mathfrak{f}(x) = 1$, $x \geqslant \frac{1}{2}$. Nor need a continuous function be of bounded variation.

Worked examples

(i) *If* $\mathfrak{f}\uparrow$ *then the set of points* x *in* (a, b) *at which* $\lim_{x \to c+} \mathfrak{f}(x) > \lim_{x \to c-} \mathfrak{f}(x)$ *is enumerable.*

$\forall \delta > 0$ suppose that there exist k points $x_1, \ldots, x_k$ such that if $y = x_i$ then $\lim_{x \to y+} \mathfrak{f}(x) > \lim_{x \to y-} \mathfrak{f}(x) + \delta$. Then if $z > y$ and $w < y$ we have $\mathfrak{f}(z) > \mathfrak{f}(w) + \delta$. Let $\mathscr{D}$ be the dissection

$$a = y_0 < y_1 < \ldots < y_{k-1} < y_k = b,$$

where $y_i < x_i < y_{i-1}$ $(i = 0, \ldots, k)$. Then

$$V^{\mathfrak{f}}_{\mathscr{D}} = \sum_{i=0}^{k-1} |\mathfrak{f}(y_{i+1}) - \mathfrak{f}(y_i)| > k\delta.$$

But $V^{\mathfrak{f}}_{\mathscr{D}} = \mathfrak{f}(b) - \mathfrak{f}(a)$ and thus $k < (\mathfrak{f}(b) - \mathfrak{f}(a))/\delta$. Thus for every $n = 1, 2, \ldots$ the set of points at which $\lim_{x \to c+} \mathfrak{f}(x) > \lim_{x \to c-} \mathfrak{f}(x) + 1/n$ is finite. Hence the union of these sets is enumerable and this union contains all points c at which $\lim_{x \to c+} \mathfrak{f}(x) > \lim_{x \to c-} \mathfrak{f}(x)$.

(ii) $\mathfrak{f}$ *is real and continuous over* $[0, 1]$. *If* $\mathfrak{f}$ *takes each of its values at most once then show that either* $\mathfrak{f}\Uparrow$ *or* $\mathfrak{f}\Downarrow$.

We assume that $\mathfrak{f}(0) < \mathfrak{f}(1)$. If for two numbers x, y with $x < y$ we had $\mathfrak{f}(x) < \mathfrak{f}(y)$ then for any z satisfying $x < z < y$ we should have $\mathfrak{f}(x) < \mathfrak{f}(z) < \mathfrak{f}(y)$. For if say $\mathfrak{f}(z) > \mathfrak{f}(y)$ then since $\mathfrak{f}$ is continuous and therefore Darboux continuous $\exists\, w$ such that $x < w < z$ and $\mathfrak{f}(w) = \mathfrak{f}(y)$. This is impossible since $\mathfrak{f}$ takes each value at most once and $w \neq y$. Similarly we cannot have $\mathfrak{f}(z) < \mathfrak{f}(x)$. Thus $\mathfrak{f}(x) \leqslant \mathfrak{f}(z) \leqslant \mathfrak{f}(y)$ and since again equality is impossible

$$\mathfrak{f}(x) < \mathfrak{f}(z) < \mathfrak{f}(y).$$

Consider then any two numbers a, b with $0 < a < b < 1$. Then by the above argument applied to 0, a, 1 in place of x, z, y we have $\mathfrak{f}(0) < \mathfrak{f}(a) < \mathfrak{f}(1)$. By a repetition of this argument applied to a, b, 1 in place of x, z, y we have $\mathfrak{f}(a) < \mathfrak{f}(b) < \mathfrak{f}(1)$. Thus if $0 < a < b < 1$, then $\mathfrak{f}(a) < \mathfrak{f}(b)$, i.e. $\mathfrak{f} \Uparrow$.

If $\mathfrak{f}(0) > \mathfrak{f}(1)$ a similar argument shows that $\mathfrak{f} \Downarrow$.

(iii) *$\mathfrak{f}$ is defined and uniformly continuous over the rational numbers. Show that $\lim_{y \to x} \mathfrak{f}(y)$, y rational exists for all real x and is a uniformly continuous function of x.*

Since $\mathfrak{f}$ is uniformly continuous over the rationals $\forall \epsilon > 0 \, \exists \delta > 0$ such that $|\mathfrak{f}(y_1) - \mathfrak{f}(y_2)| < \epsilon$ if $|y_1 - y_2| < \delta$, y_1, y_2 rational. Now let x be any real number then if y_1, y_2 are rational and satisfy

$$|x - y_1| < \tfrac{1}{2}\delta, \quad |x - y_2| < \tfrac{1}{2}\delta$$

it follows that
$$\overline{\lim_{y \to x}}\, \mathfrak{f}(y) - \underline{\lim_{y \to x}}\, \mathfrak{f}(y) \leqslant \epsilon.$$

Since this is true for every $\epsilon > 0$ it follows that $\overline{\lim}_{y \to x} \mathfrak{f}(y)$ coincides with $\underline{\lim}_{y \to x} \mathfrak{f}(y)$, i.e. $\lim_{y \to x} \mathfrak{f}(x)$ exists. Also if x_1, x_2 are any two real numbers satisfying $|x_1 - x_2| < \delta$, then $\exists$ sequences of rational numbers $\{y_n^{(1)}\}$, $\{y_n^{(2)}\}$ such that $y_n^{(1)} \to x_1$, $y_n^{(2)} \to x_2$ as $n \to \infty$, and $|y_n^{(1)} - y_n^{(2)}| < \delta$. Then
$$|\lim_{y \to x_1} \mathfrak{f}(y) - \lim_{y \to x_2} \mathfrak{f}(y)| = \lim_{n \to \infty} |\mathfrak{f}(y_n^{(1)}) - \mathfrak{f}(y_n^{(2)})| \leqslant \epsilon.$$

Thus the limit is itself a uniformly continuous function of x.

Exercises

1. $\mathfrak{f}_n(x) \uparrow$ for $0 \leqslant x \leqslant 1$ $(n = 1, 2, \ldots)$ and $0 \leqslant \mathfrak{f}_n(x) \leqslant 1$. Show that $\overline{\lim}_{n \to \infty} \mathfrak{f}_n(x) \uparrow$ and $\underline{\lim}_{n \to \infty} \mathfrak{f}_n(x) \uparrow$.

2. u is a fixed positive number. Show that u^x for real x is (i) a monotonic function of x, (ii) a continuous function of x.

3. $\Sigma a_{n-1} z^{n-1}$ is a real power series with a positive radius of convergence R. If $-R < a < b < R$ show that $\sum_0^\infty a_n z^n \in \mathsf{B}(a, b)$.

4. $\mathfrak{f}$ is monotonic over $[a, b]$ and $\mathfrak{g}$ is monotonic over a domain that includes the range of $\mathfrak{f}$. Show that $\mathfrak{g}(\mathfrak{f}(x))$ is monotonic.

5. $\mathfrak{f} \in \mathsf{B}(0,1)$. Show that the subset of points of $(0,1)$ at which $\mathfrak{f}$ is not continuous is enumerable.

6. $\mathfrak{f}$ is Darboux continuous and $\mathfrak{f} \in \mathsf{B}(0,1)$. Show that $\mathfrak{f}$ is continuous over $[0,1]$.

7. $\mathfrak{f} \in \mathsf{B}(0,1)$ and $\mathfrak{f}$ is continuous over $[0,1]$. Show that the total variation of $\mathfrak{f}$ over $[0,x]$ is continuous.

8. Give an example of a convergent sequence of functions each belonging to $\mathsf{B}(0,1)$ and bounded by the same bound such that the limit function does not belong to $\mathsf{B}(0,1)$.

9. A sequence of functions $\{\mathfrak{f}_n(x)\}$ defined over $[a,b]$ are such that $V\mathfrak{f}_n \leqslant 1$. The sequence converges to a limit function $\mathfrak{f}(x)$. Show that $V\mathfrak{f} \leqslant 1$.

10. Give an example of a function $\mathfrak{f}$ such that $\mathfrak{f}\uparrow$ over $[0,1]$ and $\forall x_1, x_2$, $0 \leqslant x_1 \leqslant x_2 \leqslant 1$ $\exists x_3$ such that $\mathfrak{f}(x_1) < \mathfrak{f}(x_3) < \mathfrak{f}(x_2)$ and $\mathfrak{f}$ is discontinuous at x_3.

11. $\mathfrak{f}\uparrow$ over $[0,1]$ and α is a fixed number satisfying $0 < \alpha < \frac{1}{2}$ such that if x_1, x_2 are any two numbers for which $0 \leqslant x_1 \leqslant x_2 \leqslant 1$ then $\exists$ a number x_3 such that

$$(1-\alpha)\,\mathfrak{f}(x_1)+\alpha\mathfrak{f}(x_2) < \mathfrak{f}(x_3) < \alpha\mathfrak{f}(x_1)+(1-\alpha)\,\mathfrak{f}(x_2).$$

Show that $\mathfrak{f}$ is continuous.

12. $\mathfrak{f}$ is continuous over $[a,b]$ and $\mathfrak{g}$ is defined by $\mathfrak{g}(x) = \sup_{a \leqslant y \leqslant x} \mathfrak{f}(y)$. Show that $\mathfrak{g}$ is continuous and that $\mathfrak{f}\Uparrow$ if and only if $\mathfrak{g}\Uparrow$.

13. $\mathfrak{f}$ is defined and bounded over $[a,b]$ and $\exists\, c, d$ such that $a < c < d < b$ and $\mathfrak{f} \notin \mathsf{B}(c,d)$. Show that $\exists$ a point $k \in [a,b]$ such that $\mathfrak{f}$ is not of bounded variation over any interval that contains k as an interior point and is contained in $[a,b]$.

17

DIFFERENTIATION: MEAN-VALUE THEOREMS

It is important to be able to assess the rate of change of a function $\mathfrak{f}$ and it is natural to use the ratio $(\mathfrak{f}(x_1)-\mathfrak{f}(x_2))/(x_1-x_2)$ for this purpose. The difficulty is that this ratio is a function of two variables x_1, x_2. Instead we replace it by $\lim_{x_2\to x_1} (\mathfrak{f}(x_1)-\mathfrak{f}(x_2))/(x_1-x_2)$ when this limit exists. This limit is a function of one variable and thus no more difficult to handle than $\mathfrak{f}$ itself and in fact in many cases it is actually simpler. Of course it does not provide immediately all the information that the ratio $(\mathfrak{f}(x_1)-\mathfrak{f}(x_2))/(x_1-x_2)$ provides but we shall show how to obtain information from a knowledge of the limit function.

All functions in this chapter are real functions of a real variable.

Definition 43. *If* $\mathfrak{f}$ *is defined in a neighbourhood of* x *and* $\lim\limits_{h\to 0}(\mathfrak{f}(x+h)-\mathfrak{f}(x))/h$ *exists then we say that* $\mathfrak{f}$ *is differentiable at* x. *The value of this limit is called the differential coefficient of* $\mathfrak{f}$ *at* x *or the derivative of* $\mathfrak{f}$ *at* x *and is denoted by* $D\mathfrak{f}(x)$, $d\mathfrak{f}/dx$ or $\mathfrak{f}'(x)$.

The definition 43 applies only when the limit is a number. Thus $D\mathfrak{f}(x)$ if it exists is itself a function of x which we denote by $D\mathfrak{f}$.

If $D\mathfrak{f}$ is differentiable we denote its differential coefficient by $D^2\mathfrak{f}$ or $d^2\mathfrak{f}/dx^2$ or $\mathfrak{f}''(x)$ and call it the *second derivative* of $\mathfrak{f}$. Similarly the nth derivative, if it exists, of $\mathfrak{f}$ is the derivative of the $n-1$th derivative and is denoted by $D^n\mathfrak{f}$, $d^n\mathfrak{f}/dx^n$ or $\mathfrak{f}^{(n)}(x)$.

The definition of differentiability may also be written

$$\mathfrak{f}(x+h)-\mathfrak{f}(x) = h.D\mathfrak{f}(x)+o\,|h| \quad \text{as} \quad h\to 0.$$

If $\mathfrak{f}$ is not differentiable at x then the upper and lower one-sided limits as $h\to 0+$ or $h\to 0-$ of $(\mathfrak{f}(x+h)-\mathfrak{f}(x))/h$ will exist (possibly ∞ or $-\infty$). These are denoted by $D^+\mathfrak{f}$, $D_+\mathfrak{f}$, $D^-\mathfrak{f}$, $D_-\mathfrak{f}$ respectively and are called the Dini derivates at x. For example

$$D^+\mathfrak{f}(x_0) = \overline{\lim_{h\to 0+}}\,\frac{\mathfrak{f}(x_0+h)-\mathfrak{f}(x_0)}{h},$$

$$D_+\mathfrak{f}(x_0) = \underline{\lim_{h\to 0+}}\,\frac{\mathfrak{f}(x_0+h)-\mathfrak{f}(x_0)}{h}, \quad \text{etc.}$$

If $D^+\mathfrak{f}(x) = D_+\mathfrak{f}(x) = \lambda$ then we say that the *right-hand derivative* at x exists and its value is λ. If $D^-\mathfrak{f}(x) = D_-\mathfrak{f}(x) = \mu$ then we say that the *left-hand derivate* exists and is equal to μ.

If $\mathfrak{f}$ and $\mathfrak{g}$ are differentiable at x then so are $\mathfrak{f}+\mathfrak{g}$, $\mathfrak{f}.\mathfrak{g}$, $1/\mathfrak{f}$ if $\mathfrak{f}(x)\neq 0$. A constant function is differentiable and has derivative zero. If $\mathfrak{f}$ is differentiable at x then it is continuous at x.

We have the following formulae

$$D(\mathfrak{f}+\mathfrak{g}) = D\mathfrak{f}+D\mathfrak{g},$$

$$D(\mathfrak{f}.\mathfrak{g}) = D\mathfrak{f}.\mathfrak{g}+\mathfrak{f}.D\mathfrak{g},$$

$$D(1/\mathfrak{f}) = -D\mathfrak{f}/(\mathfrak{f})^2 \quad \text{if} \quad \mathfrak{f}(x)\neq 0,$$

$$D^n(\mathfrak{f}.\mathfrak{g}) = \sum_{r=0}^{n}\binom{n}{r}D^r\mathfrak{f}.D^{n-r}\mathfrak{g} \quad \text{(Leibnitz's formula).}$$

The verification of these formulae is left to the reader. They depend upon the basic properties of limits (see chapter 4). Leibnitz's formula may be proved by induction using the binomial coefficient identity

$$\binom{n+1}{r} = \binom{n}{r-1} + \binom{n}{r}.$$

From the second relation above we have $D(x^n) = nx^{n-1}$ for $n = 1, 2, \ldots$, by an inductive argument.

Exercise. Show that $|x|$ is continuous at $x = 0$ but not differentiable at $x = 0$.

Theorem 30. *The function $y = \mathfrak{f}(x)$ is strictly monotonic and continuous in a neighbourhood of x_0, $D\mathfrak{f}(x_0)$ exists and $D\mathfrak{f}(x_0) \neq 0$. Let $x = \mathfrak{g}(y)$ be the inverse function defined as in theorem 28. Then if $y_0 = \mathfrak{f}(x_0)$ it follows that $D\mathfrak{g}(y_0)$ exists and is equal to $1/D\mathfrak{f}(x_0)$.*

Since $\mathfrak{g}(y)$ is continuous at y_0 (by theorem 28) it follows that $x \to x_0$ as $y \to y_0$. Hence

$$\frac{\mathfrak{g}(y)-\mathfrak{g}(y_0)}{y-y_0} = \frac{x-x_0}{\mathfrak{f}(x)-\mathfrak{f}(x_0)} \to \frac{1}{D\mathfrak{f}(x_0)} \quad \text{as} \quad y \to y_0.$$

This is the required result.

Theorem 31. *If $\mathfrak{f}$ is differentiable at u_0 and $\mathfrak{g}$ at x_0 where $u_0 = \mathfrak{g}(x_0)$ then the composite function $\mathfrak{f}(\mathfrak{g}(x))$ is differentiable at x_0 and*

$$D\mathfrak{f}(\mathfrak{g}(x_0)) = D\mathfrak{f}(u_0).D\mathfrak{g}(x_0).$$

Write u_0+k for $\mathfrak{g}(x_0+h)$. Since $\mathfrak{g}$ is differentiable at x_0,

$$k = \mathfrak{g}(x_0+h)-\mathfrak{g}(x_0) = hD\mathfrak{g}(x_0)+o(|h|) \quad \text{as} \quad h \to 0.$$

i.e. $\forall \epsilon > 0 \, \exists \, \delta > 0$ such that

$$|k-h\,D\mathfrak{g}(x_0)| \leqslant \epsilon\,|h| \quad \text{if} \quad |h| < \delta.$$

Thus $k = O\,|h|$ and $k \to 0$ as $h \to 0$. By the differentiability of $\mathfrak{f}$ at $u_0 \, \exists \, \delta_1$ such that

$$|\mathfrak{f}(u_0+k)-\mathfrak{f}(u_0)-kD\mathfrak{f}(u_0)| \leqslant \epsilon\,|k| \quad \text{if} \quad |k| < \delta_1.$$

Thus substituting $k = hD\mathfrak{g}(x_0)+\theta\epsilon\,|h|$ where $|\theta| < 1$ we have

$$|\mathfrak{f}(\mathfrak{g}(x_0+h))-\mathfrak{f}(\mathfrak{g}(x_0))-hD\mathfrak{g}(x_0)\,D\mathfrak{f}(u_0)| \leqslant \epsilon\,|k|+|\theta|\,\epsilon\,|h|\,|D\mathfrak{f}(u_0)|$$

provided that $|h| < \delta$ and $|h|\,|D\mathfrak{g}(x_0)|+\theta\epsilon\,|h| < \delta_1$. Moreover, the right-hand side of the above inequality is less than

$$\epsilon((|D\mathfrak{g}(x_0)|+\epsilon)+|\theta|\,|D\mathfrak{f}(u_0)|)\,|h| \leqslant \epsilon K\,|h|, \quad \text{say}$$

and the inequality is true provided

$$|h| < \min\{\delta, \delta_1/(\epsilon + |D\mathfrak{g}(x_0)|)\}.$$

But this means that $\mathfrak{f}\mathfrak{g}(x)$ is a differentiable function of x at x_0 and that

$$D\mathfrak{f}(\mathfrak{g}(x_0)) = D\mathfrak{f}(u_0) . D\mathfrak{g}(x_0).$$

Exercise. By considering $\mathfrak{g}$, $\mathfrak{f}$ defined by

$$\mathfrak{g}(x) = |x|, \quad \mathfrak{f}(u) = u^2 \quad (u \geqslant 0), \quad \mathfrak{f}(u) = -1 \quad (u < 0)$$

show that it is possible for $\mathfrak{f}(\mathfrak{g}(x))$ to be differentiable at 0 even though $\mathfrak{g}(x)$ is not differentiable at 0 and $\mathfrak{f}(u)$ is not differentiable at $\mathfrak{g}(0)$.

Mean-value theorems

Theorem 32. *Rolle's theorem. If* $\mathfrak{f}$ *is continuous over* $[a, b]$ *and differentiable in* $a < x < b$ *and* $\mathfrak{f}(a) = \mathfrak{f}(b)$, *then* $\exists\, c$ *such that* $a < c < b$ *and* $D\mathfrak{f}(c) = 0$.

Since $\mathfrak{f}(a) = \mathfrak{f}(b)$, either
(i) $\mathfrak{f}$ is a constant function, or
(ii) $\sup \mathfrak{f}(x) > \mathfrak{f}(a) = \mathfrak{f}(b)$, or
(iii) $\inf \mathfrak{f}(x) < \mathfrak{f}(a) = \mathfrak{f}(b)$.

In case (i) any point of (a, b) will do for c.

In case (ii) by theorem 26 (ii), p. 90, $\exists\, c$ such that $\mathfrak{f}(c) = \sup \mathfrak{f}(x)$. Thus $\mathfrak{f}(c) \geqslant \mathfrak{f}(x)$ $a \leqslant x \leqslant b$, also since $\sup \mathfrak{f}(x) > \mathfrak{f}(a) = \mathfrak{f}(b)$ the point c is neither a nor b. Hence $a < c < b$ and $\mathfrak{f}$ is differentiable at c. But $(\mathfrak{f}(c+h) - \mathfrak{f}(c))/h$ is non-positive or non-negative according as h is positive or negative and thus if $\lim_{h \to 0} (\mathfrak{f}(c+h) - \mathfrak{f}(c))/h$ exists its value can only be zero. Hence $D\mathfrak{f}(c) = 0$ and the theorem is proved.

The proof of case (iii) is similar to that of (ii) and is omitted.

Theorem 33. *The mean-value theorem. If* $\mathfrak{f}$ *is continuous over* $[a, b]$ *and differentiable in* (a, b), *then* $\exists\, c$ *such that* $a < c < b$ *and* $\mathfrak{f}(b) - \mathfrak{f}(a) = (b-a) D\mathfrak{f}(c)$.

Consider $\mathfrak{g}$ defined by

$$\mathfrak{g}(x) = \mathfrak{f}(x) - \mathfrak{f}(a) - (x-a)(\mathfrak{f}(b) - \mathfrak{f}(a))/(b-a).$$

Then $\mathfrak{g}(a) = \mathfrak{g}(b)$ and $\mathfrak{g}$ satisfies the conditions of continuity and differentiability specified in theorem 32. Thus $\exists\, c$ such that $a < c < b$ and $D\mathfrak{g}(c) = 0$, i.e. $D\mathfrak{f}(c) - (\mathfrak{f}(b) - \mathfrak{f}(a))/(b-a) = 0$.

The theorem is proved.

Exercise. $D\mathfrak{f}(x)$ exists at c and $D\mathfrak{f}(x) \to l$ as $x \to c$. Show that $l = D\mathfrak{f}(c)$.

Applications of the mean-value theorem

In (1), (2) below it is assumed that $\mathfrak{f}$, $\mathfrak{g}$ are continuous over $[a, b]$ and differentiable in $a < x < b$.

(1) (*a*) *If* $D\mathfrak{f}(x) = 0$ *for* $a < x < b$ *then* $\mathfrak{f}$ *is a constant.*

(*b*) *If* $D\mathfrak{f}(x) \geqslant 0$ *for* $a < x < b$ *then* $\mathfrak{f}\uparrow$ *and if* $D\mathfrak{f}(x) > 0$ *then* $\mathfrak{f}\Uparrow$.

(*a*) Take x, $a < x \leqslant b$ and apply the mean-value theorem. Then $\exists c$ such that $a < c < x$ and $\mathfrak{f}(x) - \mathfrak{f}(a) = (x-a)\, D\mathfrak{f}(c) = 0$. Thus $\mathfrak{f}(x) = \mathfrak{f}(a)$, $a < x \leqslant b$.

(*b*) Take x, y in $a \leqslant x < y \leqslant b$. By the mean-value theorem $\exists c$ such that $x < c < y$ and $\mathfrak{f}(y) - \mathfrak{f}(x) = (y-x)\, D\mathfrak{f}(c) \geqslant 0$. Thus $\mathfrak{f}\uparrow$. Further if $D\mathfrak{f}(c) > 0$ then $\mathfrak{f}(y) - \mathfrak{f}(x) > 0$ and this means that $\mathfrak{f}\Uparrow$.

Remark. The converse of the last result above is false. For example, $x^3\Uparrow$, but $Dx^3 = 3x^2 = 0$ at $x = 0$.

Exercise. $\mathfrak{f}\Uparrow$ over $[a, b]$. Show that $\exists c$ such that $D\mathfrak{f}(c) > 0$ and $a < c < b$.

(2) *If* $\mathfrak{f}(a) = \mathfrak{g}(a)$ *and* $D\mathfrak{f}(x) \geqslant D\mathfrak{g}(x)$ *for* $a \leqslant x \leqslant b$ *then* $\mathfrak{f}(b) \geqslant \mathfrak{g}(b)$.

Apply the mean-value theorem to $\mathfrak{f} - \mathfrak{g}$ in $[a, b]$. Then $\exists c$ such that $a < c < b$ and

$$(\mathfrak{f}(b) - \mathfrak{g}(b)) - (\mathfrak{f}(a) - \mathfrak{g}(a)) = (b-a)\,(D\mathfrak{f}(c) - D\mathfrak{g}(c)) \geqslant 0.$$

Thus $\mathfrak{f}(b) - \mathfrak{g}(b) \geqslant \mathfrak{f}(a) - \mathfrak{g}(a) = 0$.

Remark. When comparing the size of two functions $\mathfrak{f}$, $\mathfrak{g}$ it is possible to consider either $\mathfrak{f}/\mathfrak{g}$ or $\mathfrak{f} - \mathfrak{g}$. The second alternative is often the simpler.

(3) *If* α *is real and* x *real and positive then* $Dx^\alpha = \alpha x^{\alpha-1}$.

If $\alpha = 1$, then by definition

$$Dx^\alpha = \lim_{h \to 0} \frac{(x+h)^\alpha - x^\alpha}{h} = 1 = \alpha x^{\alpha-1}.$$

If α is a positive integer n and we assume inductively that $Dx^{\alpha-1} = (\alpha-1)\, x^{\alpha-2}$, then

$$Dx^\alpha = D(x^{\alpha-1}.x) = x^{\alpha-1}.1 + (\alpha-1)\, x^{\alpha-2}.x = \alpha x^{\alpha-1}.$$

Thus the formula is true for all positive integers.

If α is a positive rational of the form $1/q$, q an integer, then $x^{1/q}\Uparrow$ and $x^{1/q} \to 0$ as $x \to 0+$, $x^{1/q} \to \infty$ as $x \to \infty$. If $y = x^{1/q}$ then the inverse function (cf. theorem 28, p. 97) is $x = y^q$ defined for $y > 0$. By theorem 30 and the case above

$$Dx^{1/q} = \frac{1}{Dy^q} = \frac{1}{qy^{q-1}} = \frac{1}{q} x^{1/q-1}$$

and the formula is proved in this case.

If $\alpha = p/q$ where p, q are positive integers, then with $u = x^{1/q}$,

$$Dx^{p/q} = Du^p . Dx^{1/q} = pu^{p-1}(1/q)\,x^{1/q-1} = p/q\,x^{p/q-1}$$

and the formula is proved.

If α is a positive real irrational $\exists \{\alpha_n\}\downarrow$, α_n rational and $\alpha_n \to \alpha$ as $n \to \infty$ and $x^\alpha = \lim_{n\to\infty} x^{\alpha_n}$ (p. 25). By the mean-value theorem

$$\frac{(x+h)^{\alpha_n} - x^{\alpha_n}}{h} = \alpha_n y_n^{\alpha_n - 1},$$

where y_n lies between x and $x+h$. Thus $y_n^{\alpha_n-1}$ lies between x^{α_n-1} and $(x+h)^{\alpha_n-1}$ and $((x+h)^\alpha - x^\alpha)/h$ lies between $\alpha x^{\alpha-1}$ and $\alpha(x+h)^{\alpha-1}$. $(1+h/x)^{\alpha-1}$ is a monotonic function of α if h/x is fixed. If α is an integer $(1+h/x)^{\alpha-1} \to 1$ as $h \to 0$. Hence for all α fixed, $(x+h)^{\alpha-1} \to x^{\alpha-1}$ as $h \to 0$. Thus $\lim_{h\to 0}((x+h)^\alpha - x^\alpha)/h = \alpha x^{\alpha-1}$ and the theorem is proved in this case.

Finally if α is a negative real number

$$D(x^\alpha) = D\left(\frac{1}{x^{-\alpha}}\right) = \frac{-D(x^{-\alpha})}{x^{-2\alpha}} = \frac{\alpha x^{-\alpha-1}}{x^{-2\alpha}} = \alpha x^{\alpha-1}.$$

This establishes the result in all cases.

(4) *The Descartes–Ursell rule of signs. If $\alpha_1 > \alpha_2 > \ldots > \alpha_n$, then the number of positive roots of the equation $c_1 x^{\alpha_1} + \ldots + c_n x^{\alpha_n} = 0$, c_i real, is less than or equal to the number of changes of sign in the ordered set $c_1, c_2, \ldots, c_n$.*

Let p be the number of positive roots and q be the number of changes of sign. If $q = 0$ then clearly $p = 0$. Assume inductively that any such equation with $q-1$ changes of sign in the sequence of coefficients has at most $q-1$ roots. Of the given equation with q changes of sign in $c_1, c_2, \ldots, c_n$ suppose that the first change of sign occurs between c_k and c_{k+1}. Choose β so that $\alpha_k > \beta > \alpha_{k+1}$. Then the equation

$$(*) \quad c_1(\alpha_1 - \beta)\,x^{\alpha_1} + c_2(\alpha_2 - \beta)\,x^{\alpha_2} + \ldots + c_m(\alpha_m - \beta)\,x^{\alpha_m} = 0$$

has at most $q-1$ positive roots (by the induction hypothesis), since its coefficients $c_1(\alpha_1 - \beta), \ldots, c_m(\alpha_m - \beta)$ have exactly $q-1$ changes of sign. But

$$c_1(\alpha_1 - \beta)\,x^{\alpha_1} + \ldots + c_m(\alpha_m - \beta)\,x^{\alpha_m} = x^{\beta+1}D(c_1 x^{\alpha_1-\beta} + \ldots + c_m x^{\alpha_m - \beta})$$

and it follows from Rolle's theorem, p. 104, that (*) has at least $p-1$ positive roots. Hence $p-1 \leqslant q-1$, i.e. $p \leqslant q$ and the next stage of the induction argument is complete.

The rule is proved.

(5) *The mean-value theorem is useful in establishing elementary inequalities For example:*

$$\text{(i)}\ 1-x^{\alpha} < \alpha(1-x) \quad (0 < x < 1,\ \alpha > 1);$$

$$\text{(ii)}\ q(x^{p}-1) \geqslant p(x^{q}-1) \quad (0 < x,\ 0 < q < p).$$

(i) Write $\mathfrak{f}(x) = 1-x^{\alpha}-\alpha(1-x)$. Then $D\mathfrak{f}(x) = -\alpha x^{\alpha-1}+\alpha > 0$ since $\alpha > 1$, $0 < x < 1$. Thus $\mathfrak{f}\Uparrow$. But $\mathfrak{f}(1) = 0$ and thus $\mathfrak{f}(x) < 0$ if $0 < x < 1$. This is the required inequality.

(ii) Write $\mathfrak{f}(x) = q(x^{p}-1)-p(x^{q}-1)$. Then

$$D\mathfrak{f}(x) = pq(x^{p-1}-x^{q-1}).$$

Thus $D\mathfrak{f}(1) = 0$; $D\mathfrak{f}(x) > 0$ if $x > 1$; $D\mathfrak{f}(x) < 0$ if $x < 1$. But $\mathfrak{f}(1) = 0$. Thus $\mathfrak{f}(x) \geqslant 0$ if $x > 0$ (since $\mathfrak{f}\Downarrow$ $0 < x < 1$ and $\mathfrak{f}\Uparrow$ $x > 1$). This is the required inequality.

(6) *The Hadamard–Littlewood lemma. If* $\mathfrak{f}$ *is continuous and twice differentiable for* $x \geqslant 0$ *and if further* $\mathfrak{f}(x) = o(1)$, $D^{2}\mathfrak{f}(x) = O(x^{-2})$ *as* $x \to \infty$ *then* $D\mathfrak{f}(x) = o(x^{-1})$ *as* $x \to \infty$.

The values of $D\mathfrak{f}$ can be related by the mean-value theorem either to values of $\mathfrak{f}$ or to those of $D^{2}\mathfrak{f}$. The idea is first to use the relation with $\mathfrak{f}$ to show that $xD\mathfrak{f}(x)$ is small for *some* large x, and then to use the relation with $D^{2}\mathfrak{f}$ to show that $xD\mathfrak{f}(x)$ is small for *all* large x.

Let λ be a fixed number greater than 1. By the mean-value theorem

$$\mathfrak{f}(\lambda^{n})-\mathfrak{f}(\lambda^{n-1}) = \lambda^{n-1}(\lambda-1)\,D\mathfrak{f}(c_{n}) \quad (n = 1, 2, \ldots),$$

$\lambda^{n-1} < c_{n} < \lambda^{n}$. But $\mathfrak{f}(x) = o(1)$ as $x \to \infty$. Thus $\lambda^{n-1}.D\mathfrak{f}(c_{n}) = o(1)$ as $n \to \infty$. But $c_{n} < \lambda.\lambda^{n-1}$ and thus $c_{n}.D\mathfrak{f}(c_{n}) = o(1)$ as $n \to \infty$.

Next for x in $[\lambda^{n-1}, \lambda^{n}]$ by the mean-value theorem

$$D\mathfrak{f}(x)-D\mathfrak{f}(c_{n}) = (x-c_{n})\,D^{2}\mathfrak{f}(d_{n}),$$

where d_{n} lies between x and c_{n} (d_{n} depends on x in general). Thus

$$xD\mathfrak{f}(x) = (x/c_{n}).c_{n}D\mathfrak{f}(c_{n})+x(x-c_{n})\,D^{2}\mathfrak{f}(d_{n}).$$

Now $D^{2}\mathfrak{f}(x) = O(x^{-2})$ as $x \to \infty$, i.e. $\exists\, K, x_{0}$ such that

$$|x^{2}D^{2}\mathfrak{f}(x)| < K \quad \text{all} \quad x > x_{0}.$$

Hence $\exists\, N$ such that

$$\begin{aligned} |x(x-c_{n})\,D^{2}\mathfrak{f}(d_{n})| &< K\,|x(x-c_{n})\,d_{n}^{-2}| && (n > N) \\ &< K\,|\lambda^{n}.\lambda^{n-1}(\lambda-1).\lambda^{-2n+2}| && (n > N) \\ &= K\lambda(\lambda-1) && (n > N). \end{aligned}$$

Also $c_n D\mathfrak{f}(c_n) = o(1)$ as $n \to \infty$. Thus $\forall \epsilon > 0 \, \exists M$ such that

$$|c_n D\mathfrak{f}(c_n)| < \epsilon \quad (n > M).$$

If $n > \max(N, M)$ and $x > \lambda^n$ then

$$|xD\mathfrak{f}(x)| < \epsilon + K\lambda(\lambda - 1).$$

λ could be chosen initially to be any number greater than 1. The value of K does not depend on λ in any way and we can therefore choose λ so that $K\lambda(\lambda-1) < \epsilon$. The value of M does depend on λ but this does not matter, since having fixed λ we can still choose M so that $|c_n D\mathfrak{f}(c_n)| < \epsilon$ $(n > M)$.

Then
$$|xD\mathfrak{f}(x)| < 2\epsilon \quad \text{for} \quad x > \lambda^n \quad \text{and} \quad n > \max(N, M),$$

i.e. $xD\mathfrak{f}(x) \to 0$ as $x \to \infty$. This is the result which had to be proved.

Theorem 34. *Cauchy's formula. If* $\mathfrak{f}$, $\mathfrak{g}$ *are continuous over* $[a, b]$ *and differentiable in* (a, b) *then* $\exists c$ *such that* $a < c < b$ *and* $D\mathfrak{f}(c)\,(\mathfrak{g}(a) - \mathfrak{g}(b)) = D\mathfrak{g}(c)\,(\mathfrak{f}(a) - \mathfrak{f}(b))$.

Write $\mathfrak{h}(x) = \mathfrak{f}(x)\,(\mathfrak{g}(a) - \mathfrak{g}(b)) - \mathfrak{g}(x)\,(\mathfrak{f}(a) - \mathfrak{f}(b))$. Then $\mathfrak{h}(a) = \mathfrak{h}(b)$ and $\mathfrak{h}$ satisfies the other conditions of Rolle's theorem, p. 104. Hence $\exists c$ such that $a < c < b$ and $D\mathfrak{h}(c) = 0$. This is the required result.

Remark. The usefulness of this formula depends upon the fact that the values of $D\mathfrak{f}$, $D\mathfrak{g}$ are calculated at the *same* point c.

Corollary. *If in addition to the conditions of theorem* 34 $D\mathfrak{g}(x) \neq 0$ *for* $a < c < b$, *then*

$$\frac{\mathfrak{f}(b) - \mathfrak{f}(a)}{\mathfrak{g}(b) - \mathfrak{g}(a)} = \frac{D\mathfrak{f}(c)}{D\mathfrak{g}(c)}.$$

For by the mean-value theorem applied to $\mathfrak{g}$ in $[a, b]$,

$$\mathfrak{g}(b) - \mathfrak{g}(a) = (b - a)\,D\mathfrak{g}(\xi)$$

for some ξ in (a, b). Thus $\mathfrak{g}(b) - \mathfrak{g}(a) \neq 0$ and in the result of the theorem we may divide by $D\mathfrak{g}(c) . (\mathfrak{g}(b) - \mathfrak{g}(a))$ to give the required formula.

Theorem 35. (i) *If* $\mathfrak{f}(x) \to 0$, $\mathfrak{g}(x) \to 0$ *and* $D\mathfrak{f}(x)/D\mathfrak{g}(x) \to l$ *as* $x \to a$, *then* $\mathfrak{f}(x)/\mathfrak{g}(x) \to l$ *as* $x \to a$.

(ii) *If* $\mathfrak{g}(x) \to \infty$ *and* $D\mathfrak{f}(x)/D\mathfrak{g}(x) \to l$ *as* $x \to a$, *then* $\mathfrak{f}(x)/\mathfrak{g}(x) \to l$ *as* $x \to a$.

(i) Define new functions $\mathfrak{f}_1$, $\mathfrak{g}_1$ by

$$\mathfrak{f}_1(x) = \mathfrak{f}(x) \quad (x \neq a), \qquad \mathfrak{g}_1(x) = \mathfrak{g}(x) \quad (x \neq a),$$
$$\mathfrak{f}_1(a) = 0, \qquad \mathfrak{g}_1(a) = 0.$$

Choose $x_0 < a$ so that $D\mathfrak{g}(x) \neq 0$ in $[x_0, a]$. The conditions of theorem 34 and its corollary are satisfied. Thus

$$\frac{\mathfrak{f}(x_0)}{\mathfrak{g}(x_0)} = \frac{\mathfrak{f}_1(x_0)-\mathfrak{f}_1(a)}{\mathfrak{g}_1(x_0)-\mathfrak{g}_1(a)} = \frac{D\mathfrak{f}(c)}{D\mathfrak{g}(c)}$$

for some c satisfying $x_0 < c < a$. Let $x_0 \to a-$, then $c \to a-$ and $D\mathfrak{f}(c)/D\mathfrak{g}(c) \to l$. Hence $\mathfrak{f}(x_0)/\mathfrak{g}(x_0) \to l$ as $x_0 \to a-$. A similar result holds if $x_0 \to a+$ and the theorem is proved.

(ii) Since $D\mathfrak{f}(x)/D\mathfrak{g}(x) \to l$ as $x \to a$ $\forall \epsilon > 0 \exists \delta > 0$ such that

(1) $$\left|\frac{D\mathfrak{f}(x)}{D\mathfrak{g}(x)} - l\right| < \epsilon \quad \text{if} \quad 0 < |x-a| < \delta.$$

If x_1, x_2 satisfy $0 < x_1 - a < \delta$, $0 < x_2 - a < \delta$ we have from Cauchy's formula

$$(\mathfrak{f}(x_1)-\mathfrak{f}(x_2))\, D\mathfrak{g}(c) = (\mathfrak{g}(x_1)-\mathfrak{g}(x_2))\, D\mathfrak{f}(c),$$

i.e. dividing by $D\mathfrak{g}(c)$ and using (1).

(2) $$\mathfrak{f}(x_1)-\mathfrak{f}(x_2) = (\mathfrak{g}(x_1)-\mathfrak{g}(x_2)).\chi$$

where $|\chi - l| < \epsilon$. Now keep x_2 fixed and let $x_1 \to a+$. Then c varies and therefore χ varies but always we have $|\chi - l| < \epsilon$. If we divide (2) by $\mathfrak{g}(x_1)$ and observe that $\mathfrak{f}(x_2)/\mathfrak{g}(x_1) \to 0$, $\mathfrak{g}(x_2)/\mathfrak{g}(x_1) \to 0$ as $x_1 \to a+$ we obtain

$$l - \epsilon \leqslant \varliminf_{x_1 \to a+} \mathfrak{f}(x_1)/\mathfrak{g}(x_1) \leqslant \varlimsup_{x_1 \to a+} \mathfrak{f}(x_1)/\mathfrak{g}(x_1) \leqslant l + \epsilon.$$

These inequalities hold for every $\epsilon > 0$; thus $\lim_{x \to a+} \mathfrak{f}(x)/\mathfrak{g}(x)$ exists and is equal to l. Similarly $\lim_{x \to a-} \mathfrak{f}(x)/\mathfrak{g}(x)$ exists and is equal to l. Thus $\mathfrak{f}(x)/\mathfrak{g}(x) \to l$ as $x \to a$.

Remark. In application one often needs to consider higher-order derivatives. For example if

$$D^n\mathfrak{f}(x)/D^n\mathfrak{g}(x) \to l \quad \text{as} \quad x \to a, \quad D^r\mathfrak{f}(a) = D^r\mathfrak{g}(a) = \mathfrak{f}(a) = \mathfrak{g}(a) = 0 \quad (1 \leqslant r \leqslant n)$$

and $D^r\mathfrak{g}(x) \neq 0$ for $1 \leqslant r \leqslant n$ except at $x = a$, then $\mathfrak{f}(x)/\mathfrak{g}(x) \to l$ as $x \to a$.

Exercises. (1) $D\mathfrak{f}(a)$, $D\mathfrak{g}(a)$ exist and $D\mathfrak{g}(a) \neq 0$. Show that

$$\mathfrak{f}(x)/\mathfrak{g}(x) \to D\mathfrak{f}(a)/D\mathfrak{g}(a) \quad \text{as} \quad x \to a.$$

(2) $D\mathfrak{f}$, $D\mathfrak{g}$ exist for $x > 0$;

$$\mathfrak{f}(x) \to 0, \quad \mathfrak{g}(x) \to 0 \quad \text{as} \quad x \to \infty \quad \text{and} \quad D\mathfrak{f}(x)/D\mathfrak{g}(x) \to l \quad \text{as} \quad x \to \infty.$$

Show that $\mathfrak{f}(x)/\mathfrak{g}(x) \to l$ as $x \to \infty$.

(3) $D\mathfrak{f}$, $D\mathfrak{g}$ are defined for $x > 0$. $D\mathfrak{f}(x)/D\mathfrak{g}(x) \to l$ and $\mathfrak{g}(x) \to \infty$ as $x \to \infty$. Show that $\mathfrak{f}(x)/\mathfrak{g}(x) \to l$ as $x \to \infty$.

Theorem 36. *If $\mathfrak{f}$ is differentiable for $a \leqslant x \leqslant b$ then $D\mathfrak{f}$ is Darboux continuous over $[a, b]$.*

Suppose that $a \leqslant x_1 < x_2 \leqslant b$, $D\mathfrak{f}(x_1) = \alpha$, $D\mathfrak{f}(x_2) = \beta$ and that γ is a number lying between α and β. Assume for definiteness that $\alpha > \gamma > \beta$; the argument in the alternative case is similar and is omitted. Write $\mathfrak{g}(x) = \mathfrak{f}(x) - \gamma x$. Then

$$D\mathfrak{g}(x_1) = \alpha - \gamma > 0, \quad D\mathfrak{g}(x_2) = \beta - \gamma < 0.$$

Thus $\mathfrak{g}$ does not attain its least upper bound over $[x_1, x_2]$ either at x_1 or at x_2. But $\mathfrak{g}$ being differentiable at each point of $[x_1, x_2]$ is continuous over $[x_1, x_2]$, and by theorem 26 attains its upper bound at say c where by the above remark $x_1 < c < x_2$. But then $D\mathfrak{g}(c) = 0$, i.e. $D\mathfrak{f}(c) = \gamma$ and it follows that $D\mathfrak{f}$ is Darboux continuous.

Worked examples

(i) *The real-valued function $\mathfrak{g}$ is defined and Darboux continuous in $[0, 1]$, $D\mathfrak{g}$ exists and is bounded over $(0, 1)$. Show that $\mathfrak{g}$ is continuous over $[0, 1]$.*

We shall show that $\mathfrak{g}$ is continuous on the left at 1. Suppose that $|D\mathfrak{g}(y)| < M$ for $0 < y < 1$. By the mean-value theorem $\forall \epsilon > 0$ if $1 - (\epsilon/M) < y_1 < y_2 < 1$ then $\exists\, \eta$, $y_1 < \eta < y_2$ such that

$$|\mathfrak{g}(y_1) - \mathfrak{g}(y_2)| = |(y_1 - y_2)\, D\mathfrak{g}(\eta)| < \epsilon.$$

Thus by the general principle of convergence for limits $y \to 1-$ (see, for example, p. 89) it follows that $\lim_{y \to 1-} \mathfrak{g}(y)$ exists. Denote this limit by l. If $l < \mathfrak{g}(1)$ then $\exists\, \delta > 0$ such that if $x \in [1 - \delta, 1)$ we have $\mathfrak{g}(x) < \frac{1}{2}(l + \mathfrak{g}(1))$ and thus $\mathfrak{g}(x)$ does not take any value between $\frac{1}{2}(l + \mathfrak{g}(1))$ and $\mathfrak{g}(1)$ for $x \in [1 - \delta, 1)$. But $\mathfrak{g}(1 - \delta) \leqslant \frac{1}{2}(l + \mathfrak{g}(1))$ and thus we are led to a contradiction with the Darboux continuity condition for the interval $1 - \delta \leqslant x \leqslant 1$.

Thus $l \geqslant \mathfrak{g}(1)$. Similarly $l \leqslant \mathfrak{g}(1)$. Hence $l = \mathfrak{g}(1)$ and $\mathfrak{g}$ is continuous on the left at 1. By a similar argument $\mathfrak{g}$ is continuous on the right at 0.

The result is established.

(ii) *$D\mathfrak{f}$ is defined in a neighbourhood of $x = a$ and $D^2\mathfrak{f}(a)$ exists. Show that*

$$\lim_{h \to 0} \frac{\mathfrak{f}(a + 2h) - 2\mathfrak{f}(a + h) + \mathfrak{f}(a)}{h^2}$$

exists and is equal to $D^2\mathfrak{f}(a)$. Give an example to show that the second limit may exist even though $D\mathfrak{f}(a)$ does not exist.

Since $D^2\mathfrak{f}(a)$ exists we have

$$\lim_{h\to 0}\frac{D\mathfrak{f}(a+2h)-D\mathfrak{f}(a)}{2h} = \lim_{h\to 0}\frac{D\mathfrak{f}(a+h)-D\mathfrak{f}(a)}{h} = D^2\mathfrak{f}(a).$$

Hence

$$\lim_{h\to 0}\frac{2D\mathfrak{f}(a+2h)-2D\mathfrak{f}(a+h)}{2h}$$

$$= \lim_{h\to 0}\left(2\left(\frac{D\mathfrak{f}(a+2h)-D\mathfrak{f}(a)}{2h}\right)-\frac{D\mathfrak{f}(a+h)-D\mathfrak{f}(a)}{h}\right) = D^2\mathfrak{f}(a).$$

By theorem 35 (i)

$$\lim_{h\to 0}\frac{\mathfrak{f}(a+2h)-2\mathfrak{f}(a+h)+\mathfrak{f}(a)}{h^2}$$

exists and is equal to $D^2\mathfrak{f}(a)$.

An example of the required kind is $\mathfrak{f}(x) = |x|$, $a = 0$.

Exercises

1. $\mathfrak{f}$ is differentiable in a neighbourhood of $x = a$. Show that there exists $\{x_n\}$ such that $x_n \to a$ and $D\mathfrak{f}(x_n) \to D\mathfrak{f}(a)$ as $n \to \infty$.

2. $\mathfrak{f}$ is continuous in (a, b) and differentiable except possibly at $x = c$. If $\lim_{x\to c} D\mathfrak{f}(x)$ exists and is equal to l ($l \neq \pm\infty$) then show that $D\mathfrak{f}(c)$ exists and is equal to l.

3. $\mathfrak{f}$ is Darboux continuous over $[a, b]$ and differentiable at all points of (a, b). Prove that the conclusion of Rolle's theorem is still valid, i.e. if $\mathfrak{f}(a) = \mathfrak{f}(b)$ then $\exists\, c$ such that $a < c < b$ and $D\mathfrak{f}(c) = 0$.

4. Prove that if $\dfrac{a_0}{n+1}+\dfrac{a_1}{n}+\ldots+\dfrac{a_{n-1}}{2}+a_n = 0$, then the equation

$$a_0x^n+a_1x^{n-1}+\ldots+a_{n-1}x+a_n = 0$$

has at least one root between 0 and 1.

5. $\mathfrak{f}$ is non-negative and possesses a derivative of the third order for $0 < x < 1$. Prove that if $\mathfrak{f}$ is zero for at least two values of x in $(0, 1)$ then $D^3\mathfrak{f}$ takes the value zero for at least one value of x in $(0, 1)$.

6. $\mathfrak{f}$ has a continuous derivative in $[a-h, a+2h]$, $h > 0$ and its second derivative exists in $(a-h, a+2h)$. Show that $\exists\, c$ such that $a-h < c < a+2h$ and

$$\big(\mathfrak{f}(a+2h)-3\mathfrak{f}(a)+2\mathfrak{f}(a-h)\big)/h^2 = 3D^2\mathfrak{f}(c).$$

7. $D\mathfrak{f}$ exists for $a \leqslant x \leqslant b+h$, $h > 0$, $b > a$ and $D\mathfrak{f} > 0$, $D\mathfrak{f}\Uparrow$. Show that $\mathfrak{f}(x+h)-\mathfrak{f}(x) = hD\mathfrak{f}(c)$ and $\mathfrak{c}(x) = c$ defines $\mathfrak{c}$ as a single-valued continuous increasing function of x.

8. $\mathfrak{f}$ is continuous in $[a, b]$ and $D^2\mathfrak{f}$ exists and is positive at all points of (a, b). Show that $(\mathfrak{f}(x)-\mathfrak{f}(a))/(x-a)$ is a strictly increasing function of x in (a, b).

9. Show that

(i) if $D\mathfrak{f}(x) \to a$ as $x \to \infty$, $a \neq 0$ then $\mathfrak{f}(x) \sim xa$;

(ii) if $D\mathfrak{f}(x) \to 0$ as $x \to \infty$ then $\mathfrak{f}(x) = o(x)$;

(iii) if $D\mathfrak{f}(x) \to \infty$ as $x \to \infty$ then $\mathfrak{f}(x) \to \infty$;

(iv) if $\mathfrak{f}(x) \to a$ as $x \to \infty$ then $D\mathfrak{f}(x)$ cannot tend to any limit other than zero as $x \to \infty$.

10. If $\mathfrak{f}(x)+D\mathfrak{f}(x) \to a$ as $x \to \infty$ then show that $\mathfrak{f}(x) \to a$ and $D\mathfrak{f}(x) \to 0$ as $x \to \infty$.

11. $\mathfrak{f}$ and $\mathfrak{g}$ are differentiable for all large x and $D\mathfrak{g}(x) > 0$, $\mathfrak{g}(x) \to \infty$ as $x \to \infty$. Prove that

$$\underline{\lim}_{x\to\infty} \frac{D\mathfrak{f}(x)}{D\mathfrak{g}(x)} \leqslant \underline{\lim}_{x\to\infty} \frac{\mathfrak{f}(x)}{\mathfrak{g}(x)} \leqslant \overline{\lim}_{x\to\infty} \frac{\mathfrak{f}(x)}{\mathfrak{g}(x)} \leqslant \overline{\lim}_{x\to\infty} \frac{D\mathfrak{f}(x)}{D\mathfrak{g}(x)}.$$

12. Prove that if the nth derivative of $\mathfrak{f}$ exists in some neighbourhood of the point $x = 0$ and the $n+1$th derivative exists at $x = 0$ then

$$\mathfrak{f}(h)-\mathfrak{f}(0)-\frac{h}{1!}D\mathfrak{f}(0)-\frac{h^2}{2!}D^2\mathfrak{f}(0)-\dots-\frac{h^n}{n!}D^n\mathfrak{f}(0) = \frac{h^{n+1}}{(n+1)!}(D^{n+1}\mathfrak{f}(0)+o(1))$$

as $h \to 0$.

13. $\mathfrak{f}$ is differentiable in (a, b). Prove that the closure of the set of values taken by the expression $(\mathfrak{f}(x)-\mathfrak{f}(y))/(x-y)$, $a < x < y < b$ is identical with the closure of the set of values taken by $D\mathfrak{f}$ in $a < x < b$.

14. (i) $\mathfrak{f}$ is defined over $[a, b]$ and its derivative at each point of this interval is zero. Prove that $\mathfrak{f}$ is a constant.

(ii) $\mathfrak{f}$ is defined and continuous over $[a,b]$ and $D^+\mathfrak{f} = D_+\mathfrak{f} = 0$ for $a \leqslant x \leqslant b$. Prove that $\mathfrak{f}$ is a constant. (Hint: use the Heine–Borel theorem. Given $\epsilon > 0$ a pre-assigned positive number we can find a finite set of intervals covering $[a, b]$ such that if (l, m) is one of these intervals and $x_1, x_2 \in (l, m)$ then $|\mathfrak{f}(x_1)-\mathfrak{f}(x_2)| < 2\epsilon\,|l-m|$ and then use the fact that any such covering contains a subcovering such that no point of $[a, b]$ belongs to more than two intervals of the subcovering.)

15. $\mathfrak{f}$ is defined for $0 \leqslant x \leqslant a$, $\mathfrak{f}(0) = 0$ and $D\mathfrak{f}$ is positive and increasing. Show that $\mathfrak{f}(x)/x \uparrow$ for $0 < x \leqslant a$.

18

THE *n*th MEAN-VALUE THEOREM: TAYLOR'S THEOREM

The mean-value theorem can be extended for suitable functions to give a formula relating the value of a function at an arbitrary point with its value and those of its first $n-1$ derivatives at a fixed point, together with its nth derivative at a suitable point depending on the arbitrary point but not explicitly defined. In many cases one can take the limit as $n \to \infty$ and obtain an expression for the value of the function as a power series.

Theorem 37. *The n-th mean-value theorem. The n-th differential coefficient of* $\mathfrak{f}$ *exists for* $a < x < b$ *and the* $(n-1)$*-th differential coefficient is continuous for* $a \leqslant x \leqslant b$. *Then if* p *is a fixed positive number* $\exists\, \xi$ *such that* $a < \xi < b$ *and*

$$\mathfrak{f}(b) = \mathfrak{f}(a) + (b-a)\,D\mathfrak{f}(a) + \frac{(b-a)^2}{2!}D^2\mathfrak{f}(a) + \dots$$
$$+ \frac{(b-a)^{n-1}}{(n-1)!}D^{n-1}\mathfrak{f}(a) + \frac{(b-\xi)^{n-p}\,(b-a)^p}{p(n-1)!}D^n\mathfrak{f}(\xi).$$

Write

$$\mathfrak{F}(x) = \mathfrak{f}(b) - \mathfrak{f}(x) - (b-x)\,D\mathfrak{f}(x) - \dots - \frac{(b-x)^{n-1}}{(n-1)!}D^{n-1}\mathfrak{f}(x)$$

and
$$\mathfrak{g}(x) = \mathfrak{F}(x) - \left(\frac{b-x}{b-a}\right)^p \mathfrak{F}(a).$$

Then $\mathfrak{g}(a) = \mathfrak{g}(b) = 0$ and $\mathfrak{g}$ satisfies all the conditions of Rolle's theorem (theorem 32). Thus $\exists\, \xi$ such that $a < \xi < b$ and $D\mathfrak{g}(\xi) = 0$, i.e.

$$0 = D\mathfrak{F}(\xi) + \frac{p(b-\xi)^{p-1}}{(b-a)^p}\mathfrak{F}(a).$$

On rewriting this equation and substituting for $\mathfrak{F}$ we have

$$\mathfrak{f}(b) = \mathfrak{f}(a) + (b-a)\,D\mathfrak{f}(a) + \dots + \frac{(b-a)^{n-1}}{(n-1)!}D^{n-1}\mathfrak{f}(a) + R_n,$$

where
$$R_n = \frac{(b-\xi)^{n-p}}{p(n-1)!}(b-a)^p\,D^n\mathfrak{f}(\xi).$$

The theorem is proved.

R_n is called the *remainder* and this form of it is due to Schlömilch. Particular cases are

$(p = n)$ $$R_n = \frac{(b-a)^n}{n!}D^n\mathfrak{f}(\xi) \quad \text{(Lagrange's form)},$$

$(p = 1)$ $$R_n = \frac{(b-a)^n}{(n-1)!}(1-\theta)^{n-1}D^n\mathfrak{f}(a+\theta(b-a))$$

(Cauchy's form) where $0 < \theta < 1$.

Remark. In applications of this formula we usually (but not always) regard a as fixed and b as variable. In the proof we adopt the opposite point of view and replace a by x as an explicit recognition of this fact. We could consider other functions instead of $\mathfrak{g}(x)$, for example, if $\mathfrak{g}(x) = \mathfrak{F}(x) - \mathfrak{G}(x).\mathfrak{F}(a)$ where $\mathfrak{G}$ is any function such that $\mathfrak{G}(b) = 0$, $\mathfrak{G}(a) = 1$, $\mathfrak{G}$ is continuous over $[a, b]$ and differentiable over (a, b) then we obtain the following form for the remainder (assuming $D\mathfrak{G}(x) \neq 0, a \leqslant x \leqslant b$)

$$\frac{D\mathfrak{F}(\xi)}{D\mathfrak{G}(\xi)} = \frac{-(b-\xi)^{n-1}}{(n-1)!} \frac{D^n\mathfrak{f}(\xi)}{D\mathfrak{G}(\xi)}$$

for an appropriate ξ, $a < \xi < b$.

Corollary 1. *If $\mathfrak{f}$ has derivatives of all orders for x in $[a, b]$ and the remainder R_n in its n-th mean-value expansion tends to zero as $n \to \infty$, then*

$$\mathfrak{f}(b) = \sum_{n=0}^{\infty} \frac{(b-a)^n}{n!} D^n\mathfrak{f}(a).$$

This result is known as *Taylor's theorem* and the series is called the *Taylor's series* of the function $\mathfrak{f}(x)$ at b expanded about the point a. The fact that $R_n \to 0$ implies both that the series

$$\Sigma \frac{(b-a)^{n-1}}{(n-1)!} D^{n-1}\mathfrak{f}(a)$$

converges and that its sum is $\mathfrak{f}(b)$. The series may, however, converge without its sum being equal to $\mathfrak{f}(b)$. We can construct an example by using power series but we need another result first.

If $\mathfrak{f}$ is defined over say $[0, b]$ and we know that it is equal to its Taylor series expansion $\sum_{0}^{\infty} D^n\mathfrak{f}(0) . x^n/n!$ in $0 \leqslant x \leqslant b$ then the series $\Sigma D^{n-1}\mathfrak{f}(0)\, x^{n-1}/(n-1)!$ is a power series whose radius of convergence is at least b; in particular it is convergent for $-b < x \leqslant 0$. (It need not converge for $x = -b$.) Thus the sum of this series exists for $-b < x \leqslant 0$. Denote it by $\mathfrak{g}(x)$. A natural question is whether $\mathfrak{g}(x)$ has a Taylor series expansion and if it is equal to the sum of its Taylor series for $-b < x \leqslant 0$. This is in fact the case and if we define $\mathfrak{h}(x)$ by $\mathfrak{h}(x) = \mathfrak{f}(x)$, $0 \leqslant x \leqslant b$, $\mathfrak{h}(x) = \mathfrak{g}(x)$, $-b < x \leqslant 0$, then $\mathfrak{h}(x)$ has a Taylor series expansion that is valid throughout $-b < x \leqslant b$. The proof depends upon the following theorem in which a_n $(n = 0, 1 \ldots)$ are real.

Theorem 38. *If the real power series $\Sigma a_{n-1} x^{n-1}$ converges for $|x| < R$ to the sum $\mathfrak{f}(x)$, then $\mathfrak{f}(x)$ is differentiable for $|x| < R$ and $D\mathfrak{f}(x) = \sum_{1}^{\infty} n a_n x^{n-1}$.*

By (v), p. 66 we have

$$\mathfrak{f}(x) = \sum_{0}^{\infty} c_n (x - x_1)^n \quad \text{if} \quad |x_1| < R \quad \text{and} \quad |x - x_1| < R - |x_1|,$$

where
$$c_n = \sum_{k=0}^{\infty} \binom{n+k}{n} a_{n+k} x_1^k.$$

By the corollary to theorem 15

$$\mathfrak{f}(x)-c_0-c_1(x-x_1) = O(x-x_1)^2 \quad \text{as} \quad x \to x_1,$$

i.e.

$$\frac{\mathfrak{f}(x)-\mathfrak{f}(x_1)}{x-x_1}-c_1 \to 0 \quad \text{as} \quad x \to x_1.$$

Thus $D\mathfrak{f}(x_1)$ exists and is equal to $c_1 = \sum_1^\infty n a_n x_1^{n-1}$.

The theorem is proved.

Corollary. *Any power series with a positive radius of convergence is the Taylor series of its sum inside its circle of convergence.*

For if

$$\mathfrak{f}(x) = \sum_0^\infty a_n x^n \quad \text{in} \quad |x| < R$$

then $D\mathfrak{f}(0) = a_1$ and $D\mathfrak{f}(x) = \sum_1^\infty n a_n x^{n-1}$ in $|x| < R$.

Thus by a simple induction argument $D^n\mathfrak{f}(0) = n!\,a_n$ and the Taylor series of $\mathfrak{f}$ namely

$$\Sigma \frac{D^{n-1}\mathfrak{f}(0)}{(n-1)!} x^{n-1}$$

coincides with the power series $\Sigma a_{n-1} x^{n-1}$.

An important example. We next give an example of a function whose Taylor series converges but does not converge to the value of the function. Consider first the series $\Sigma x^{n-1}/(n-1)!$. This series converges for all x and by theorem 38 if $\mathfrak{f}(x)$ is its sum then $D\mathfrak{f} = \mathfrak{f}$. Also for any number k, $\mathfrak{f}(x)/x^k \to \infty$ as $x \to \infty$; for $\exists\, N$ such that N is an integer $> k$. Then $\mathfrak{f}(x)/x^k > x^{N-k}/N! \to \infty$ as $x \to \infty$. Thus $x^k/\mathfrak{f}(x) \to 0$ as $x \to \infty$. Similarly $x^k/\mathfrak{f}(x^2) \to 0$ as $x \to -\infty$. Now the function of the example is defined by

$$\mathfrak{g}(0) = 0,$$

$$\mathfrak{g}(x) = 1/\mathfrak{f}(x^{-2}) \quad (x \neq 0).$$

Then

$$\frac{\mathfrak{g}(x)-\mathfrak{g}(0)}{x} = \frac{1/x}{\mathfrak{f}(1/x^2)} \to 0 \quad \text{as} \quad x \to 0.$$

For if we put $x = 1/y$ this reduces to $y/\mathfrak{f}(y^2) \to 0$ as $y \to \infty$, or as $y \to -\infty$. Thus $D\mathfrak{g}(0)$ exists and is equal to zero. Further if $x \neq 0$

$$D\mathfrak{g}(x) = \frac{-D\mathfrak{f}(x^{-2})}{(\mathfrak{f}(x^{-2}))^2} = \frac{2}{x^3}\frac{1}{\mathfrak{f}(x^{-2})} = \mathfrak{g}(x)\,.\,\mathfrak{P}_1(x),$$

where $\mathfrak{P}_1(x)$ is a polynomial in $1/x$. Next

$$\frac{D\mathfrak{g}(x)-D\mathfrak{g}(0)}{x} = \frac{\mathfrak{g}(x)\,.\,\mathfrak{P}_1(x)}{x} \to 0 \quad \text{as} \quad x \to 0$$

as above. Thus $D^2\mathfrak{g}(0)$ exists and is equal to zero. Further if $x \neq 0$

$$D^2\mathfrak{g}(x) = D\mathfrak{g}(x).\mathfrak{P}_1(x) + \mathfrak{g}(x).\mathfrak{Q}(x) = \mathfrak{g}(x).\mathfrak{P}_2(x),$$

where $\mathfrak{P}_2(x)$ is a polynomial in $1/x$.

By an induction argument $D^n\mathfrak{g}(0) = 0$, and $D^n\mathfrak{g}(x) = \mathfrak{g}(x).\mathfrak{P}_n(x)$, where $\mathfrak{P}_n(x)$ is a polynomial in x. Thus the Taylor series of $\mathfrak{g}(x)$ exists and is $\Sigma 0.x^{n-1}$. But $\mathfrak{g}(x) \neq 0$ for $x \neq 0$ and thus although the Taylor series of $\mathfrak{g}(x)$ exists it does not converge to $\mathfrak{g}(x)$ for any $x \neq 0$.

The Binomial series

An important Taylor series is that for $(1-x)^{-m}$ when $|x| < 1$. We have

$$\frac{1}{(1-x)^m} = 1 + mx + \frac{m(m+1)}{1.2}x^2 + \dots$$
$$+ \frac{m(m+1)\dots(m+n-1)}{1.2.\dots.n}x^n + \dots,$$

where $|x| < 1$ and m is any real number. This identity may be established by using the nth mean-value theorem with Cauchy's form of the remainder and showing that the remainder tends to zero as $n \to \infty$. Alternatively it may be verified as follows. Write

$$a_n = \frac{m(m+1)\dots(m+n-1)}{1.2.\dots.n}.$$

Then $a_{n+1}/a_n = (m+n)/(n+1)$ and thus $a_{n+1}/a_n \to 1$ as $n \to \infty$. Hence $\Sigma a_{n+1}x^{n-1}$ converges absolutely for $|x| < 1$. Using theorem 38 we have

$$D\left((1-x)^m \sum_0^\infty a_n x^n\right)$$
$$= -m(1-x)^{m-1}\sum_0^\infty a_n x^n + (1-x)^m \sum_0^\infty n a_n x^{n-1}$$
$$= (1-x)^{m-1}\left(\sum_0^\infty -m a_n x^n + \sum_0^\infty n a_n x^{n-1} - \sum_0^\infty n a_n x^n\right)$$
$$= (1-x)^{m-1}\sum_0^\infty ((n+1)a_{n+1} - (m+n)a_n)x^n$$
$$= 0,$$

where we have used the fact that all the series involved converge for $|x| < 1$.

Thus $(1-x)^m \sum_0^\infty a_n x^n$ is a constant for $|x| < 1$. But its value when $x = 0$ is 1. Thus it is equal to 1 for $|x| < 1$, i.e.

$$\frac{1}{(1-x)^m} = \sum_0^\infty a_n x^n = 1 + mx + \frac{m(m+1)}{1.2} x^2 + \ldots \quad \text{for} \quad |x| < 1.$$

Maxima and minima

The nth mean-value theorem can be used to determine the local behaviour of suitable functions.

If $D^n\mathfrak{f}(x)$ exists in a neighbourhood of a and is continuous at a and if further $D^r\mathfrak{f}(a) = 0$, $1 \leqslant r \leqslant n-1$, $D^n\mathfrak{f}(a) \neq 0$, then

$$\mathfrak{f}(x) - \mathfrak{f}(a) \sim \frac{(x-a)^n}{n!} D^n\mathfrak{f}(a) \quad \text{as} \quad x \to a.$$

Thus if n is even a is a *maximum* if $D^n\mathfrak{f}(a) < 0$,

a is a *minimum* if $D^n\mathfrak{f}(a) > 0$;

if n is odd a is neither maximum nor minimum. It is a *point of inflexion.*

Worked examples

(i) *$\mathfrak{f}$ has a fifth derivative in $[-h, h]$ whose modulus is bounded by M and* $\mathfrak{g}(x) = \mathfrak{f}(x) - \mathfrak{f}(-x) - \frac{1}{3}x(D\mathfrak{f}(x) + D\mathfrak{f}(-x) + 4D\mathfrak{f}(0))$. *Show that*

$$|D^3\mathfrak{g}(x)| \leqslant \tfrac{2}{3}Mx^2 \quad \textit{if} \quad |x| \leqslant h.$$

Deduce that

$$|\mathfrak{g}(h)| \leqslant \tfrac{1}{9}Mh^5.$$

By direct calculation

$$D\mathfrak{g}(x) = \tfrac{2}{3}(D\mathfrak{f}(x) + D\mathfrak{f}(-x)) - \tfrac{4}{3}D\mathfrak{f}(0) - \tfrac{1}{3}x(D^2\mathfrak{f}(x) - D^2\mathfrak{f}(-x)),$$

$$D^2\mathfrak{g}(x) = \tfrac{1}{3}D^2\mathfrak{f}(x) - \tfrac{1}{3}D^2\mathfrak{f}(-x) - \tfrac{1}{3}x(D^3\mathfrak{f}(x) + D^3\mathfrak{f}(-x)),$$

$$D^3\mathfrak{g}(x) = -\tfrac{1}{3}x(D^4\mathfrak{f}(x) - D^4\mathfrak{f}(-x)).$$

From this last relation by the mean-value theorem applied to $D^4\mathfrak{f}$ we obtain $D^3\mathfrak{g}(x) = -\frac{2}{3}x^2D^5\mathfrak{f}(\xi)$ for some ξ lying between x and $-x$. Thus

$$|D^3\mathfrak{g}(x)| \leqslant \tfrac{2}{3}Mx^2.$$

Next by the third mean-value theorem applied to $\mathfrak{g}(x)$

$$\mathfrak{g}(h) = \mathfrak{g}(0) + hD\mathfrak{g}(0) + \frac{h^2}{2!}D^2\mathfrak{g}(0) + \frac{h^3}{3!}D^3\mathfrak{g}(\xi_1)$$

for some ξ_1, $0 < \xi_1 < h$. Also $\mathfrak{g}(0) = D\mathfrak{g}(0) = D^2\mathfrak{g}(0) = 0$. Thus

$$|\mathfrak{g}(h)| \leqslant \frac{h^3}{3!}\tfrac{2}{3}M\xi_1^2 \leqslant \tfrac{1}{9}Mh^5.$$

(ii) $\mathfrak{f}$ *and its first* $n+2$ *derivatives are continuous in a neighbourhood of* $x = 0$ *and* $D^{n+1}\mathfrak{f}(0) \neq 0$. *The number* θ_n *is such that Lagrange's form of the n-th mean-value theorem is*

$$\mathfrak{f}(x) = \mathfrak{f}(0) + xD\mathfrak{f}(0) + \frac{x^2}{2!}D^2\mathfrak{f}(0) + \ldots + \frac{x^{n-1}}{(n-1)!}D^{n-1}\mathfrak{f}(0) + \frac{x^n}{n!}D^n\mathfrak{f}(\theta_n x),$$

where $0 < \theta_n < 1$. *Show that*

$$\theta_n = \frac{1}{n+1} + \frac{nx}{2(n+1)^2(n+2)}\frac{D^{n+2}\mathfrak{f}(0)}{D^{n+1}\mathfrak{f}(0)} + o(|x|) \quad \text{as} \quad x \to 0.$$

The $n+2$nd mean-value theorem is

$$\mathfrak{f}(x) = \mathfrak{f}(0) + xD\mathfrak{f}(0) + \ldots + \frac{x^n}{n!}D^n\mathfrak{f}(0) + \frac{x^{n+1}}{(n+1)!}D^{n+1}\mathfrak{f}(0) + \frac{x^{n+2}}{(n+2)!}D^{n+2}\mathfrak{f}(\theta x),$$

where $0 < \theta < 1$. Combine this with the nth mean-value expression for $\mathfrak{f}(x)$ given above. We get

$$D^n\mathfrak{f}(\theta_n x) = \frac{x}{n+1}D^{n+1}\mathfrak{f}(0) + \frac{x^2}{(n+1)(n+2)}D^{n+2}\mathfrak{f}(\theta x) + D^n\mathfrak{f}(0).$$

The second mean-value theorem applied to $D^n\mathfrak{f}(\theta_n x)$ gives

$$D^n\mathfrak{f}(\theta_n x) = D^n\mathfrak{f}(0) + \theta_n x D^{n+1}\mathfrak{f}(0) + \tfrac{1}{2}\theta_n^2 x^2 D^{n+2}\mathfrak{f}(\theta'\theta_n x),$$

where $0 < \theta' < 1$. Hence on combining these two equations

$$(*) \quad \theta_n x D^{n+1}\mathfrak{f}(0) + \tfrac{1}{2}\theta_n^2 x^2 D^{n+2}\mathfrak{f}(\theta'\theta_n x) = \frac{x}{n+1}D^{n+1}\mathfrak{f}(0) + \frac{x^2}{(n+1)(n+2)}D^{n+2}\mathfrak{f}(\theta x).$$

Now $D^{n+2}\mathfrak{f}(x)$ is continuous in a neighbourhood of $x = 0$ and is therefore bounded in some smaller neighbourhood of $x = 0$. Thus if we divide (*) by $x \, . \, D^{n+1}\mathfrak{f}(0)$ and use the fact that $|\theta_n| < 1$ we obtain

$$\theta_n = 1/(n+1) + O(|x|) \quad \text{as} \quad x \to 0.$$

In (*) replace θ_n in the second term by $1/(n+1)+O(|x|)$ and divide by $xD^{n+1}\mathfrak{f}(0)$ again. We obtain

$$\theta_n + \tfrac{1}{2}xD^{n+2}\mathfrak{f}(\theta'\theta_n x)\left(\frac{1}{n+1}+O(x)\right)^2 \Big/ D^{n+1}\mathfrak{f}(0)$$
$$= \frac{1}{n+1}+\frac{x}{(n+1)(n+2)}\frac{D^{n+2}\mathfrak{f}(\theta x)}{D^{n+1}\mathfrak{f}(0)}.$$

By the continuity of $D^{n+2}\mathfrak{f}(x)$ at $x = 0$,

$$D^{n+2}\mathfrak{f}(\theta'\theta_n x) = D^{n+2}\mathfrak{f}(0)+o(1) \quad \text{as} \quad x \to 0,$$
$$D^{n+2}\mathfrak{f}(\theta x) = D^{n+2}\mathfrak{f}(0)+o(1) \quad \text{as} \quad x \to 0.$$

Thus

$$\theta_n = \frac{1}{n+1}+\frac{nx}{2(n+1)^2(n+2)}\frac{D^{n+2}\mathfrak{f}(0)}{D^{n+1}\mathfrak{f}(0)}+o(|x|) \quad \text{as} \quad x \to 0.$$

Exercises

1. $\mathfrak{F}(x)$ has a continuous nth derivative in $[a,b]$ with upper and lower bounds H and h. $\mathfrak{F}$ and its first $n-1$ derivatives are zero at $x = a$. Show that

$$\frac{h(x-a)^n}{n!} \leqslant \mathfrak{F}(x) \leqslant \frac{H(x-a)^n}{n!} \quad (a \leqslant x \leqslant b).$$

By taking $$\mathfrak{F}(x) = \mathfrak{f}(x)-\mathfrak{f}(a)-\sum_1^{n-1}\frac{D^r\mathfrak{f}(a)}{r!}(x-a)^r,$$
deduce that if $D^n\mathfrak{f}(x)$ is continuous in $[a,b]$, then $\exists\, \xi$ such that $a \leqslant \xi \leqslant b$ and

$$\mathfrak{f}(x) = \mathfrak{f}(a)+\sum_1^{n-1}\frac{D^r\mathfrak{f}(a)}{r!}(x-a)^r+\frac{D^n\mathfrak{f}(\xi)}{n!}(x-a)^n.$$

2. If $D^2\mathfrak{f}$ is continuous and not zero at $x = a$ and

$$\mathfrak{f}(a+h)-\mathfrak{f}(a) = hD\mathfrak{f}(a+\theta h) \quad (0 < \theta < 1),$$

show that $\lim_{h\to 0}\theta = \frac{1}{2}$. Further if $D^n\mathfrak{f}$ is continuous and not zero at $x = a$ and $D\mathfrak{f}(a) = D^2\mathfrak{f}(a) = \ldots = D^{n-1}\mathfrak{f}(a) = 0$, then

$$\mathfrak{f}(a+h)-\mathfrak{f}(a) = \frac{h^{n-1}}{(n-1)!}D^{n-1}\mathfrak{f}(a+\theta h),$$

where $0 < \theta < 1$ and $\lim\theta = 1/n$.

3. If $D^2\mathfrak{f}$ exists for $x \geqslant 0$ show that for $x > 0\, \exists\, \xi$ such that $0 < \xi < x$ and $D\mathfrak{f}(x)-\mathfrak{f}(x)/x = \frac{1}{2}xD^2\mathfrak{f}(\xi)-\mathfrak{f}(0)/x$. Hence show that
(a) if $\mathfrak{f}(0) = 0$ and $D^2\mathfrak{f}(x) > 0$ for $x > 0$ then $\mathfrak{f}(x)/x \Uparrow$,
(b) if $\mathfrak{f}(0) \geqslant 0$ and $D^2\mathfrak{f}(x) < 0$ for $x > 0$ then $\mathfrak{f}(x)/x \Downarrow$.

4. $\mathfrak{f}$ has a continuous $n-1$th derivative for $a \leqslant x \leqslant a+nh$ and its nth derivative exists for $a < x < a+nh$ where $h > 0$, n a positive integer. Show that $\exists\, \xi$ such that

$$a < \xi < a+nh \quad \text{and} \quad \sum_{r=0}^{n}(-1)^r\binom{n}{r}\mathfrak{f}(x+rh) = (-h)^n D^n\mathfrak{f}(\xi).$$

Show that if $m > n$

$$x^{n-m} \sum_{r=0}^{n} (-1)^r \binom{n}{r} (x+rh)^m \to m(m-1)\dots(m-n+1)(-h)^n$$

as $x \to \infty$.

5. Write $$\mathfrak{f}(m,x) = \sum_{0}^{\infty} \frac{m(m-1)\dots(m-r+1)}{r!} x^r \quad (|x| < 1)$$

and show that

(a) $\mathfrak{f}(m+1,x) - \mathfrak{f}(m,x) = x\mathfrak{f}(m,x)$;
(b) $D\mathfrak{f}(m,x) = m\mathfrak{f}(m-1,x)$;
(c) $D(\mathfrak{f}(m,x)/(1+x)^m) = 0$;
(d) $\mathfrak{f}(m,x) = (1+x)^m$.

19

CONVEX AND CONCAVE FUNCTIONS

All functions in this chapter are real-valued functions of a real variable. The convex or concave functions are generalizations of functions with a second derivative of constant sign in the same way that monotonic functions are generalizations of functions with a first derivative of constant sign.

Definition 44. *$\mathfrak{f}$ defined over (a,b) is convex if for any pair of real numbers λ, μ where $\lambda \geqslant 0$, $\mu \geqslant 0$, $\lambda+\mu = 1$ and any pair of real numbers x, y satisfying $a < x < y < b$ we have*

$$\mathfrak{f}(\lambda x + \mu y) \leqslant \lambda \mathfrak{f}(x) + \mu \mathfrak{f}(y).$$

If $-\mathfrak{f}$ is a convex function then $\mathfrak{f}$ is said to be a *concave* function.

If $\mathfrak{f}$ is defined over $[a,b]$ and satisfies the inequality above for all x, y satisfying $a \leqslant x < y \leqslant b$, then we say that $\mathfrak{f}$ is convex over $[a,b]$.

Exercises. (1) Show that if $\mathfrak{f}$ and $\mathfrak{g}$ are convex over (a,b) then so is $\mathfrak{f}+\mathfrak{g}$.

(2) $\mathfrak{f}_i$ is convex over (a,b) and $\mathfrak{f}_i(x) \leqslant M$ for $i \in I$ and $a < x < b$. Show that $\sup \mathfrak{f}_i(x)$ is convex over (a,b).

Theorem 39. *If $\mathfrak{f}$ is convex over (a,b), then*

(i) *$\mathfrak{f}$ is bounded over any closed interval contained in (a,b) and is continuous at each point of (a,b).*

(ii) *The right-hand and left-hand derivatives of $\mathfrak{f}$ exist and both are increasing functions of x.*

(i) By the convexity condition if $a < x_1 \leqslant x_2 \leqslant x_3 < b$, $x_1 < x_3$ with $\lambda = (x_3 - x_2)/(x_3 - x_1)$, $\mu = (x_2 - x_1)/(x_3 - x_1)$, x replaced by x_1 and y by x_3 we have on multiplication by $(x_3 - x_1)$,

$$(1) \qquad (x_3 - x_1)\,\mathfrak{f}(x_2) \leqslant (x_3 - x_2)\,\mathfrak{f}(x_1) + (x_2 - x_1)\,\mathfrak{f}(x_3).$$

Thus
$$\mathfrak{f}(x_2) \leqslant \max(\mathfrak{f}(x_1), \mathfrak{f}(x_3))$$

and $\mathfrak{f}$ is bounded above in $[x_1, x_3]$.

Next if $x_1 < x_2 < x_3$ and $x_2 \leqslant x \leqslant x_3$, then with x_3 in (1) replaced by x we have

$$(x - x_1)\,\mathfrak{f}(x_2) \leqslant (x - x_2)\,\mathfrak{f}(x_1) + (x_2 - x_1)\,\mathfrak{f}(x),$$
$$\mathfrak{f}(x) \geqslant \big((x - x_1)\,\mathfrak{f}(x_2) - (x - x_2)\,\mathfrak{f}(x_1)\big)/(x_2 - x_1)$$
$$\geqslant -\frac{|x_3 - x_1|}{x_2 - x_1}\,(|\mathfrak{f}(x_1)| + |\mathfrak{f}(x_2)|)$$

and $\mathfrak{f}$ is bounded below in $[x_2, x_3]$. Similarly it is bounded below in $[x_1, x_2]$ and thus it is bounded below in $[x_1, x_3]$.

Hence $\mathfrak{f}$ is bounded in $[x_1, x_3]$.

If $a < x < b$ and $|\delta|$ is sufficiently small then $a < x + 2\delta < b$, $a < x + \delta < b$, $a < x - \delta < b$. By the convexity condition

$$\mathfrak{f}(x + \delta) \leqslant \tfrac{1}{2}(\mathfrak{f}(x) + \mathfrak{f}(x + 2\delta)).$$

Taking upper limits as $\delta \to 0$ and denoting $\overline{\lim\limits_{y \to x}}\,\mathfrak{f}(y)$ by L, we have

$$L \leqslant \tfrac{1}{2}(\mathfrak{f}(x) + L),$$

i.e.
$$\mathfrak{f}(x) \geqslant L.$$

By a second application of the convexity condition

$$\mathfrak{f}(x) \leqslant \tfrac{1}{2}(\mathfrak{f}(x + \delta) + \mathfrak{f}(x - \delta))$$

taking lower limits as $\delta \to 0$ (see worked example (iv), chapter 15) and writing l for $\underline{\lim\limits_{y \to x}}\,\mathfrak{f}(y)$ we have

$$\mathfrak{f}(x) \leqslant \tfrac{1}{2}\underline{\lim_{\delta \to 0}}\,(\mathfrak{f}(x + \delta) + \mathfrak{f}(x - \delta))$$
$$\leqslant \tfrac{1}{2}(l + L).$$

But $\mathfrak{f}(x) \geqslant L$. Hence $\mathfrak{f}(x) \leqslant \tfrac{1}{2}(l + \mathfrak{f}(x))$. Thus $\mathfrak{f}(x) \leqslant l$, i.e. $L \leqslant \mathfrak{f}(x) \leqslant l$. Since in any case $L \geqslant l$ we have $\mathfrak{f}(x) = L = l$ and $\mathfrak{f}$ is continuous at x.

(ii) If $\eta > \delta > 0$ and x, $x + \eta$ both belong to (a, b), then

$$\mathfrak{f}(x + \delta) \leqslant \frac{\delta}{\eta}\,\mathfrak{f}(x + \eta) + \frac{\eta - \delta}{\eta}\,\mathfrak{f}(x)$$

by the convexity condition. Write $\mathfrak{g}(\delta) = (\mathfrak{f}(x+\delta)-\mathfrak{f}(x))/\delta$, then the above inequality gives $\mathfrak{g}(\delta) \leqslant \mathfrak{g}(\eta)$. Hence $\mathfrak{g}(x)\uparrow$ function of x defined for $x > 0$. Thus $\lim_{\delta\to 0+} \mathfrak{g}(\delta)$ exists, possibly $-\infty$. But the convexity condition applied again gives, for $\zeta > 0$, $a < x-\zeta < b$,

$$(\mathfrak{f}(x)-\mathfrak{f}(x-\zeta))/\zeta \leqslant \mathfrak{g}(\delta). \tag{2}$$

Thus $\lim_{\delta\to 0+} \mathfrak{g}(\delta)$ exists and is finite. That is to say the right-hand derivate of $\mathfrak{f}$ exists at x.

Similarly the left-hand derivate exists and it follows from (2) that it is less than or equal to the right-hand derivate at the same point.

Further if $a < x_2 < x_1 < b$ and n is so large that $x_1+(x_1-x_2)/n < b$ then by the convexity condition

$$\mathfrak{f}\left(x_2+\frac{r(x_1-x_2)}{n}\right)-\mathfrak{f}\left(x_2+\frac{(r-1)(x_1-x_2)}{n}\right)$$
$$\leqslant \mathfrak{f}\left(x_2+\frac{(r+1)(x_1-x_2)}{n}\right)-\mathfrak{f}\left(x_2+\frac{r(x_1-x_2)}{n}\right)$$

for $r = 1, 2, \ldots, n$. By combining these inequalities we have

$$\mathfrak{f}\left(x_2+\frac{x_1-x_2}{n}\right)-\mathfrak{f}(x_2) \leqslant \mathfrak{f}\left(x_1+\frac{x_1-x_2}{n}\right)-\mathfrak{f}(x_1)$$

and on letting $n\to\infty$ we deduce that the right-hand derivate of $\mathfrak{f}$ is increasing.

Similarly the left-hand derivate is also increasing.

Jensen's inequality. If $\mathfrak{f}$ *is convex,*

$$\alpha_i \geqslant 0, \alpha_1+\alpha_2+\ldots+\alpha_n = 1 \quad \textit{and} \quad a < x_i < b, i = 1, 2, \ldots, n,$$

then we have

$$\mathfrak{f}(\alpha_1 x_1+\alpha_2 x_2+\ldots+\alpha_n x_n) \leqslant \alpha_1\mathfrak{f}(x_1)+\ldots+\alpha_n\mathfrak{f}(x_n).$$

If $\alpha_n = 1$ then $\alpha_i = 0$ $(i = 1, 2, \ldots, n-1)$ and the inequality reduces to a trivial identity.

If $\alpha_n < 1$ then from the definition of convexity

$$\mathfrak{f}(\alpha_1 x_1+\alpha_2 x_2+\ldots+\alpha_n x_n) \tag{3}$$
$$\leqslant (1-\alpha_n)\mathfrak{f}\left(\frac{\alpha_1 x_1+\ldots+\alpha_{n-1}x_{n-1}}{1-\alpha_n}\right)+\alpha_n\mathfrak{f}(x_n).$$

The inequality to be proved is true if $n = 2$, since it is then the definition of convexity. Assume inductively that it is true with n replaced by $n-1$, that is to say

$$\mathfrak{f}(\beta_1 x_1 + \ldots + \beta_{n-1} x_{n-1}) \leqslant \beta_1 \mathfrak{f}(x_1) + \ldots + \beta_{n-1} \mathfrak{f}(x_{n-1})$$

if $\beta_i \geqslant 0$ and $\beta_1 + \ldots + \beta_{n-1} = 1$. In particular if $\beta_i = \alpha_i/(1-\alpha_n)$ then

$$\mathfrak{f}\left(\frac{\alpha_1 x_1 + \ldots + \alpha_{n-1} x_{n-1}}{1-\alpha_n}\right) \leqslant \frac{\alpha_1}{1-\alpha_n} \mathfrak{f}(x_1) + \ldots + \frac{\alpha_{n-1}}{1-\alpha_n} \mathfrak{f}(x_{n-1}).$$

If this inequality is combined with (3) we obtain the required inequality. Thus, by induction, the inequality is true for all integers n greater than or equal to 2.

$\mathfrak{f}$ *is such that* $D^2\mathfrak{f}$ *is defined over* (a, b). *Then* $\mathfrak{f}$ *is convex over* (a, b) *if and only if* $D^2\mathfrak{f} \geqslant 0$.

If $\mathfrak{f}$ is convex and $D\mathfrak{f}$ exists then it follows from theorem 39 (ii) that $D\mathfrak{f}\uparrow$ and thus $D^2\mathfrak{f} \geqslant 0$.

Conversely if $D^2\mathfrak{f} \geqslant 0$ then $D\mathfrak{f}\uparrow$. Given x, y, λ, μ such that $a < x < y < b$, $\lambda \geqslant 0$, $\mu \geqslant 0$, and $\lambda + \mu = 1$ we have by two applications of the mean-value theorem

$$\mathfrak{f}(\lambda x + \mu y) - \mathfrak{f}(x) = ((\lambda - 1) x + \mu y) D\mathfrak{f}(\xi) = \mu(y-x) D\mathfrak{f}(\xi),$$

$$\mathfrak{f}(y) - \mathfrak{f}(\lambda x + \mu y) = (-\lambda x + (1-\mu) y) D\mathfrak{f}(\eta) = \lambda(y-x) D\mathfrak{f}(\eta),$$

where $x \leqslant \xi \leqslant \lambda x + \mu y$, $\lambda x + \mu y \leqslant \eta \leqslant y$. Since $D\mathfrak{f}\uparrow$ and these inequalities imply that $\xi \leqslant \eta$ it follows that

$$\lambda\big(\mathfrak{f}(\lambda x + \mu y) - \mathfrak{f}(x)\big) \leqslant \mu\big(\mathfrak{f}(y) - \mathfrak{f}(\lambda x + \mu y)\big),$$

i.e. that

$$\mathfrak{f}(\lambda x + \mu y) \leqslant \lambda \mathfrak{f}(x) + \mu \mathfrak{f}(y).$$

Thus $\mathfrak{f}$ is convex over (a, b).

Exercises. (1) Show that x^u regarded as a function of x is convex over $x > 0$ when $u \geqslant 1$, concave when $0 \leqslant u \leqslant 1$ and convex when $u \leqslant 0$.

(2) $\Sigma a_{n-1} x^{n-1}$ converges to $\mathfrak{f}(x)$ for $0 \leqslant x \leqslant R$ and $a_n \geqslant 0$. Show that $\mathfrak{f}$ is convex over $(0, R)$.

If in the definition 44 the case of equality can arise only when $\lambda = 0$ or $\mu = 0$, then $\mathfrak{f}$ is said to be strictly convex over (a, b) or $[a, b]$ as the case may be. By an argument similar to that above $\mathfrak{f}$ is strictly convex over (a, b) if $D^2\mathfrak{f}(x) > 0$ for $x \in (a, b)$.

If $\mathfrak{f}$ is strictly convex then equality holds in Jensen's inequality if and only if all the points a_i for which $\alpha_i \neq 0$ coincide.

Worked examples

(i) $\mathfrak{f}$ *is bounded over* (a, b) *and satisfies*

$$\mathfrak{f}(\tfrac{1}{2}(x+y)) \leqslant \tfrac{1}{2}(\mathfrak{f}(x)+\mathfrak{f}(y))$$

for x, y *satisfying* $a < x < y < b$. *Show that* (*a*) $\mathfrak{f}$ *is continuous,* (*b*) $\mathfrak{f}$ *is convex.*

(*a*) The proof of continuity is the same as that given in theorem 39 (i).

(*b*) From the given condition by an inductive argument it follows that if $a < x < b$ and $x < x+y < b$

$$\mathfrak{f}(x+py/2^r) \leqslant (1-p/2^r)\,\mathfrak{f}(x)+(p/2^r)\,\mathfrak{f}(x+y),$$

where $p = 0, 1, 2, \ldots, 2^r$ and $r = 1, 2, \ldots$. By continuity, for any λ with $0 \leqslant \lambda \leqslant 1$ it follows that

$$\mathfrak{f}(x+\lambda y) \leqslant (1-\lambda)\,\mathfrak{f}(x)+\lambda\mathfrak{f}(x+y).$$

Replacing $1-\lambda$ by μ and y by $y-x$ we obtain

$$\mathfrak{f}(\mu x+\lambda y) \leqslant \mu\mathfrak{f}(x)+\lambda\mathfrak{f}(y),$$

where $\lambda \geqslant 0$, $\mu \geqslant 0$, $\lambda+\mu = 1$, $a < x < b$ and $x < y < b$. Since it is merely a matter of notation that $x < y$ it follows that $\mathfrak{f}$ is convex.

(ii) *If* $\mathfrak{f}$ *is convex over* $a < x < b$ *then* $\mathfrak{f}$ *is differentiable except at an enumerable set of points.*

Write $D^+(x)$ and $D^-(x)$ for the right and left derivates of $\mathfrak{f}$ at x. By the argument on p. 122 $D^+(x) \geqslant D^-(x)$ and if $x_1 < x_2$ then $D^+(x_1) \leqslant (\mathfrak{f}(x_2)-\mathfrak{f}(x_1))/(x_2-x_1) \leqslant D^-(x_2)$. Thus if D^+ and D^- are continuous at x they are equal there. By monotoneity (p. 99 (i)) this is so except for an enumerable set of points. The result follows.

Exercises

1. $\mathfrak{f}$ is a strictly increasing convex function defined and bounded over $[a, b]$. Show that its inverse function is concave.

2. $\mathfrak{f}$ is convex over (a, b) and $\mathfrak{g}$ is increasing and convex over an interval which includes the range of $\mathfrak{f}$. Show that $\mathfrak{g}(\mathfrak{f}(x))$ is convex over (a, b).

3. $\mathfrak{f}$ is defined and convex over $[a, b]$. Show that if for some c with $a < c < b$ we have $\mathfrak{f}(a) = \mathfrak{f}(b) = \mathfrak{f}(c)$ then $\mathfrak{f}$ is a constant.

4. $\mathfrak{f}$ is convex and increasing for $x > 0$ and $\mathfrak{f}(x) = o(x)$ as $x \to \infty$. Show that $\mathfrak{f}$ is a constant.

5. Show that if $\lambda_i \geqslant 0$, $x_i \geqslant 0$ $(i = 1, 2, \ldots, n)$, then for $\alpha > 1$

$$(\lambda_1 x_1^\alpha + \lambda_2 x_2^\alpha + \ldots + \lambda_n x_n^\alpha) \geqslant (\lambda_1 + \lambda_2 + \ldots + \lambda_n)^{1-\alpha} (\lambda_1 x_1 + \lambda_2 x_2 + \ldots + \lambda_n x_n)^\alpha.$$

6. $\{\mathfrak{f}_n(x)\}$ is a sequence of functions convex and defined over (a, b) and $|\mathfrak{f}_n(x)| \leqslant M$ for $a < x < b$, $n = 1, 2, \ldots$. Show that $\overline{\lim_{n\to\infty}}\, \mathfrak{f}_n(x)$ is convex over (a, b). Give an example to show that $\underline{\lim_{n\to\infty}}\, \mathfrak{f}_n(x)$ need not be convex over (a, b).

7. Show that a convex function is of bounded variation.

20

THE ELEMENTARY TRANSCENDENTAL FUNCTIONS

There are certain functions which have proved to be of fundamental importance in analysis. We introduce the simplest of these functions as the sums of power series.

(1) *The exponential function* $\exp z$.

$\exp z$ is defined for all complex or all real z as the sum of the convergent power series $1 + z + (z^2/2!) + \ldots$. Thus

$$\exp z = \sum_0^\infty \frac{z^n}{n!}.$$

From chapter 11, theorem 15, corollary 1 and property (v) of power series, $\exp z$ is a continuous function of z. From theorem 38, chapter 18, if z is real, say $z = x$, then $\exp x$ is a differentiable function of x and $D(\exp x) = \exp x$.

Write

$$\begin{aligned}\mathfrak{f}_N(z) &= (1 + z/N)^N \\ &= 1 + z + 1\left(1 - \frac{1}{N}\right)\frac{z^2}{2!} + \ldots + 1\left(1 - \frac{1}{N}\right)\ldots\left(1 - \frac{N-1}{N}\right)\frac{z^N}{N!}.\end{aligned}$$

Now
$$\left(1 - \frac{1}{N}\right)\ldots\left(1 - \frac{n-1}{N}\right)\frac{1}{n!} \to \frac{1}{n!} \quad \text{as} \quad N \to \infty$$

and
$$\left|\left(1 - \frac{1}{N}\right)\ldots\left(1 - \frac{n-1}{N}\right)\frac{1}{n!}\right| < \frac{1}{n!}.$$

Thus by the dominated convergence of power series (p. 63) it follows that $\mathfrak{f}_N(z) \to \exp z$ as $N \to \infty$. In particular

$$e = \lim (1 + 1/N)^N = \exp 1 = 1 + \frac{1}{1!} + \frac{1}{2!} + \ldots.$$

By the multiplication of two absolutely convergent series (p. 50) we deduce that $\exp z_1 . \exp z_2 = \exp(z_1+z_2)$ and by repeated application, for any integer m, $\exp m = (\exp 1)^m = e^m$. If p, q are positive integers then $(\exp p/q)^q = \exp p = e^p$. Thus $\exp p/q = e^{p/q}$. By the continuity of the two functions $\exp x$, e^x for real x it follows that $\exp x = e^x$ for any real positive number x. But for real negative x, $\exp x = 1/\exp(-x)$, $e^x = 1/e^{-x}$. Thus for all real x, $e^x = \exp x$.

For complex z we *define* e^z to be $\exp z$.

Exercise. Show that for any complex number z $\exp z \neq 0$.

(2) *The logarithmic function* $\log x$ *for real positive* x.

In this section the variable x is always real. $\exp x$ is continuous; if $x \geqslant 0$ by definition $\exp x > 0$, if $x < 0$, $\exp x = 1/(\exp -x) > 0$. But $D \exp x = \exp x$ and thus $\exp x \Uparrow$. Further $\exp x \to \infty$ as $x \to \infty$ and $\exp x = 1/(\exp -x) \to 0$ as $x \to -\infty$. By theorem 28, p. 97, $\exists$ a function $\mathfrak{g}(y)$ defined over $0 < y$ such that

$$\exp \mathfrak{g}(y) = y, \quad \mathfrak{g}(\exp x) = x$$

and $\mathfrak{g}(y)$ is continuous and strictly increasing. The function $\mathfrak{g}$ is called the *logarithmic function* and we denote $\mathfrak{g}(y)$ by $\log y$. By theorem 30

$$D \log y = 1/\exp x = 1/y \quad (0 < y).$$

By theorem 37, since $D^n \log y = (-1)^{n-1}(n-1)!/y^n$, we have

$$\log(1+y) = \log 1 + y - \frac{y^2}{2} + \ldots + (-1)^{n-2}\frac{y^{n-1}}{n-1} + R_n,$$

where
$$R_n = \frac{(-1)^{n-1}}{n}\frac{y^n}{\xi^n}$$

and ξ lies between 1 and $1+y$. If $0 \leqslant y \leqslant 1$ then $y \leqslant 1 \leqslant \xi$ and $|R_n| \leqslant 1/n \to 0$ as $n \to \infty$. Also $\log 1 = \log(\exp 0) = 0$. Thus

$$\log(1+y) = \sum_1^\infty \frac{(-1)^{n-1}y^n}{n} \quad \text{for} \quad 0 \leqslant y \leqslant 1.$$

This expansion is also valid if $-1 < y < 0$. For the radius of convergence of $\Sigma(-1)^{n-2}y^{n-1}/(n-1)$ is 1 and if $|y| < 1$ then

$$D\left(\log(1+y) - \sum_1^\infty \frac{(-1)^{n-1}y^n}{n}\right) = \frac{1}{1+y} - \sum_1^\infty (-1)^{n-1}y^{n-1} = 0.$$

Thus $\log(1+y)$ differs from $\sum_1^\infty (-1)^{n-1}y^n/n$ by a constant, and since the two functions are equal when $y = 0$, this constant is zero. Thus

$$\log(1+y) = \sum_1^\infty (-1)^{n-1}y^n/n \quad \text{for} \quad -1 < y \leqslant 1.$$

Since (i) $\exp(a+b) = \exp a . \exp b$

and (ii) $(\exp a)^b = e^{ab} = \exp ab$,

we also have (i) $\log AB = \log A + \log B$

and (ii) $\log(A^B) = B \log A$.

Remark. No attempt is made here to define $\log z$ for complex numbers z. Such a definition requires a careful analysis involving the complex variable calculus.

(3) *The trigonometric functions* $\sin z$ *and* $\cos z$.
These are defined for all z, real or complex by

$$\sin z = z - \frac{z^3}{3!} + \frac{z^5}{5!} - \ldots = \sum_0^\infty \frac{(-1)^n z^{2n+1}}{(2n+1)!},$$

$$\cos z = 1 - \frac{z^2}{2!} + \frac{z^4}{4!} - \ldots = \sum_0^\infty \frac{(-1)^n z^{2n}}{(2n)!}.$$

These functions are continuous for all z and if z is real they are differentiable. Also $D \sin x = \cos x$, $D \cos x = -\sin x$. From the definitions and the properties of the multiplication of power series it follows that

(*a*) $\sin z = -\sin(-z)$, $\cos z = \cos(-z)$;

(*b*) $\cos z \cos w \pm \sin z \sin w = \cos(z \mp w)$;

(*c*) $\cos^2 z + \sin^2 z = 1$, $\cos^2 z - \sin^2 z = \cos 2z$;

(*d*) $\sin z \cos w \pm \cos z \sin w = \sin(z \pm w)$;

(*e*) $\sin 2z = 2 \cos z \sin z$.

If x is real then $\cos x$, $\sin x$ are also real and thus the identity $\sin^2 x + \cos^2 x = 1$ implies that

$$-1 \leqslant \sin x \leqslant 1, \quad -1 \leqslant \cos x \leqslant 1.$$

Denote by i that complex number whose real part is 0 and whose imaginary part is 1. Then

$$\cos z + i \sin z = \exp(iz), \quad \cos z - i \sin z = \exp(-iz),$$

$$\cos z = \tfrac{1}{2}(\exp iz + \exp(-iz)), \quad \sin z = \frac{1}{2i}(\exp(iz) - \exp(-iz)).$$

If $\sin\frac{1}{2}z \neq 0$ then for positive integers M, N

$$\sum_N^M \cos nz = \sum_N^M (2\cos nz . \sin\tfrac{1}{2}z)/(2\sin\tfrac{1}{2}z)$$

$$= \sum_N^M (\sin(n+\tfrac{1}{2})z - \sin(n-\tfrac{1}{2})z)/(2\sin\tfrac{1}{2}z)$$

$$= (\sin(M+\tfrac{1}{2})z - \sin(N-\tfrac{1}{2})z)/(2\sin\tfrac{1}{2}z).$$

Similarly

$$\sum_N^M \sin nz = (\cos(N-\tfrac{1}{2})z - \cos(M+\tfrac{1}{2})z)/(2\sin\tfrac{1}{2}z).$$

We show next that the trigonometric functions are periodic, i.e. $\exists$ a number α such that $\sin(z+\alpha) = \sin z$, $\cos(z+\alpha) = \cos z$.

It is sufficient to establish the existence of a real positive number 2π such that $\cos 2\pi = 1$, for if such a number exists then since $\cos^2 x + \sin^2 x = 1$ it follows that $\sin 2\pi = 0$ and the required periodicity is established since by the formulae (*b*) and (*d*) given above

$$\sin(z+2\pi) = \sin z \cos 2\pi + \cos z \sin 2\pi,$$

$$\cos(z+2\pi) = \cos z \cos 2\pi - \sin z \sin 2\pi.$$

Let X be the set of all positive real numbers x such that $\cos x = 1$. We show first that $X \neq \phi$. If $\exists\,\beta$ such that $\cos\beta = -1$ then $\cos 2\beta = 2\cos^2\beta - 1$ (from the two identities (*c*)) and $\cos 2\beta = 1$, i.e. $X \neq \phi$. If $\exists\,\gamma$ such that $\cos\gamma = 0$ then $\cos 2\gamma = -1$ and $\cos 4\gamma = 1$. Either such a γ exists or, since $\cos x$ is continuous and $\cos 0 = 1$, $\cos x > 0$ for all $x > 0$. But in this second case

$$D\sin x = \cos x > 0 \quad \text{and} \quad \sin x \Uparrow.$$

Hence $\sin x$ converges to a finite limit say m as $x \to \infty$. As $\sin 0 = 0$ $m > 0$, $D\cos x = -\sin x < 0$ ($\sin x > 0$ for $x > 0$ since $\sin x \Uparrow$, $\sin 0 = 0$). Thus $\cos x \Downarrow$ and $\cos x \to$ a limit say l as $x \to \infty$. Also $l < 1$. Let $x \to \infty$ in (*e*) then $m = 2ml$ or, as $m > 0$, $l = \frac{1}{2}$. But from (*c*) $\cos 2x = 2\cos^2 x - 1$ and letting $x \to \infty$, $l = 2l^2 - 1$. However, when $l = \frac{1}{2}$, $l \neq 2l^2 - 1$ and this contradiction shows that $X \neq \phi$. The function $\cos x$ is continuous for real x and thus either 0 is a limit point of X or X is a closed set. But by the mean-value theorem

$$\cos x = 1 - \frac{x^2}{2!} + \frac{x^4}{4!}\cos\theta_1 x, \quad \text{where} \quad 0 < \theta_1 < 1,\ x > 0$$

$$\leqslant 1 - \frac{x^2}{2!} + \frac{x^4}{4!} < 1 \quad \text{if} \quad 0 < x < 1.$$

Thus 0 is not a limit point of X and X is closed. We denote $\inf X$ by 2π and this real positive number has the properties stated.

Actually 2π is the least positive number with the periodicity properties. For suppose that α is such that

$$\sin(z+\alpha) = \sin z \quad \text{and} \quad \cos(z+\alpha) = \cos z.$$

Then
$$\sin z \cos\alpha + \cos z \sin\alpha = \sin z,$$

$$\cos z \cos\alpha - \sin z \sin\alpha = \cos z$$

and on multiplying the first equation by $\sin z$ and the second by $\cos z$ and adding we have (using (c)) $\cos\alpha = 1$. Thus if α is real and positive then $\alpha \in X$. Thus $\alpha \geqslant 2\pi$.

Moreover, the only values of z for which $\cos z = 1$ are precisely the members of X, the number 0 and the negatives of the members of X. For if $z = x+iy$, then

$$\cos z = \cos x \cos iy - \sin x \sin iy,$$

$\mathscr{I}(\cos z) = -\sin x.\{y+(y^3/3!)+\ldots\}$. This if $\cos z = 1$ then $y = 0$ or $\sin x = 0$. If $\sin x = 0$ then $\cos x = \pm 1$ and since

$$\cos iy = 1+\frac{y^2}{2!}+\frac{y^4}{4!}+\ldots \geqslant 1$$

we can have $\cos z = 1$ only if $\cos x = 1$, $y = 0$. Thus $\cos z = 1$ only if z is real and since $\cos z = \cos(-z)$ the statement made above is true.

If m is a positive integer then

$$\cos 2\pi m = \cos 2\pi(m-1)\cos 2\pi - \sin 2\pi(m-1)\sin 2\pi$$

$$= \cos 2\pi(m-1)$$

and by a simple induction $\cos 2\pi m = 1$, i.e. $2\pi m \in X$ for all positive integers m. These numbers comprise the whole of the set X. For if $\exists\ \alpha \in X$ let m be such that $2\pi m \leqslant \alpha < 2\pi(m+1)$. Then $0 \leqslant \alpha - 2\pi m < 2\pi$ and

$$\cos(\alpha - 2\pi m) = \cos\alpha\cos 2\pi m + \sin\alpha\sin 2\pi m = 1.$$

Since 2π is the least member of X, $\alpha - 2\pi m = 0$.

The other trigonometric functions are defined for all appropriate z by

$$\tan z = \frac{\sin z}{\cos z}, \quad \sec z = \frac{1}{\cos z}, \quad \operatorname{cosec} z = \frac{1}{\sin z}, \quad \cot z = \frac{1}{\tan z}.$$

(4) *The hyperbolic functions.*

These are

$$\cosh z = \tfrac{1}{2}(\exp z + \exp(-z)), \quad \sinh z = \tfrac{1}{2}(\exp z - \exp(-z)),$$

$$\tanh z = \frac{\sinh z}{\cosh z}, \quad \operatorname{sech} z = \frac{1}{\cosh z},$$

$$\operatorname{cosech} z = \frac{1}{\sinh z}, \quad \coth z = \frac{1}{\tanh z}.$$

There are relations between the hyperbolic functions analogous to these between the trigonometric functions and which can be deduced from them by observing that

$$\cosh z = \cos iz, \quad \sinh z = (\sin iz)/i.$$

(5) *The power of a real positive number.*

For any real positive number a and any number z real or complex we define a^z to be $e^{z \log a}$.

If z is real then $z \log a = \log a^z$ and this definition agrees with that given previously (p. 25).

Worked examples

(i) *$\{v_n\}$ is a real decreasing null sequence. Show that*

(*a*) *$\Sigma v_n \sin nx$ converges for all real x and*

(*b*) *$\Sigma v_n \cos nx$ converges for all real x for which $\sin \tfrac{1}{2}x \neq 0$.*

If $\sin \tfrac{1}{2}x \neq 0$ it follows from identities in (3) above that

$$\left| \sum_N^M \sin nx \right| \leqslant |\sin \tfrac{1}{2}x|^{-1}, \quad \left| \sum_N^M \cos nx \right| \leqslant |\sin \tfrac{1}{2}x|^{-1}.$$

By theorem 8, p. 46, both $\Sigma v_n \sin nx$ and $\Sigma v_n \cos nx$ converge. Thus (*b*) is proved. To complete the proof of (*a*) we observe that if $\sin \tfrac{1}{2}x = 0$, then $\sin nx = 0$ for every integer n and the series is simply a series of zeros.

(ii) *Show that for any real k, $\exp n/n^k \to \infty$ as $n \to \infty$.*

Let R be an integer greater than k. Then since

$$\exp n = 1 + n + \ldots + \frac{n^R}{R!} + \ldots$$

$$\frac{\exp n}{n^k} > \frac{n^R}{R!\, n^k} = \frac{n^{R-k}}{R!} \to \infty \quad \text{as} \quad n \to \infty.$$

(iii) *The partial sums of the series $\Sigma a_n/n^z$ are bounded and w is such that $\mathscr{R}(w) > \mathscr{R}(z)$. Show that $\Sigma a_n/n^w$ converges.*

Write $\Sigma a_n/n^w = \Sigma a_n/n^z . 1/n^{w-z}$. If we can show that

$$(a)\ \frac{1}{n^{w-z}} \to 0 \quad \text{as} \quad n \to \infty, \quad \text{and} \quad (b)\ \Sigma\left|\frac{1}{n^{w-z}} - \frac{1}{(n+1)^{w-z}}\right|$$

converges, then the convergence of $\Sigma a_n/n^w$ is a consequence of theorem 8 (d), p. 46.

For (a) $|1/n^{w-z}| = 1/n^\delta$, where $\delta = \mathscr{R}(w-z) > 0$. Thus ($a$) is true. For ($b$) write $w-z = u$ and

$$\frac{1}{(n+1)^u} = \frac{1}{n^u\left(1+\frac{1}{n}\right)^u} = \frac{1}{n^u} e^{-u \log(1+1/n)}$$

$$= \frac{1}{n^u}\left\{1 - u \log\left(1+\frac{1}{n}\right) + O\left(\log\left(1+\frac{1}{n}\right)\right)^2\right\} \quad \text{as} \quad n \to \infty$$

$$= \frac{1}{n^u}\left(1 - \frac{u}{n} + O\left(\frac{1}{n^2}\right)\right) \quad \text{as} \quad n \to \infty,$$

since $\log(1+1/n) = 1/n + O(1/n^2)$ by the power series expansion given above. Thus

$$\left|\frac{1}{n^u} - \frac{1}{(n+1)^u}\right| = \left|\frac{1}{n^u}\left(-\frac{u}{n} + O\left(\frac{1}{n^2}\right)\right)\right| = O\left(\frac{1}{n^{1+\mathscr{R}u}}\right)$$

and since $\mathscr{R}(u) > 0$ it follows from the comparison principle that

$$\Sigma\left|\frac{1}{n^u} - \frac{1}{(n+1)^u}\right|$$

converges. Hence (b) is proved.

Exercises

1. (a) Show that $x - \sin x \uparrow$ for all real x and that $\tan x - x \uparrow$ for real x in $(-\frac{1}{2}\pi, \frac{1}{2}\pi)$.
(b) Show that $x/\sin x \uparrow$ for x in $(0, \pi)$.

2. Deduce from exercise 1 (a) that $\sin x < x$ if $x > 0$ and hence that $\cos x - 1 + \frac{1}{2}x^2 > 0$ if $x > 0$.

Show that if m is odd and $x > 0$

$$\cos x - 1 + \frac{x^2}{2!} - \dots (-1)^{m+1}\frac{x^{2m}}{2m!} > 0,$$

$$\sin x - x + \frac{x^3}{3!} - \dots (-1)^{m+1}\frac{x^{2m+1}}{(2m+1)!} > 0,$$

and that these inequalities are reversed if m is even.

3. Prove that if m is a positive integer and $x < 0$ then

$$1 + x + \frac{x^2}{2!} + \dots + \frac{x^{2m+1}}{(2m+1)!} < e^x < 1 + x + \frac{x^2}{2!} + \dots + \frac{x^{2m}}{(2m)!}.$$

4. Prove that if $0 < x < 1$ then

$$\pi < \frac{\sin \pi x}{x(1-x)}.$$

5. If x, λ are real show that

$$|\sin \lambda x| \leqslant |\sin (\lambda - 1) x| + |\sin x|$$

and that for any positive integer m

$$|\sin mx| < m\,|\sin x|.$$

6. Evaluate, if possible (cf. theorem 35), where x is real.

(i) $\lim\limits_{x \to 0} (\cos ax)^{\operatorname{cosec}^2 bx} \quad (a \neq 0,\ b \neq 0).$

(ii) $\lim\limits_{x \to 0} \dfrac{x^2 \sin 1/x}{e^x - 1}.$

(iii) $\lim\limits_{x \to 0} \dfrac{\sin^4 x}{2 \sin (e^x - 1) - x(x+2)}.$

(iv) $\lim\limits_{x \to 0} \dfrac{1}{x^3}\left(\cot x - \dfrac{1}{x} + \dfrac{x}{3}\right).$

7. Show that, x being real,

$$\lim_{x \to 1} \frac{x - (n+1)\,x^{n+1} + nx^{n+2}}{(1-x)^2} = \frac{n(n+1)}{2},$$

$$\lim_{x \to 1} \frac{1 - 4 \sin^2 \pi x/6}{1 - x^2} = \frac{\pi}{2\sqrt{3}},$$

$$\lim_{x \to 0} \frac{\tan x - x}{x - \sin x} = 2,$$

$$\lim_{x \to 0} \frac{\tan nx - n \tan x}{n \sin x - \sin nx} = 2.$$

8. For what real values of x are the series

$$\sum_{n=1}^{\infty} n^{i-x}, \quad \sum_{n=1}^{\infty} (-1)^n n^{i-x},$$

(*a*) absolutely convergent, (*b*) convergent

9. Show that $\Sigma\, e^{\sqrt{n}} z^n$ converges abolutely if $|z| < 1$ and diverges if $|z| \geqslant 1$. Show that $\Sigma\, e^{-\sqrt{n}} z^n$ converges absolutely if $|z| \leqslant 1$ and diverges if $|z| > 1$.

10. p and q are real numbers. What additional conditions must they satisfy in order that the function $\mathfrak{f}$ defined by

$$\mathfrak{f}(0) = 0, \quad \mathfrak{f}(x) = x^p \cos (1/x^q) \quad (x \neq 0)$$

be (*a*) continuous, (*b*) of bounded variation, or (*c*) Darboux continuous in $0 \leqslant x \leqslant 1$?

21

INEQUALITIES

There are certain inequalities that are often useful in analysis. Four of the most important of these are given in this chapter.

The Arithmetic Mean–Geometric Mean inequality.

If the real numbers a_i $(i = 1, 2, ..., n)$ are non-negative, then

$$(a_1+a_2+\ldots+a_n)/n \geqslant (a_1 a_2 \ldots a_n)^{1/n}$$

and equality holds if and only if all the a_i are equal.

Since for $x > 0$, $D^2(-\log x) = 1/x^2 > 0$ the function $-\log x$ is strictly convex and by Jensen's inequality (p. 122)

$$\begin{aligned}-\log((a_1+\ldots+a_n)/n) &\leqslant -(\log a_1+\ldots+\log a_n)/n \\ &= -\log(a_1 \ldots a_n)^{1/n}\end{aligned}$$

with equality if and only if all the a_i are equal. Since $-\log x \Downarrow$ this implies that

$$(a_1+a_2+\ldots+a_n)/n \geqslant (a_1 . a_2 . \ldots . a_n)^{1/n}$$

with equality if and only if all the a_i are equal.

Exercises. (1) $a_i > 0$ and $p_i > 0$ are real numbers and $p_1+p_2+\ldots+p_n = 1$. Prove that $p_1 a_1+\ldots+p_n a_n \geqslant a_1^{p_1} . a_2^{p_2} . \ldots . a_n^{p_n}$.

(2) Prove that $((n+1)/2)^n \geqslant 1.2.\ldots.n$.

(3) x, y, z are positive real numbers such that $x+y+z = 1$. Show that $x^2yz \leqslant \frac{1}{64}$. (Hint: consider the two pairs of numbers y, z and x, $y+z$.)

Hölder's inequality. α, β are positive real numbers and $\alpha+\beta = 1$. If $a_i \geqslant 0$ $(i = 1, 2, ..., n)$ then

$$\left(\sum_{i=1}^{n} a_i\right)^{\alpha} \left(\sum_{i=1}^{n} b_i\right)^{\beta} \geqslant \sum_{i=1}^{n} a_i^{\alpha} b_i^{\beta},$$

with equality if and only if $\exists$ fixed numbers p, q not both zero and such that $pa_i+qb_i = 0$ $(i = 1, 2, ..., n)$.

Write $A = \sum_{i=1}^{n} a_i$. If $A = 0$ then $a_i = 0$ $(i = 1, 2, ..., n)$, both sides of the inequality are zero and the criterion for equality is satisfied with $p = 1$, $q = 0$.

If $A \neq 0 \; \exists$ some i for which $a_i \neq 0$. The function $-x^\beta$ is strictly convex for $x > 0$, $0 < \beta < 1$ and from Jensen's inequality (p. 122)

$$-\Sigma \frac{a_i}{A}\left(\frac{b_i}{a_i}\right)^\beta \geqslant -\left(\Sigma \frac{a_i}{A}\frac{b_i}{a_i}\right)^\beta = -\left(\Sigma \frac{b_i}{A}\right)^\beta,$$

where the summation is over integers i for which $a_i \neq 0$. Thus

$$\Sigma a_i^\alpha b_i^\beta \leqslant (\Sigma b_i)^\beta \left(\sum_{i=1}^n a_i\right)^\alpha \tag{1}$$

with the same range of summation. Moreover, there is strict inequality unless b_i/a_i has the same value for all i for which $a_i \neq 0$. Now we add to both sides of (1) the terms corresponding to integers i for which $a_i = 0$. This does not affect the left-hand side, it does not decrease the right-hand side and will actually increase the right-hand side unless $a_i = 0$ implies $b_i = 0$. Thus

$$\sum_{i=1}^n a_i^\alpha b_i^\beta \leqslant \left(\sum_{i=1}^n b_i\right)^\beta \left(\sum_{i=1}^n a_i\right)^\alpha$$

and equality holds only if (i) $a_i = 0$ implies $b_i = 0$, and (ii) b_i/a_i has the same value, say k, for all integers i for which $a_i \neq 0$. Thus if (i) and (ii) hold we have $pa_i + qb_i = 0$, where $p = k$, $q = -1$.

Finally, suppose that $pa_i + qb_i = 0$ $(i = 1, 2, \ldots, n)$, where not both p, q are zero. If $p = 0$ then $q \neq 0$ and therefore $b_i = 0$ $(i = 1, 2, \ldots, n)$ and both sides of the inequality are zero. Similarly if $q = 0$ the inequality also reduces to $0 = 0$. If neither p nor q is zero then the two conditions above hold, i.e. (i) $a_i = 0$ implies $b_i = 0$, and (ii) b_i/a_i has the same value k for all integers for which $a_i \neq 0$, where $k = -p/q$. But then if $A = 0$ we have equality as above and if $A \neq 0$, substitution for b_i in terms of a_i and k establishes the equality.

Cauchy's inequality. If

$$a_i \geqslant 0, \quad b_i \geqslant 0 \quad (i = 1, 2, \ldots, n)$$

then

$$\left(\sum_{i=1}^n a_i^2\right)\left(\sum_{i=1}^n b_i^2\right) \geqslant \left(\sum_{i=1}^n a_i b_i\right)^2,$$

with equality if and only if $\exists$ fixed numbers p, q not both zero and such that $pa_i + qb_i = 0$ $(i = 1, 2, \ldots, n)$.

This follows from Hölder's inequality by writing $\alpha = \beta = \frac{1}{2}$ and replacing a_i, b_i by a_i^2, b_i^2 respectively.

Exercises. (1) $m > 1$ and $x > 0$, $x \neq 1$. Show that $x^m - 1 > m(x-1)$ and deduce that if $x_i > 0$ and $\sum_{i=1}^{n} x_i = n$ then $\sum_{i=1}^{n} x_i^m \geqslant n$.

(2) Prove that

$$\sum_{i=1}^{n} i(n+1-i) \leqslant \sum_{i=1}^{n} i^2 \quad \text{and} \quad \sum_{i=1}^{n} i(n-i)^2 \leqslant \sum_{i=1}^{n} i^3.$$

Minkowski's inequality. If $a_i \geqslant 0$, $b_i \geqslant 0$ and $k \geqslant 1$ then

$$\left(\sum_{i=1}^{n} (a_i+b_i)^k\right)^{1/k} \leqslant \left(\sum_{i=1}^{n} a_i^k\right)^{1/k} + \left(\sum_{i=1}^{n} b_i^k\right)^{1/k}.$$

If $k = 1$ both sides of the inequality are equal. If $k > 1$ we have on applying Hölder's inequality twice after rewriting the sum $\sum_{i=1}^{n} (a_i+b_i)^k$,

$$\begin{aligned}\sum_{i=1}^{n} (a_i+b_i)^k &= \sum_{i=1}^{n} (a_i+b_i)^{k-1} a_i + \sum_{i=1}^{n} (a_i+b_i)^{k-1} b_i \\ &\leqslant \left(\sum_{i=1}^{n} (a_i+b_i)^k\right)^{(k-1)/k} \left(\sum_{i=1}^{n} a_i^k\right)^{1/k} \\ &\qquad + \left(\sum_{i=1}^{n} (a_i+b_i)^k\right)^{(k-1)/k} \left(\sum_{i=1}^{n} b_i^k\right)^{1/k}\end{aligned}$$

and thus $$\left(\sum_{i=1}^{n} (a_i+b_i)^k\right)^{1/k} \leqslant \left(\sum_{i=1}^{n} a_i^k\right)^{1/k} + \left(\sum_{i=1}^{n} b_i^k\right)^{1/k}.$$

Minkowski's inequality is proved.

Worked example

Give a direct proof of Cauchy's inequality.

We have

$$(2) \qquad \left(\sum_{i=1}^{n} a_i^2\right)\left(\sum_{i=1}^{n} b_i^2\right) = \sum_{i,j=1}^{n} a_i^2 b_j^2.$$

$$(3) \qquad \left(\sum_{i=1}^{n} a_i b_i\right)^2 = \sum_{i,j=1}^{n} a_i b_i a_j b_j.$$

Consider two integers say k, l, $1 \leqslant k < l \leqslant n$. Corresponding to these integers are two terms in (2), namely, $a_k^2 b_l^2$ and $a_l^2 b_k^2$ and two terms in (3), namely, $a_k b_k a_l b_l$ and $a_l b_l a_k b_k$. Now

$$a_k^2 b_l^2 + a_l^2 b_k^2 - 2a_k a_l b_k b_l = (a_k b_l - a_l b_k)^2 \geqslant 0.$$

Since similar inequalities hold for every pair of integers k, l we have

$$\left(\sum_{i=1}^{n} a_i^2\right)\left(\sum_{i=1}^{n} b_i^2\right) \geqslant \left(\sum_{i=1}^{n} a_i b_i\right)^2,$$

i.e. Cauchy's inequality. Moreover, equality occurs if and only if

$$a_k b_l - a_l b_k = 0 \quad \text{for every pair of integers } k, l,$$

and this is the criterion for equality if we choose k so that $a_k \neq 0$ or $b_k \neq 0$ and let l take all values $1, \ldots, n$. If such a k cannot be chosen then $a_k = b_k = 0$ for all k.

Exercises

1. z_1, z_2 are complex and $\lambda > 0$. Prove that

$$|z_1+z_2|^2 \leqslant (1+\lambda)\,|z_1|^2+(1+1/\lambda)\,|z_2|^2.$$

2. a, b, c are real. Prove that

$$|a|+|b|+|c|+|a+b+c| \geqslant |a+b|+|b+c|+|c+a|.$$

3. Real numbers $a_1, a_2, \ldots, a_n$ are such that $a_i(a_1+\ldots+a_{i-1}) > 0$ $(i = 2, \ldots, n)$. Prove that

$$(1+a_1)(1+a_2)\ldots(1+a_n) > 1+a_1+a_2+\ldots+a_n.$$

4. $a_i > 0$ $(i = 1, 2, \ldots, n)$. Prove that

$$(a_1+a_2+\ldots+a_n)(a_1^{-1}+a_2^{-1}+\ldots+a_n^{-1}) \geqslant n^2.$$

5. $a_i > 0$, $b_i > 0$ $(i = 1, 2, \ldots, n)$. Prove that

$$\left(\sum_{i=1}^{n} a_i\right)\left(\sum_{i=1}^{n} b_i\right) \geqslant \sum_{i=1}^{n} (a_i+b_i) \sum_{i=1}^{n} a_i b_i/(a_i+b_i).$$

6. $a_i > 0$, $p_i > 0$ $(i = 1, 2, \ldots, n)$ and $r > 0$. Prove that

$$\left(\sum_{i=1}^{n} p_i a_i^r\right)^2 \leqslant \left(\sum_{i=1}^{n} p_i\right)\left(\sum_{i=1}^{n} p_i a_i^{2r}\right).$$

7. $a_i > 0$ $(i = 1, 2, \ldots, n)$ and p_i is a positive integer $(i = 1, 2, \ldots, n)$. Show that

$$a_1^{p_1} a_2^{p_2} \ldots a_n^{p_n} \leqslant \left(\frac{p_1 a_1+p_2 a_2+\ldots+p_n a_n}{p_1+p_2+\ldots+p_n}\right)^{p_1+p_2+\ldots+p_n}.$$

Deduce that $$1.2^2.3^3.\ldots.n^n \leqslant (\tfrac{1}{3}(2n+1))^{\frac{1}{2}n(n+1)}.$$

8. q_i, a_i, b_i, c_i $(i = 1, 2, 3)$ are positive real numbers and $q_1+q_2+q_3 = 1$. Prove that

$$a_1^{q_1}a_2^{q_2}a_3^{q_3}+b_1^{q_1}b_2^{q_2}b_3^{q_3}+c_1^{q_1}c_2^{q_2}c_3^{q_3} \leqslant (a_1+b_1+c_1)^{q_1}(a_2+b_2+c_2)^{q_2}(a_3+b_3+c_3)^{q_3}.$$

9. a_i, b_i, c_i $(i = 1, 2, \ldots, n)$ are non-negative real numbers and $k > 1$. Prove that

$$\left(\sum_{i=1}^{n} (a_i+b_i+c_i)^k\right)^{1/k} \leqslant \left(\sum_{i=1}^{n} a_i^k\right)^{1/k} + \left(\sum_{i=1}^{n} b_i^k\right)^{1/k} + \left(\sum_{i=1}^{n} c_i^k\right)^{1/k}.$$

10. $a_{ij} \geqslant 0$ for $1 \leqslant i \leqslant n$ and $1 \leqslant j \leqslant m$, $k > 1$. Prove that

$$\left(\sum_{i=1}^{n} \left(\sum_{j=1}^{m} a_{ij}\right)^k\right)^{1/k} \leqslant \sum_{j=1}^{m} \left(\sum_{i=1}^{n} a_{ij}^k\right)^{1/k}.$$

22

THE RIEMANN INTEGRAL

We have defined the derivative of a function $\mathfrak{f}(x)$ and denoted it by $D\mathfrak{f}(x)$. It is often necessary to reverse this process and given a function $\mathfrak{g}$ to find, if possible, a function $\mathfrak{f}$ such that $D\mathfrak{f} = \mathfrak{g}$. When the calculus was first formulated such a function $\mathfrak{f}$ was called the integral of $\mathfrak{g}$ (or the indefinite integral). $\mathfrak{f}$ is of course not uniquely defined since it could be altered by the addition of a constant function without changing its derivative. Thus if D is regarded as a mapping of a certain class of functions into another class then integration was the inverse mapping D^{-1}. However, this definition is inconveniently restrictive; for example, the integral of $\mathfrak{g}^2$ need not be defined even though that of $\mathfrak{g}$ is defined. The Riemann integral is an extension of this idea in which these restrictions are removed. (Actually we define the Riemann integral over a fixed interval $[a, b]$ and it is a number. If we allow b to vary this number varies and as a function of b coincides with $D^{-1}\mathfrak{g}$ when this last is defined.)

Throughout this chapter $\mathfrak{f}$ is a *bounded* real-valued function of a real variable.

We suppose that $\mathfrak{f}$ is defined over $[a, b]$. A *dissection* of $[a, b]$ denoted by $\mathscr{D}$ is a finite set of points contained in $[a, b]$ and containing a, b (see also chapter 16). Any point of $\mathscr{D}$ is called a *dissection point*. If $\mathscr{D}$ is the set of points $x_0, x_1, \ldots, x_k$, where

$$a = x_0 < x_1 < x_2 < \ldots < x_k = b,$$

then $\max(x_{i+1} - x_i)$, $0 \leqslant i \leqslant k-1$, is called the *mesh* of $\mathscr{D}$ and is denoted by $\tau(\mathscr{D})$. We shall use w_i for the interval (x_i, x_{i+1}). Such an interval is called an interval of the dissection. We shall also write

$$M_i(\mathfrak{f}) = \sup_{x \in w_i} \mathfrak{f}(x), \quad m_i(\mathfrak{f}) = \inf_{x \in w_i} \mathfrak{f}(x),$$

or sometimes simply M_i, m_i when it is clear from the context which function is being considered.

If two dissections $\mathscr{D}_1$, $\mathscr{D}_2$ of a, b are such that $\mathscr{D}_1 \supset \mathscr{D}_2$, then we say that $\mathscr{D}_1$ is a *refinement* of $\mathscr{D}_2$.

$\mathfrak{f}$ is said to be a *step-function on* $\mathscr{D}$ if $\mathfrak{f}$ is constant for $x \in w_i$ $(i = 0, 1, \ldots, k-1)$. If $\mathfrak{f}$ is a step-function on a particular dissection it is one also on every refinement of $\mathscr{D}$.

Any function which is such that $\exists$ a dissection on which it is a step-function is referred to simply as a step-function.

Definition 45. *$\mathfrak{f}$ is Riemann integrable over $[a, b]$ if $\forall \epsilon > 0 \, \exists \delta > 0$ and a number l such that*

$$\left| l - \sum_{i=0}^{k-1} \mathfrak{f}(\xi_i)(x_{i+1} - x_i) \right| < \epsilon,$$

where $a = x_0 < x_1 < \ldots < x_k = b$, $x_i < \xi_i < x_{i+1}$ and the inequality is satisfied for any such sets of numbers x_i, ξ_i for which $x_{i+1} - x_i < \delta$ $(i = 0, 1, \ldots, k-1)$.

The number l is called the *Riemann integral of* $\mathfrak{f}$ *over* $[a, b]$ and will be denoted by

$$\int_a^b \mathfrak{f}(x)\,dx \quad \text{or} \quad \int_a^b \mathfrak{f}.$$

When the Riemann integral of $\mathfrak{f}$ exists it is unique. The interval $[a, b]$ is the *range of integration* and $\mathfrak{f}$ is the *integrand*. We shall also say that $\mathfrak{f}$ is R-integrable and denote the class of functions R-integrable over $[a, b]$ by $\mathsf{R}(a, b)$.

The definition of $\int_a^b \mathfrak{f}$ is extended to the case $b \leqslant a$ by making the two conventions

$$\text{(i)} \int_a^a \mathfrak{f} = 0, \quad \text{(ii)} \int_a^b \mathfrak{f} = -\int_b^a \mathfrak{f} \quad \text{if} \quad b < a.$$

The definition is valid whether $\mathfrak{f}$ is real-valued or complex-valued. But x must be a real variable and in fact we use the definition only when $\mathfrak{f}$ is real-valued.

Exercise. Show that the constant function $k \in \mathsf{R}(a, b)$ and $\int_a^b k = k(b-a)$.

Definition 46. *For any dissection $\mathscr{D}$ of $[a, b]$ the upper and lower Riemann sums of $\mathfrak{f}$ are $S^{\mathfrak{f}}_{\mathscr{D}}$, $s^{\mathfrak{f}}_{\mathscr{D}}$ defined by*

$$S^{\mathfrak{f}}_{\mathscr{D}} = \sum_{0}^{k-1} M_i(\mathfrak{f})\,(x_{i+1}-x_i), \quad s^{\mathfrak{f}}_{\mathscr{D}} = \sum_{0}^{k-1} m_i(\mathfrak{f})\,(x_{i+1}-x_i).$$

Lemma 1. *For any dissection $\mathscr{D}$ of $[a, b]$*

$$(b-a)\sup \mathfrak{f} \geqslant S^{\mathfrak{f}}_{\mathscr{D}} \geqslant s^{\mathfrak{f}}_{\mathscr{D}} \geqslant (b-a)\inf \mathfrak{f}.$$

This follows immediately from the fact that

$$\sup \mathfrak{f} \geqslant M_i(\mathfrak{f}) \geqslant m_i(\mathfrak{f}) \geqslant \inf \mathfrak{f}$$

for each interval of $\mathscr{D}$.

It follows that the numbers $S^{\mathfrak{f}}_{\mathscr{D}}$ are bounded below by a number that does not depend on $\mathscr{D}$. Thus $\inf_{\mathscr{D}} S^{\mathfrak{f}}_{\mathscr{D}}$ exists and is a number. We write it as $S^{\mathfrak{f}}$. Similarly we define $s^{\mathfrak{f}}$ to be $\sup_{\mathscr{D}} s^{\mathfrak{f}}_{\mathscr{D}}$.

Lemma 2. $S^{\mathfrak{f}} \geqslant s^{\mathfrak{f}}$.

Given $\epsilon > 0 \; \exists$ dissections say $\mathscr{D}$, $\mathscr{E}$ such that

$$S^{\mathfrak{f}}_{\mathscr{D}} < S^{\mathfrak{f}} + \epsilon, \quad s^{\mathfrak{f}}_{\mathscr{E}} > s^{\mathfrak{f}} - \epsilon.$$

We shall show that $S^{\mathfrak{f}}_{\mathscr{D}} \geqslant s^{\mathfrak{f}}_{\mathscr{E}}$. Consider the dissection $\mathscr{F}$ defined by $\mathscr{F} = \mathscr{D} \cup \mathscr{E}$, and compare the two Riemann sums $S^{\mathfrak{f}}_{\mathscr{D}}$, $S^{\mathfrak{f}}_{\mathscr{F}}$. In calculating $S^{\mathfrak{f}}_{\mathscr{D}}$ we multiply the length of each interval of $\mathscr{D}$ by the value of the least upper bound of $\mathfrak{f}$ in that interval and then form the sum of all these products and we calculate $S^{\mathfrak{f}}_{\mathscr{F}}$ similarly. Now each interval of $\mathscr{D}$ is formed from one or more intervals of $\mathscr{F}$ and the value of the least upper bound of $\mathfrak{f}$ in each of these intervals of $\mathscr{F}$ is less than or equal to the value of the least upper bound of $\mathfrak{f}$ in the corresponding interval of $\mathscr{D}$. It follows that $S^{\mathfrak{f}}_{\mathscr{F}} \leqslant S^{\mathfrak{f}}_{\mathscr{D}}$.

Similarly $s^{\mathfrak{f}}_{\mathscr{F}} \geqslant s^{\mathfrak{f}}_{\mathscr{E}}$. Thus

$$S^{\mathfrak{f}}_{\mathscr{F}} < S^{\mathfrak{f}} + \epsilon, \quad S^{\mathfrak{f}}_{\mathscr{F}} > s^{\mathfrak{f}} - \epsilon$$

and since, by lemma 1, $S^{\mathfrak{f}}_{\mathscr{F}} \geqslant s^{\mathfrak{f}}_{\mathscr{F}}$ it follows that

$$S^{\mathfrak{f}} + \epsilon > s^{\mathfrak{f}} - \epsilon.$$

But this is true for every $\epsilon > 0$, hence $S^{\mathfrak{f}} \geqslant s^{\mathfrak{f}}$ and the lemma is proved.

Definition 47. *$S^{\mathfrak{f}}$, $s^{\mathfrak{f}}$ are called respectively the upper and the lower Riemann integrals of $\mathfrak{f}$ over $[a, b]$.*

Theorem 40. *$\mathfrak{f} \in \mathsf{R}(a, b)$ if and only if $\forall \epsilon > 0 \; \exists \delta > 0$ such that if $\tau(\mathscr{D}) < \delta$ then $S^{\mathfrak{f}}_{\mathscr{D}} - s^{\mathfrak{f}}_{\mathscr{D}} < \epsilon$.*

If $\mathfrak{f} \in \mathsf{R}(a,b)$ then $\forall \epsilon > 0 \, \exists \, \delta > 0$ such that if $\tau(\mathscr{D}) < \delta$ then

$$\left| \int_a^b \mathfrak{f} - \sum_0^{k-1} \mathfrak{f}(\xi_i)(x_{i+1} - x_i) \right| < \tfrac{1}{3}\epsilon,$$

where $\mathscr{D}$ is

$$x_0, x_1, \ldots, x_k, \quad a = x_0 < x_1 < \ldots < x_k = b \quad \text{and} \quad x_i < \xi_i < x_{i+1}.$$

Hence
$$S^{\mathfrak{f}}_{\mathscr{D}} \leqslant \int_a^b \mathfrak{f} + \tfrac{1}{3}\epsilon, \quad s^{\mathfrak{f}}_{\mathscr{D}} \geqslant \int_a^b \mathfrak{f} - \tfrac{1}{3}\epsilon$$

and thus $S^{\mathfrak{f}}_{\mathscr{D}} - s^{\mathfrak{f}}_{\mathscr{D}} \leqslant \tfrac{2}{3}\epsilon < \epsilon$. The condition is proved.

Conversely if $\mathfrak{f}$ satisfies the given condition then $S^{\mathfrak{f}} = s^{\mathfrak{f}}$. Using the condition $\forall \epsilon > 0 \, \exists \, \delta > 0$ such that if $\mathscr{D}$ is a dissection with $\tau(\mathscr{D}) < \delta$ then $s^{\mathfrak{f}}_{\mathscr{D}} + \epsilon > S^{\mathfrak{f}}_{\mathscr{D}}$ we have

$$S^{\mathfrak{f}} + \epsilon = s^{\mathfrak{f}} + \epsilon \geqslant s^{\mathfrak{f}}_{\mathscr{D}} + \epsilon > S^{\mathfrak{f}}_{\mathscr{D}} \geqslant \sum_{i=0}^{k-1} \mathfrak{f}(\xi_i)(x_{i+1} - x_i)$$
$$\geqslant s^{\mathfrak{f}}_{\mathscr{D}} > S^{\mathfrak{f}}_{\mathscr{D}} - \epsilon \geqslant S^{\mathfrak{f}} - \epsilon.$$

Thus
$$\left| S^{\mathfrak{f}} - \sum_{i=0}^{k-1} \mathfrak{f}(\xi_i)(x_{i+1} - x_i) \right| < \epsilon \quad \text{and} \quad \int_a^b \mathfrak{f}$$

exists and is equal to the common value of $S^{\mathfrak{f}}$ and $s^{\mathfrak{f}}$.

Exercise. $\mathfrak{f} \in \mathsf{R}(a,b)$ and $\mathscr{D}$ is any dissection. Show that

$$S^{\mathfrak{f}}_{\mathscr{D}} \geqslant \int_a^b \mathfrak{f} \geqslant s^{\mathfrak{f}}_{\mathscr{D}}.$$

Theorem 41. *$\mathfrak{f} \in \mathsf{R}(a,b)$ if and only if $\forall \epsilon > 0 \, \exists$ a dissection $\mathscr{D}$ such that*

$$S^{\mathfrak{f}}_{\mathscr{D}} - s^{\mathfrak{f}}_{\mathscr{D}} < \epsilon.$$

If $\mathfrak{f} \in \mathsf{R}(a,b)$ then by theorem 40 the given condition is satisfied.

Conversely to prove that the given condition implies R-integrability it is sufficient, by theorem 40, to show that $\forall \epsilon > 0 \, \exists \, \delta > 0$ such that if $\mathscr{E}$ is a dissection with $\tau(\mathscr{E}) < \delta$ then $S^{\mathfrak{f}}_{\mathscr{E}} - s^{\mathfrak{f}}_{\mathscr{E}} < \epsilon$.

Write $M = \sup |\mathfrak{f}(x)|$, $a \leqslant x \leqslant b$; and choose $\mathscr{D}$ so that

$$S^{\mathfrak{f}}_{\mathscr{D}} - s^{\mathfrak{f}}_{\mathscr{D}} < \tfrac{1}{2}\epsilon.$$

Let N be the number of dissection points of $\mathscr{D}$ and $\mathscr{E}$ a dissection such that $\tau(\mathscr{E}) < \epsilon/8N(M+1)$. At most $N-2$ intervals of $\mathscr{E}$ contain points of $\mathscr{D}$ and each such interval is of length less than $\tau(\mathscr{E})$. Thus

$$S^{\mathfrak{f}}_{\mathscr{E}} \leqslant S^{\mathfrak{f}}_{\mathscr{E} \cup \mathscr{D}} + (N-2) \,.\, 2M \,.\, \tau(\mathscr{E}).$$

$\mathscr{E} \cup \mathscr{D}$ is a refinement of $\mathscr{D}$. Thus $S^{\mathfrak{f}}_{\mathscr{E} \cup \mathscr{D}} \leqslant S^{\mathfrak{f}}_{\mathscr{D}}$ and finally

$$S^{\mathfrak{f}}_{\mathscr{E}} \leqslant S^{\mathfrak{f}}_{\mathscr{D}} + 2(N-2)\, M\tau(\mathscr{E}) \leqslant S^{\mathfrak{f}}_{\mathscr{D}} + \tfrac{1}{4}\epsilon.$$

Similarly $$s^{\mathfrak{f}}_{\mathscr{E}} \geqslant s^{\mathfrak{f}}_{\mathscr{D}} - \tfrac{1}{4}\epsilon$$
and thus $$S^{\mathfrak{f}}_{\mathscr{E}} - s^{\mathfrak{f}}_{\mathscr{E}} \leqslant S^{\mathfrak{f}}_{\mathscr{D}} - s^{\mathfrak{f}}_{\mathscr{D}} + \tfrac{1}{2}\epsilon < \epsilon$$
and thus $\mathfrak{f}$ is R-integrable over $[a, b]$.

Corollary 1. *$\mathfrak{f} \in \mathsf{R}(a, b)$ if and only if $S^{\mathfrak{f}} = s^{\mathfrak{f}}$.*

If $S^{\mathfrak{f}} = s^{\mathfrak{f}}$ and $\epsilon > 0$ is given then, first, $\exists$ dissections $\mathscr{E}$, $\mathscr{F}$ such that
$$S^{\mathfrak{f}}_{\mathscr{E}} < S^{\mathfrak{f}} + \tfrac{1}{2}\epsilon, \quad s^{\mathfrak{f}}_{\mathscr{F}} > s^{\mathfrak{f}} - \tfrac{1}{2}\epsilon.$$
Write $\mathscr{D} = \mathscr{E} \cup \mathscr{F}$ then since $S^{\mathfrak{f}}_{\mathscr{D}} \leqslant S^{\mathfrak{f}}_{\mathscr{E}}$ and $s^{\mathfrak{f}}_{\mathscr{D}} \geqslant s^{\mathfrak{f}}_{\mathscr{F}}$, it follows that $S^{\mathfrak{f}}_{\mathscr{D}} - s^{\mathfrak{f}}_{\mathscr{D}} < \epsilon$. By the theorem $\mathfrak{f} \in \mathsf{R}(a, b)$.

If $\mathfrak{f} \in \mathsf{R}(a, b)$ then $\forall \epsilon > 0 \, \exists \, \delta > 0$ such that for any dissection $\mathscr{D}$ with $\tau(\mathscr{D}) < \delta$
$$\left|\int_a^b \mathfrak{f} - S^{\mathfrak{f}}_{\mathscr{D}}\right| \leqslant \epsilon, \quad \left|\int_a^b \mathfrak{f} - s^{\mathfrak{f}}_{\mathscr{D}}\right| \leqslant \epsilon.$$
Hence using lemma 2,
$$0 \leqslant S^{\mathfrak{f}} - s^{\mathfrak{f}} \leqslant S^{\mathfrak{f}}_{\mathscr{D}} - s^{\mathfrak{f}}_{\mathscr{D}} \leqslant 2\epsilon.$$
This is true for every $\epsilon > 0$. Thus $S^{\mathfrak{f}} = s^{\mathfrak{f}}$.

Corollary 2. *If $\mathfrak{g}$ is a step-function over $\mathscr{D}$ then $\mathfrak{g} \in \mathsf{R}(a, b)$ and*
$$\int_a^b \mathfrak{g} = S^{\mathfrak{g}}_{\mathscr{D}}.$$

For if $\mathfrak{g}$ is a step-function over $\mathscr{D}$ then $S^{\mathfrak{g}}_{\mathscr{D}} = s^{\mathfrak{g}}_{\mathscr{D}}$. By the theorem $\mathfrak{g} \in \mathsf{R}(a, b)$. Also for any $\mathscr{D}$, $S^{\mathfrak{g}}_{\mathscr{D}} \geqslant \int_a^b \mathfrak{g} \geqslant s^{\mathfrak{g}}_{\mathscr{D}}$.

Corollary 3. *$\mathfrak{f} \in \mathsf{R}(a, b)$ if and only if $\forall \epsilon > 0 \, \exists$ step-functions $\mathfrak{g}$, $\mathfrak{G}$ such that*
$$\mathfrak{g}(x) \leqslant \mathfrak{f}(x) \leqslant \mathfrak{G}(x), \quad \sup|\mathfrak{g}(x)| \leqslant \sup|\mathfrak{f}(x)|, \quad a \leqslant x \leqslant b,$$
$$\sup|\mathfrak{G}(x)| \leqslant \sup|\mathfrak{f}(x)| \quad \text{and} \quad \int_a^b (\mathfrak{G} - \mathfrak{g}) < \epsilon.$$

If $\mathfrak{f} \in \mathsf{R}(a, b)$ and $\epsilon > 0$ then $\exists$ a dissection $\mathscr{D}$ such that $S^{\mathfrak{f}}_{\mathscr{D}} < s^{\mathfrak{f}}_{\mathscr{D}} + \epsilon$ (by the theorem). Define $\mathfrak{g}$, $\mathfrak{G}$ by
$$\mathfrak{g}(x) = m_i(\mathfrak{f}), \quad \mathfrak{G}(x) = M_i(\mathfrak{f}), \quad \text{where} \quad x_i < x < x_{i+1},$$
$$\mathfrak{g}(x_i) = \mathfrak{G}(x_i) = \mathfrak{f}(x_i) \quad (i = 0, 1, \ldots, k).$$
Then $\mathfrak{G}$, $\mathfrak{g}$ are step-functions and so is $\mathfrak{G} - \mathfrak{g}$. Thus
$$\int_a^b (\mathfrak{G} - \mathfrak{g}) = S^{\mathfrak{G}}_{\mathscr{D}} - s^{\mathfrak{g}}_{\mathscr{D}} = S^{\mathfrak{f}}_{\mathscr{D}} - s^{\mathfrak{f}}_{\mathscr{D}} < \epsilon.$$

Also $\mathfrak{g}(x) \leqslant \mathfrak{f}(x) \leqslant \mathfrak{G}(x), \quad \sup|\mathfrak{g}(x)| \leqslant \sup|\mathfrak{f}(x)|,$

$$\sup|\mathfrak{G}(x)| \leqslant \sup|\mathfrak{f}(x)|.$$

Thus $\mathfrak{g}$, $\mathfrak{G}$ are two step-functions which satisfy all the stated conditions.

Conversely if $\mathfrak{g}$, $\mathfrak{G}$ are step-functions on $\mathscr{E}$, $\mathscr{F}$ respectively which satisfy the given conditions then write $\mathscr{D} = \mathscr{E} \cup \mathscr{F}$. Both $\mathfrak{g}$, $\mathfrak{G}$ are step-functions on $\mathscr{D}$ and

$$S_{\mathscr{D}}^{\mathfrak{g}} = s_{\mathscr{D}}^{\mathfrak{g}} = \int_a^b \mathfrak{g}, \quad S_{\mathscr{D}}^{\mathfrak{G}} = s_{\mathscr{D}}^{\mathfrak{G}} = \int_a^b \mathfrak{G}.$$

But $\mathfrak{g}(x) \leqslant \mathfrak{f}(x) \leqslant \mathfrak{G}(x)$ and thus $s_{\mathscr{D}}^{\mathfrak{g}} \leqslant s_{\mathscr{D}}^{\mathfrak{f}} \leqslant S_{\mathscr{D}}^{\mathfrak{f}} \leqslant S_{\mathscr{D}}^{\mathfrak{G}}$.

Hence $$S_{\mathscr{D}}^{\mathfrak{f}} - s_{\mathscr{D}}^{\mathfrak{f}} \leqslant S_{\mathscr{D}}^{\mathfrak{G}} - s_{\mathscr{D}}^{\mathfrak{g}} = \int_a^b (\mathfrak{G} - \mathfrak{g}) < \epsilon$$

and by theorem 41, $\mathfrak{f} \in \mathsf{R}(a, b)$.

If $\mathfrak{f} \in \mathsf{R}(a, b)$ and $a \leqslant c < d \leqslant b$ then $\mathfrak{f} \in \mathsf{R}(c, d)$. For if $\mathscr{D}$ is any dissection of $[a, b]$ which contains c, d then denote by $\mathscr{E}$ those points of $\mathscr{D}$ which belong to $[c, d]$. From theorem 40 $\forall \epsilon > 0 \, \exists$ a dissection $\mathscr{D}$ of $[a, b]$ such that $S_{\mathscr{D}}^{\mathfrak{f}} - s_{\mathscr{D}}^{\mathfrak{f}} < \epsilon$. But $S_{\mathscr{E}}^{\mathfrak{f}} - s_{\mathscr{E}}^{\mathfrak{f}} \leqslant S_{\mathscr{D}}^{\mathfrak{f}} - s_{\mathscr{D}}^{\mathfrak{f}}$ and thus by theorem 41, $\mathfrak{f} \in \mathsf{R}(c, d)$.

Further if $a < c < b$ then

$$\int_a^b \mathfrak{f} = \int_a^c \mathfrak{f} + \int_c^b \mathfrak{f}.$$

For let $\mathscr{D}(n)$ be a dissection of $[a, b]$ which contains c and for which $\tau(\mathscr{D}(n)) < 1/n$. Let $\mathscr{E}(n)$ denote the points of $\mathscr{D}(n)$ in $[a, c]$ and $\mathscr{F}(n)$ the points of $\mathscr{D}(n)$ in $[c, b]$. Then

$$S_{\mathscr{D}(n)}^{\mathfrak{f}} = S_{\mathscr{E}(n)}^{\mathfrak{f}} + S_{\mathscr{F}(n)}^{\mathfrak{f}}.$$

Let $n \to \infty$ in this relation and we obtain

$$\int_a^b \mathfrak{f} = \int_a^c \mathfrak{f} + \int_c^b \mathfrak{f}.$$

Theorem 42. *The following classes of functions belong to* $\mathsf{R}(a, b)$.

(i) *monotonic functions,*

(ii) *bounded functions which are continuous except at a finite number of points.*

(i) Suppose that $\mathfrak{f} \uparrow$. Let N be a positive integer and $\mathscr{D}$ be the dissection of $[a, b]$ all of whose intervals are of length $(b-a)/N$. Then if $\mathscr{D}$ is $x_0, x_1, \ldots, x_k$

$$M_i(\mathfrak{f}) \leqslant \mathfrak{f}(x_{i+1}), \quad m_i(\mathfrak{f}) \geqslant \mathfrak{f}(x_i)$$

and thus
$$S_{\mathscr{D}}^{\mathfrak{f}} - s_{\mathscr{D}}^{\mathfrak{f}} = \sum_{i=0}^{k-1} (M_i(\mathfrak{f}) - m_i(\mathfrak{f}))\,(b-a)/N$$
$$\leqslant \frac{b-a}{N} \sum_{i=0}^{k-1} (\mathfrak{f}(x_{i+1}) - \mathfrak{f}(x_i))$$
$$= \frac{b-a}{N}(\mathfrak{f}(b) - \mathfrak{f}(a)).$$

Since N can be as large as we please it follows from theorem 41 that $\mathfrak{f} \in \mathsf{R}(a, b)$.

A similar argument applies if $\mathfrak{f}\downarrow$.

(ii) Suppose that $\mathfrak{f}$ is bounded over $[a, b]$ and is continuous at all points of $[a, b]$, except the points $y_1, y_2, \ldots, y_l$, say (there may be no such points of discontinuity, the argument that follows is valid in that case also). Given $\eta > 0$ the complement of $\bigcup_{j=1}^{l} (y_j - \eta, y_j + \eta)$ in $[a, b]$ is a closed bounded set X on which $\mathfrak{f}$ is continuous and therefore, by the corollary to theorem 26, is uniformly continuous (if there are no points of discontinuity then $\mathfrak{f}$ is uniformly continuous over $[a, b]$). Denote $\sup |\mathfrak{f}(x)|$, $a \leqslant x \leqslant b$ by M. Now by the uniform continuity of $\mathfrak{f}\ \forall \epsilon > 0\ \exists \delta > 0$ such that if $|x - y| < \delta$ and $x \in X$, $y \in X$ then $|\mathfrak{f}(x) - \mathfrak{f}(y)| < \epsilon$. Consider a dissection $\mathscr{D} = x_0, x_1, \ldots, x_k$ for which $\tau(\mathscr{D}) < \delta$ and a particular interval w of $\mathscr{D}$. Either (a) w meets one of the intervals $(y_j - \eta, y_j + \eta)$, or (b) $w \subset X$. In case (a)
$$w \subset (y_j - \eta - \delta, y_j + \eta + \delta) \quad \text{and} \quad M_i(\mathfrak{f}) - m_i(\mathfrak{f}) \leqslant 2M.$$
In case (b)
$$M_i(\mathfrak{f}) - m_i(\mathfrak{f}) \leqslant \epsilon.$$
Thus
$$S_{\mathscr{D}}^{\mathfrak{f}} - s_{\mathscr{D}}^{\mathfrak{f}} = \sum_{i=0}^{k-1} (M_i(\mathfrak{f}) - m_i(\mathfrak{f}))\,(x_{i+1} - x_i)$$
$$\leqslant 2M(2\eta + 2\delta)\,l + \epsilon(b-a).$$

Since ϵ, δ, η can all be made arbitrarily small, provided that they are positive, it follows from theorem 41 that $\mathfrak{f} \in \mathsf{R}(a, b)$.

Theorem 43. *If $\mathfrak{f}$ and $\mathfrak{g}$ belong to $\mathsf{R}(a, b)$ then so also do*

(a) $A\mathfrak{f} + B\mathfrak{g}$ where A, B are constants and
$$\int_a^b A\mathfrak{f} + B\mathfrak{g} = A\int_a^b \mathfrak{f} + B\int_a^b \mathfrak{g};$$
(b) $\mathfrak{f}^+$ defined by $\mathfrak{f}^+(x) = \max(\mathfrak{f}(x), 0)$;

(c) $|\mathfrak{f}|$;

(d) $\mathfrak{f}.\mathfrak{g}$;

(e) $1/\mathfrak{f}$ provided that $\exists\, \epsilon > 0$ such that $|\mathfrak{f}(x)| \geqslant \epsilon$ for $a \leqslant x \leqslant b$.

(*a*) $\forall \epsilon > 0 \exists \delta_1 > 0$ and $\delta_2 > 0$ such that if $\mathscr{D}$ is a dissection with $\tau(\mathscr{D}) < \delta_1$ then

$$\left| A \int_a^b \mathfrak{f} - A \sum_0^{k-1} \mathfrak{f}(\xi_i)(x_{i+1} - x_i) \right| < \tfrac{1}{2}\epsilon,$$

where $\mathscr{D}$ is $x_0, x_1, \ldots, x_k$ and $x_i < \xi_i < x_{i+1}$. Similarly if $\tau(\mathscr{D}) < \delta_2$ then

$$\left| B \int_a^b \mathfrak{g} - B \sum_0^{k-1} \mathfrak{g}(\xi_i)(x_{i+1} - x_i) \right| < \tfrac{1}{2}\epsilon.$$

Thus, using the modulus inequality

$$\left| \sum_0^{k-1} (A\mathfrak{f}(\xi_i) + B\mathfrak{g}(\xi_i))(x_{i+1} - x_i) - \left(A \int_a^b \mathfrak{f} + B \int_a^b \mathfrak{g} \right) \right| < \epsilon$$

provided that $\tau(\mathscr{D}) < \min(\delta_1, \delta_2)$. Hence $A\mathfrak{f} + B\mathfrak{g} \in \mathsf{R}(a, b)$ and

$$\int_a^b (A\mathfrak{f} + B\mathfrak{g}) = A \int_a^b \mathfrak{f} + B \int_a^b \mathfrak{g}.$$

(*b*) For any dissection $\mathscr{D}$

$$M_i(\mathfrak{f}^+) = \max(M_i(\mathfrak{f}), 0), \quad m_i(\mathfrak{f}^+) = \max(m_i(\mathfrak{f}), 0).$$

Thus
$$M_i(\mathfrak{f}^+) - m_i(\mathfrak{f}^+) \leqslant M_i(\mathfrak{f}) - m_i(\mathfrak{f})$$

and
$$S_{\mathscr{D}}^{\mathfrak{f}^+} - s_{\mathscr{D}}^{\mathfrak{f}^+} \leqslant S_{\mathscr{D}}^{\mathfrak{f}} - s_{\mathscr{D}}^{\mathfrak{f}}.$$

Now $\mathfrak{f} \in \mathsf{R}(a, b)$ and thus it follows from theorem 41 that

$$\mathfrak{f}^+ \in \mathsf{R}(a, b).$$

(*c*) $|\mathfrak{f}| = 2\mathfrak{f}^+ - \mathfrak{f}$ and from (*a*), (*b*) above it follows that

$$|\mathfrak{f}| \in \mathsf{R}(a, b).$$

(*d*) Write M for $\sup |\mathfrak{f}(x)|$, $a \leqslant x \leqslant b$ and $\mathfrak{h}(x) = \mathfrak{f}^2(x)$. We shall show that $\mathfrak{h} \in \mathsf{R}(a, b)$. Since for $a \leqslant x \leqslant b$, $a \leqslant y \leqslant b$

$$|\mathfrak{h}(x) - \mathfrak{h}(y)| = |\mathfrak{f}(x) - \mathfrak{f}(y)|\,|\mathfrak{f}(x) + \mathfrak{f}(y)| \leqslant 2M\,|\mathfrak{f}(x) - \mathfrak{f}(y)|$$

for any dissection $\mathscr{D}$ we have

$$M_i(\mathfrak{h}) - m_i(\mathfrak{h}) \leqslant 2M(M_i(\mathfrak{f}) - m_i(\mathfrak{f}))$$

and thus
$$S_{\mathscr{D}}^{\mathfrak{h}} - s_{\mathscr{D}}^{\mathfrak{h}} \leqslant 2M(s_{\mathscr{D}}^{\mathfrak{f}} - s_{\mathscr{D}}^{\mathfrak{f}}).$$

But $\mathfrak{f} \in \mathsf{R}(a, b)$ and it follows from theorem 41 that $\mathfrak{h} \in \mathsf{R}(a, b)$. Write $\mathfrak{f}.\mathfrak{g} = \tfrac{1}{4}((\mathfrak{f} + \mathfrak{g})^2 - (\mathfrak{f} - \mathfrak{g})^2)$. From (*a*) and the above argument it follows that $\mathfrak{f}.\mathfrak{g} \in \mathsf{R}(a, b)$.

(e) For $a \leqslant x \leqslant b$, $a \leqslant y \leqslant b$, we have

$$\left|\frac{1}{\mathfrak{f}(x)} - \frac{1}{\mathfrak{f}(y)}\right| = \frac{|\mathfrak{f}(x) - \mathfrak{f}(y)|}{\mathfrak{f}(x) \cdot \mathfrak{f}(y)} \leqslant \frac{|\mathfrak{f}(x) - \mathfrak{f}(y)|}{\epsilon^2}.$$

Thus for any dissection $\mathscr{D}$

$$S_{\mathscr{D}}^{1/\mathfrak{f}} - s_{\mathscr{D}}^{1/\mathfrak{f}} \leqslant (S_{\mathscr{D}}^{\mathfrak{f}} - s_{\mathscr{D}}^{\mathfrak{f}})\,\epsilon^{-2}$$

and by theorem 41, $1/\mathfrak{f} \in \mathsf{R}(a, b)$.

Theorem 44. *If* $\mathfrak{f}, \mathfrak{g} \in \mathsf{R}(a, b)$ *and* $\mathfrak{f}(x) \geqslant \mathfrak{g}(x)$ *for* $a \leqslant x \leqslant b$ *then*

$$\int_a^b \mathfrak{f} \geqslant \int_a^b \mathfrak{g}.$$

For any dissection $\mathscr{D}$ $S_{\mathscr{D}}^{\mathfrak{f}} \geqslant S_{\mathscr{D}}^{\mathfrak{g}}$ and the result follows from a consideration of the limit of these numbers corresponding to a sequence of dissections whose meshes tend to zero.

Corollary 1. *If* $\mathfrak{f} \in \mathsf{R}(a, b)$ *and* $m \leqslant \mathfrak{f}(x) \leqslant M$ *then*

$$m(b-a) \leqslant \int_a^b \mathfrak{f} \leqslant M(b-a).$$

Corollary 2. *If* $\mathfrak{f}, \mathfrak{g} \in \mathsf{R}(a, b)$, $m \leqslant \mathfrak{f}(x) \leqslant M$ *and* $\mathfrak{g}(x) \geqslant 0$ *then*

$$m\int_a^b \mathfrak{g} \leqslant \int_a^b \mathfrak{f}\mathfrak{g} \leqslant M\int_a^b \mathfrak{g}.$$

Corollary 3. *The modulus inequality. If* $\mathfrak{f} \in \mathsf{R}(a, b)$ *then*

$$\int_a^b |\mathfrak{f}| \geqslant \left|\int_a^b \mathfrak{f}\right|.$$

For $|\mathfrak{f}(x)| \geqslant \mathfrak{f}(x)$ and $|\mathfrak{f}(x)| \geqslant -\mathfrak{f}(x)$.

Corollary 4. *If* $a < c < b$ *and* $\mathfrak{f} \in \mathsf{R}(a, b)$, *then* $\int_a^c \mathfrak{f}$ *is a continuous function of* c.

If $a < y < b$, then writing $M = \sup |\mathfrak{f}(x)|$, $a \leqslant x \leqslant b$ we have

$$\left|\int_a^y \mathfrak{f} - \int_a^c \mathfrak{f}\right| = \left|\int_y^c \mathfrak{f}\right| \leqslant \left|\int_y^c |\mathfrak{f}|\right| \leqslant M\,|y-c|.$$

Thus as $y \to c$,

$$\int_a^y \mathfrak{f} \to \int_a^c \mathfrak{f}.$$

If $\mathfrak{f}$ is continuous and $\mathfrak{g}(x) \geqslant 0$ for $a \leqslant x \leqslant b$ and $\mathfrak{g} \in \mathsf{R}(a, b)$, then since $\mathfrak{f}$ takes every value between m and M by corollary 2 $\exists\, \xi$ such that $a \leqslant \xi \leqslant b$ and

$$\int_a^b \mathfrak{f}\mathfrak{g} = \mathfrak{f}(\xi)\int_a^b \mathfrak{g}.$$

This result is known as the *first mean-value theorem for integrals*.

Exercises. (1) $\mathfrak{f}$ and $\mathfrak{g}$ are continuous functions such that for $a \leqslant x \leqslant b$, $\mathfrak{f}(x) \geqslant \mathfrak{g}(x)$ and $\exists\, c$, $a \leqslant c \leqslant b$ such that $\mathfrak{f}(c) > \mathfrak{g}(c)$. Show that

$$\int_a^b \mathfrak{f} > \int_a^b \mathfrak{g} \quad \text{if} \quad a < b.$$

(2) $\mathfrak{f}$ is continuous, $\mathfrak{f}(x) \geqslant 0$ for $a \leqslant x \leqslant b$ and $\int_a^b \mathfrak{f} = 0$. Show that $\mathfrak{f}(x) = 0$, $a \leqslant x \leqslant b$.

(3) $\mathfrak{f} \in \mathsf{R}(a, b)$ and λ satisfies $0 < \lambda < 1$. Show that $\exists\, \xi$ such that

$$a \leqslant \xi \leqslant b \quad \text{and} \quad \lambda \int_a^b \mathfrak{f} = \int_a^\xi \mathfrak{f}.$$

Worked examples

(i) *$\mathfrak{g}(x)$ is a strictly increasing continuous function of x defined for $x \geqslant 0$ and $\mathfrak{g}(0) = 0$. $\mathfrak{f}(y)$ is the inverse function of $\mathfrak{g}(x)$. If $a > 0$, $b > 0$ prove that*

$$ab \leqslant \int_0^a \mathfrak{g} + \int_0^b \mathfrak{f}.$$

Both $\mathfrak{g}$ and $\mathfrak{f}$ are monotonic and bounded over $[0, X]$ for any $X > 0$ and thus are R-integrable. Suppose $b \geqslant \mathfrak{g}(a)$, the argument when $b < \mathfrak{g}(a)$ is omitted because it is similar with a, $\mathfrak{g}$ interchanged with b, $\mathfrak{f}$ respectively.

Let $\mathscr{D}$ be the dissection $x_0, x_1, \ldots, x_k$ of $[0, a]$. Then since $\mathfrak{g} \Uparrow$

$$\int_0^a \mathfrak{g} \geqslant \sum_0^{k-1} \mathfrak{g}(x_i)\,(x_{i+1} - x_i)$$

$$= \sum_0^{k-1} \mathfrak{g}(x_{i+1})\,(x_{i+1} - x_i) + \sum_0^{k-1} (\mathfrak{g}(x_i) - \mathfrak{g}(x_{i+1}))\,(x_{i+1} - x_i).$$

Similarly since $\mathfrak{f} \Uparrow$ and $\mathfrak{g}(x_0)$, $\mathfrak{g}(x_1)$, $\ldots$, $\mathfrak{g}(x_k)$ is a dissection of $[0, \mathfrak{g}(a)]$ we have

$$\int_0^{\mathfrak{g}(a)} \mathfrak{f} \geqslant \sum_0^{k-1} \mathfrak{f}(\mathfrak{g}(x_i))\,(\mathfrak{g}(x_{i+1}) - \mathfrak{g}(x_i)) = \sum_0^{k-1} x_i(\mathfrak{g}(x_{i+1}) - \mathfrak{g}(x_i)).$$

Hence combining these two inequalities,

$$\int_0^a \mathfrak{g} + \int_0^{\mathfrak{g}(a)} \mathfrak{f} \geqslant \sum_0^{k-1} (x_{i+1} \cdot \mathfrak{g}(x_{i+1}) - x_i \cdot \mathfrak{g}(x_i))$$
$$+ \sum_0^{k-1} (\mathfrak{g}(x_i) - \mathfrak{g}(x_{i+1}))\,(x_{i+1} - x_i)$$

$$= x_k\,\mathfrak{g}(x_k) - x_0\,\mathfrak{g}(x_0) + \sum_0^{k-1} (\mathfrak{g}(x_i) - \mathfrak{g}(x_{i+1}))\,(x_{i+1} - x_i).$$

Now $x_0 = \mathfrak{g}(x_0) = 0$, $x_k = a$, and if all the intervals of $\mathscr{D}$ are of equal length say δ then this inequality becomes

$$\int_0^a \mathfrak{g} + \int_0^{\mathfrak{g}(a)} \mathfrak{f} \geqslant a\mathfrak{g}(a) + \delta \sum_0^{k-1} (\mathfrak{g}(x_i) - \mathfrak{g}(x_{i+1})) = (a-\delta)\,\mathfrak{g}(a).$$

But $\mathfrak{f}\Uparrow$ and thus $\mathfrak{f}(y) \geqslant \mathfrak{f}(\mathfrak{g}(a))$ if $y \geqslant \mathfrak{g}(a)$. Since $\mathfrak{f}(\mathfrak{g}(a)) = a$ this means that

$$\int_{\mathfrak{g}(a)}^b \mathfrak{f} \geqslant (b - \mathfrak{g}(a))\,a.$$

On adding to the previous inequality we obtain

$$\int_0^a \mathfrak{g} + \int_0^b \mathfrak{f} \geqslant ab - \delta\mathfrak{g}(a).$$

This inequality is true for every $\delta > 0$. Thus finally

$$\int_0^a \mathfrak{g} + \int_0^b \mathfrak{f} \geqslant ab.$$

(ii) $\mathfrak{f} \in \mathsf{R}(a,b)$ *and* c *is such that* $a < c < b$. *Prove that*

$$\lim_{h\to 0+} \int_a^c |\mathfrak{f}(x) - \mathfrak{f}(x+h)|\,dx = 0.$$

$\forall \epsilon > 0$ let $\mathfrak{g}$, $\mathfrak{G}$ be two step functions over the dissection $\mathscr{D}$ such that

$$\mathfrak{g}(x) \leqslant \mathfrak{f}(x) \leqslant \mathfrak{G}(x) \quad \text{and} \quad \int_a^b (\mathfrak{G} - \mathfrak{g}) < \epsilon.$$

Such step-functions exist by theorem 41, corollary 3, and we can always suppose that two step-functions are step-functions over the *same* dissection by replacing the two dissections over which they are given as defined by the dissection formed by their union.

For any h satisfying $0 < h < b - c$,

$$\mathfrak{g}(x+h) - \mathfrak{G}(x) \leqslant \mathfrak{f}(x+h) - \mathfrak{f}(x) \leqslant \mathfrak{G}(x+h) - \mathfrak{g}(x).$$

Thus

$$\begin{aligned}|\mathfrak{f}(x+h) - \mathfrak{f}(x)| &\leqslant \max\left(|\mathfrak{G}(x+h) - \mathfrak{g}(x)|,\ |\mathfrak{g}(x+h) - \mathfrak{G}(x)|\right) \\ &\leqslant |\mathfrak{G}(x+h) - \mathfrak{G}(x)| + |\mathfrak{G}(x) - \mathfrak{g}(x)| + |\mathfrak{g}(x+h) - \mathfrak{g}(x)|.\end{aligned}$$

If $\mathscr{D}$ is $x_0, x_1, \ldots, x_k$ and $h < \tau(\mathscr{D})$, then $\mathfrak{G}(x+h) = \mathfrak{G}(x)$ unless possibly $x_i - h \leqslant x \leqslant x_i$ $(i = 1, 2, \ldots, k-1)$. Thus

$$\int_a^c |\mathfrak{G}(x+h) - \mathfrak{G}(x)|\,dx \leqslant (k-1)\,h\,.\,2 \sup_{a\leqslant x\leqslant b} |\mathfrak{G}(x)| \to 0 \quad \text{as} \quad h \to 0+.$$

Similarly $\displaystyle\int_a^c |\mathfrak{g}(x+h)-\mathfrak{g}(x)|\,dx \to 0$ as $h \to 0+$.

Hence $\displaystyle\overline{\lim_{h\to 0+}} \int_a^c |\mathfrak{f}(x)-\mathfrak{f}(x+h)|\,dx \leqslant \int_a^c |\mathfrak{G}(x)-\mathfrak{g}(x)|\,dx \leqslant \epsilon.$

This is true for every $\epsilon > 0$. Hence

$$\overline{\lim_{h\to 0+}} \int_a^c |\mathfrak{f}(x)-\mathfrak{f}(x+h)|\,dx \leqslant 0.$$

Since the integrand is non-negative

$$\underline{\lim_{h\to 0+}} \int_a^c |\mathfrak{f}(x)-\mathfrak{f}(x+h)|\,dx \geqslant 0$$

and these two inequalities imply the required result.

(iii) *Prove that if* $\mathfrak{f} \in \mathsf{R}(0, \pi)$ *and* n *denotes a positive integer then*

$$\lim_{n\to\infty} \int_0^\pi \mathfrak{f}(x)\,|\sin nx|\,dx = \frac{2}{\pi}\int_0^\pi \mathfrak{f}(x)\,dx.$$

Consider first a function $\mathfrak{f}(x)$ of the following form

$$\begin{aligned} \mathfrak{f}(x) &= 0 \quad (x < c_1) \\ &= k \quad (c_1 \leqslant x \leqslant c_2) \\ &= 0 \quad (c_2 < x). \end{aligned}$$

If $c_1 = c_2$ the limiting relation is true since both integrals are zero. Suppose that $c_1 < c_2$ then

$$\int_0^\pi \mathfrak{f}(x)\,|\sin nx|\,dx = k\int_{c_1}^{c_2} |\sin nx|\,dx.$$

Let r_1, r_2 be integers chosen so that

$$(r_1-1)\pi/n < c_1 \leqslant r_1\pi/n, \quad r_2\pi/n \leqslant c_2 < (r_2+1)\pi/n$$

and we shall suppose n so large that $r_1 \leqslant r_2$. Then

$$\int_0^\pi \mathfrak{f}(x)\,|\sin nx|\,dx = k\int_{c_1}^{r_1\pi/n} |\sin nx|\,dx$$
$$+k\int_{r_1\pi/n}^{r_2\pi/n} |\sin nx|\,dx + k\int_{r_2\pi/n}^{c_2} |\sin nx|\,dx.$$

Now $\displaystyle\int_{c_1}^{r_1/\pi n} k\,|\sin nx|\,dx \leqslant k\left(\frac{r_1\pi}{n} - c_1\right) = O\left(\frac{1}{n}\right)$

and similarly $\displaystyle k\int_{r_2\pi/n}^{c_2} |\sin nx|\,dx = O\left(\frac{1}{n}\right).$

Also $$k\int_{r_1\pi/n}^{r_2\pi/n} |\sin nx|\,dx = \frac{k}{n}\int_{r_1\pi}^{r_2\pi} |\sin y|\,dy = \frac{2k}{n}(r_2-r_1).$$

Thus $$\int_0^\pi \mathfrak{f}(x)\,|\sin nx|\,dx = \frac{2k}{n}(r_2-r_1)+O\left(\frac{1}{n}\right)$$
$$= \frac{2k(c_2-c_1)}{\pi}+O\left(\frac{1}{n}\right)$$
$$= \frac{2}{\pi}\int_0^\pi \mathfrak{f}(x)\,dx+O\left(\frac{1}{n}\right)$$

and the stated limiting relation is true in this case.

Next if $\mathfrak{f}$ is a step-function it is the sum of a number of functions of the above form and the required result follows by addition.

Finally, if $\mathfrak{f} \in \mathsf{R}(0,\pi)$, $\forall \epsilon > 0 \, \exists$ step-functions $\mathfrak{g}$, $\mathfrak{G}$ such that

$$\mathfrak{g}(x) \leqslant \mathfrak{f}(x) \leqslant \mathfrak{G}(x) \quad \text{and} \quad \int_0^\pi (\mathfrak{G}-\mathfrak{g}) < \epsilon$$

by theorem 41, corollary 3. Thus

$$\overline{\lim_{n\to\infty}} \int_0^\pi \mathfrak{f}(x)\,|\sin nx|\,dx \leqslant \overline{\lim_{n\to\infty}} \int_0^\pi \mathfrak{G}(x)\,|\sin nx|\,dx$$
$$= \frac{2}{\pi}\int_0^\pi \mathfrak{G}(x)\,dx$$
$$\leqslant \frac{2}{\pi}\int_0^\pi \mathfrak{f}(x)\,dx+\frac{2\epsilon}{\pi}.$$

Similarly $$\underline{\lim_{n\to\infty}} \int_0^\pi \mathfrak{f}(x)\,|\sin nx|\,dx \geqslant \frac{2}{\pi}\int_0^\pi \mathfrak{f}(x)\,dx-\frac{2\epsilon}{\pi}$$

and since ϵ is any positive number it follows that

$$\lim_{n\to\infty} \int_0^\pi \mathfrak{f}(x)\,|\sin nx|\,dx \quad \text{exists and is equal to} \quad \frac{2}{\pi}\int_0^\pi \mathfrak{f}(x)\,dx.$$

Exercises

1. $\mathfrak{g}\Uparrow$ for $a \leqslant x \leqslant b$. Show that $\mathfrak{f}(\mathfrak{g}(x)) \in \mathsf{R}(a,b)$ if either $\mathfrak{f}$ is monotonic or $\mathfrak{f}$ is continuous.

2. Do the following functions defined to be zero when $x = 0$ and to be

(a) $\exp(\sin x)$; (b) $\exp(\sin 1/x)$; (c) $\exp 1/x$; (d) $\sin(\exp 1/x)$

when $x > 0$ belong to $\mathsf{R}(0,1)$?

3. $\mathfrak{f}$ is defined by $\mathfrak{f}(x) = 0$, $x \leqslant 0$, $\mathfrak{f}(x) = \sin 1/x$, $x > 0$. $\{r_n\}$ is a sequence which enumerates the rational numbers in $[0, 1]$. Show that $\Sigma \mathfrak{f}(x - r_n)/n^2$ converges and determine whether or not its sum belongs to $\mathsf{R}(0, 1)$.

4. $\{x_n\}$ is a sequence of points of $[a, b]$ and $v_N(\alpha, \beta)$ is the number of the points $x_1, x_2, \ldots, x_N$ which lie in (α, β). Prove that

$$v_N(\alpha, \beta)/N \to (\beta - \alpha)/(b - a)$$

for every subinterval (α, β) of $[a, b]$ if and only if for every $\mathfrak{f} \in \mathsf{R}(a, b)$

$$\left(\sum_1^N \mathfrak{f}(x_n)\right) \Big/ N \to \left(\int_a^b \mathfrak{f}\right)(b - a) \quad \text{as} \quad N \to \infty.$$

5. $\mathfrak{f} \Downarrow$ for $0 < x \leqslant b$ and as $x \to 0+$, $\mathfrak{f}(x) \to \infty$ and $\displaystyle\int_x^b \mathfrak{f} \to I$ where I is a number. Prove that $x\mathfrak{f}(x) \to 0$ as $x \to 0+$ and that

$$\frac{b}{n} \sum_{r=1}^{n} \mathfrak{f}\left(\frac{rb}{n}\right) \to I \quad \text{as} \quad n \to \infty.$$

6. $\mathfrak{f}$ is continuous in $-1 \leqslant x \leqslant 1$ and attains its upper bound M at the point $x = 0$ only. Prove that

$$\int_{-1}^{+1} \exp(n\mathfrak{f}(x))\, dx = o(\exp nM) \quad \text{as} \quad n \to \infty.$$

7. $\mathfrak{f}$ is such that $D^2\mathfrak{f}$ exists and is continuous for $x_0 - r \leqslant x \leqslant x_0 + r$. Show that $\exists\, \xi$ such that

$$x_0 - r < \xi < x_0 + r \quad \text{and} \quad D^2\mathfrak{f}(\xi) = \frac{3}{r^3} \int_{x_0 - r}^{x_0 + r} (\mathfrak{f}(x) - \mathfrak{f}(x_0))\, dx.$$

8. $\mathfrak{f}$ is differentiable for $a \leqslant x \leqslant b$ and $\mathfrak{f}(a) = \mathfrak{f}(b) = 0$, and the differential coefficient of $\mathfrak{f}$ is bounded. Show that $\exists\, \xi$ such that $a < \xi < b$ and

$$D\mathfrak{f}(\xi) > \frac{4}{(b-a)^2} \int_a^b \mathfrak{f}.$$

23

INTEGRATION AND DIFFERENTIATION

The Riemann integral as defined in the previous chapter has no obvious connexion with differentiation. However, this connexion is fundamentally important and we establish it in this chapter. By combining integration and differentiation we are able to find all that we need to know about $(\mathfrak{f}(x_1) - \mathfrak{f}(x_2))/(x_1 - x_2)$ without at any stage having to consider functions of more than one real variable.

Theorem 45. *If* $\mathfrak{f} \in \mathsf{R}(a,b)$ *and* $\mathfrak{f}$ *is continuous at* c, $a < c < b$, *then* $\int_a^y \mathfrak{f}$ *is a differentiable function of* y *at* c *and its differential coefficient is* $\mathfrak{f}(c)$.

Write $\mathfrak{F}(y) = \int_a^y \mathfrak{f}$. Take $h \neq 0$, $|h| < \min(c-a, b-c)$. Then

$$\left|\frac{\mathfrak{F}(c+h)-\mathfrak{F}(c)}{h} - \mathfrak{f}(c)\right| = \frac{1}{|h|}\left|\int_c^{c+h} (\mathfrak{f}(x)-\mathfrak{f}(c))\,dx\right|$$

$$\leqslant \frac{1}{|h|}\left|\int_c^{c+h} |\mathfrak{f}(x)-\mathfrak{f}(c)|\,dx\right|$$

$$= o(1) \quad \text{as} \quad h \to 0,$$

by continuity at c. The theorem is proved.

Exercise. $\mathfrak{f}$ is continuous over $[a,b]$. Show that $\exists\, c$, $a < c < b$ such that

$$\int_a^b \mathfrak{f} = (b-a)\,\mathfrak{f}(c).$$

Theorem 46. *If* $\mathfrak{f} \in \mathsf{R}(a,b)$ *then* $\int_a^x \mathfrak{f}$ *is of bounded variation over* $[a,b]$.

Write $\mathfrak{F}(x) = \int_a^x \mathfrak{f}$. For any dissection $x_0, x_1, \ldots, x_{k-1}, x_k$ we have

$$\sum_{i=0}^{k-1} |\mathfrak{F}(x_{i+1}) - \mathfrak{F}(x_i)| = \sum_{i=0}^{k-1}\left|\int_{x_i}^{x_{i+1}} \mathfrak{f}\right|$$

$$\leqslant \sum_{i=0}^{k-1}\int_{x_i}^{x_{i+1}} |\mathfrak{f}| = \int_a^b |\mathfrak{f}|$$

and since $\int_a^b |\mathfrak{f}|$ is a number independent of the dissection concerned it follows that $\mathfrak{F}$ is of bounded variation.

Theorem 47. *If* $\mathfrak{f}$, *defined over* $[a,b]$ *is such that* $D\mathfrak{f} \in \mathsf{R}(a,b)$ *then*

$$\int_a^b D\mathfrak{f} = \mathfrak{f}(b) - \mathfrak{f}(a).$$

Let $\mathscr{D}$ be the dissection $x_0, x_1, \ldots, x_k$. By the mean-value theorem $\exists\, \xi_i$ such that $x_i < \xi_i < x_{i+1}$ and

$$\mathfrak{f}(x_{i+1}) - \mathfrak{f}(x_i) = (x_{i+1} - x_i)\,D\mathfrak{f}(\xi_i) \quad (i = 0, 1, \ldots, k-1).$$

Adding these equations we obtain

$$\mathfrak{f}(b) - \mathfrak{f}(a) = \sum_{i=0}^{k-1} D\mathfrak{f}(\xi_i)\,(x_{i+1} - x_i).$$

If we now replace $\mathscr{D}$ by the elements of a sequence of dissections whose meshes tend to zero, then the right-hand side above tends to $\int_a^b D\mathfrak{f}$. Hence $\int_a^b D\mathfrak{f} = \mathfrak{f}(b)-\mathfrak{f}(a)$.

This theorem enables us to determine the integrals of those functions which are differential coefficients. For example

$$\int_a^b \sin x\,dx = \int_a^b D(-\cos x)\,dx = -\cos b + \cos a,$$

$$\int_a^b \cos x\,dx = \int_a^b D(\sin x)\,dx = \sin b - \sin a,$$

$$\int_a^b \exp x\,dx = \exp b - \exp a \quad \text{and} \quad \int_a^b \frac{dx}{x} = \log b - \log a$$
$$(b > a > 0).$$

The correspondence between integration and differentiation implied by theorems 45 and 47 extends also to formulae. In particular corresponding to the differentiation of a product we have integration by parts, and corresponding to the differentiation of a composite function we have the formula for the change of variable in an integral.

Theorem 48. *Integration by parts. If*

$$\mathfrak{f} \in \mathsf{R}(a,b),\ \mathfrak{g} \in \mathsf{R}(a,b), \quad \mathfrak{F}(x) = \int_a^x \mathfrak{f}, \quad \mathfrak{G}(x) = \int_a^x \mathfrak{g},$$

then
$$\int_a^b \mathfrak{F}\mathfrak{g} + \int_a^b \mathfrak{f}\mathfrak{G} = \mathfrak{F}(b)\,\mathfrak{G}(b).$$

If $\mathfrak{f}$ and $\mathfrak{g}$ are continuous then by theorem 45

$$D(\mathfrak{F}\mathfrak{G}) = \mathfrak{F}.D\mathfrak{G} + D\mathfrak{F}.\mathfrak{G} = \mathfrak{F}\mathfrak{g} + \mathfrak{f}\mathfrak{G}$$

and by theorem 47

$$\int_a^b D(\mathfrak{F}\mathfrak{G}) = \mathfrak{F}(b)\,\mathfrak{G}(b) - \mathfrak{F}(a)\,\mathfrak{G}(a) = \mathfrak{F}(b)\,\mathfrak{G}(b)$$

since $\mathfrak{F}(a) = 0$. Thus

$$\int_a^b \mathfrak{F}\mathfrak{g} + \int_a^b \mathfrak{f}\mathfrak{G} = \mathfrak{F}(b)\,\mathfrak{G}(b)$$

and the formula is proved in this case. For the general case we need the following lemma.

Lemma. *If* $\mathfrak{f} \in \mathsf{R}(a,b)$ *then* $\forall \epsilon > 0\ \exists$ *a continuous function* $\mathfrak{h}$ *such that*

$$\sup_{a \leqslant x \leqslant b} |\mathfrak{h}(x)| \leqslant \sup_{a \leqslant x \leqslant b} |\mathfrak{f}(x)| \quad \text{and} \quad \int_a^b |\mathfrak{f}-\mathfrak{h}| < \epsilon.$$

By corollary 3 to theorem 41 it is sufficient to consider the case when $\mathfrak{f}$ is a step-function. Suppose then that $\mathfrak{f}$ is a step-function defined over the dissection $x_0, x_1, x_2, \ldots, x_k$ such that

$$\mathfrak{f}(x) = \alpha_i \quad \text{if} \quad x_i < x < x_{i+1} \quad (i = 0, 1, \ldots, k-1).$$

Write $M = \max(|\alpha_i|, i = 0, 1, \ldots, k-1)$. Choose $\eta > 0$ such that $\eta < \frac{1}{2}\min(x_{i+1} - x_i)$ $(i = 0, 1, \ldots, k-1)$ and $\eta < \epsilon/2k(M+1)$. Define the continuous function $\mathfrak{h}$ by

$$\mathfrak{h}(x_i) = 0, \quad \mathfrak{h}(x) = \alpha_i(x - x_i)/\eta \qquad (x_i < x < x_i + \eta),$$

$$\mathfrak{h}(x) = \alpha_i \qquad (x_i + \eta \leqslant x \leqslant x_{i+1} - \eta),$$

$$\mathfrak{h}(x) = \alpha_i(x_{i+1} - x)/\eta \qquad (x_{i+1} - \eta < x < x_{i+1}).$$

Then $\mathfrak{h}(x) = \mathfrak{f}(x)$ except for x lying in $2k$ intervals each of length η. Throughout these intervals $|\mathfrak{h} - \mathfrak{f}| \leqslant M$. Thus

$$\int_a^b |\mathfrak{h} - \mathfrak{f}| \leqslant 2k\eta M < \epsilon$$

and since $\sup|\mathfrak{h}| \leqslant \sup|\mathfrak{f}|$ the lemma is proved.

Next let $\mathfrak{f}$ and $\mathfrak{g}$ be geneıal R-integrable functions. Write $M = \sup|\mathfrak{f}|$, $N = \sup|\mathfrak{g}|$ and construct sequences of continuous functions $\{\mathfrak{f}_n\}$, $\{\mathfrak{g}_n\}$ such that

$$\int_a^b |\mathfrak{f} - \mathfrak{f}_n| < 1/n, \quad \int_a^b |\mathfrak{g} - \mathfrak{g}_n| < 1/n, \quad |\mathfrak{f}_n(x)| \leqslant M, \quad |\mathfrak{g}_n(x)| \leqslant N$$

(such sequences of functions exist by the lemma). Thus

$$\left|\int_a^b \mathfrak{F}\mathfrak{g} - \int_a^b \mathfrak{F}_n\mathfrak{g}_n\right| \leqslant \int_a^b |\mathfrak{F}\mathfrak{g} - \mathfrak{F}_n\mathfrak{g}_n|$$

$$\leqslant \int_a^b |\mathfrak{F} - \mathfrak{F}_n| \cdot |\mathfrak{g}| + \int_a^b |\mathfrak{F}_n||\mathfrak{g} - \mathfrak{g}_n|$$

$$\leqslant (b-a)N/n + (b-a)M/n$$

$$\to 0 \quad \text{as} \quad n \to \infty.$$

Thus $$\int_a^b \mathfrak{F}_n\mathfrak{g}_n \to \int_a^b \mathfrak{F}\mathfrak{g} \quad \text{as} \quad n \to \infty.$$

Similarly $$\int_a^b \mathfrak{f}_n\mathfrak{G}_n \to \int_a^b \mathfrak{f}\mathfrak{G} \quad \text{as} \quad n \to \infty.$$

Also $\mathfrak{F}_n(b)\,\mathfrak{G}_n(b) \to \mathfrak{F}(b)\,\mathfrak{G}(b)$ as $n \to \infty$. Hence by the case of continuous functions proved above

$$\int_a^b (\mathfrak{F}\mathfrak{g}+\mathfrak{f}\mathfrak{G}) = \lim_{n\to\infty}\int_a^b (\mathfrak{F}_n\mathfrak{g}_n+\mathfrak{f}_n\mathfrak{G}_n)$$

$$= \lim_{n\to\infty}\mathfrak{F}_n(b)\,\mathfrak{G}_n(b) = \mathfrak{F}(b)\,\mathfrak{G}(b)$$

and the formula is proved.

Corollary 1. *The second mean-value theorem for integrals.* $\mathfrak{f}$ *and* $\mathfrak{g}$ *are defined over* $[a,b]$, $\mathfrak{f}$ *is monotonic and such that* $D\mathfrak{f} \in \mathsf{R}(a,b)$. *Also* $\mathfrak{g} \in \mathsf{R}(a,b)$. *Then* $\exists\, c$ *such that* $a \leqslant c \leqslant b$ *and*

$$\int_a^b \mathfrak{f}\mathfrak{g} = \mathfrak{f}(a)\int_a^c \mathfrak{g}+\mathfrak{f}(b)\int_c^b \mathfrak{g}.$$

Apply the theorem to $D\mathfrak{f}$, $\mathfrak{g}$ in place of $\mathfrak{f}$, $\mathfrak{g}$ and write

$$\mathfrak{H}(x) = \int_a^x D\mathfrak{f},\quad \mathfrak{G}(x) = \int_a^x \mathfrak{g}.$$

Then

$$(1)\qquad \int_a^b \mathfrak{H}\mathfrak{g}+\int_a^b D\mathfrak{f}\,.\,\mathfrak{G} = \mathfrak{H}(b)\,.\,\mathfrak{G}(b).$$

If $\mathfrak{f}\uparrow$ then $D\mathfrak{f} \geqslant 0$ and

$$\inf \mathfrak{G}\,.\int_a^b D\mathfrak{f} \leqslant \int_a^b D\mathfrak{f}\,.\,\mathfrak{G} \leqslant \sup \mathfrak{G}\,.\int_a^b D\mathfrak{f}.$$

Similar but reversed inequalities hold if $\mathfrak{f}\downarrow$. By theorem 44, corollary 4, p. 145, $\mathfrak{G}$ is continuous. Thus $\exists\, c$ such that

$$\int_a^b D\mathfrak{f}\,.\,\mathfrak{G} = \mathfrak{G}(c)\int_a^b D\mathfrak{f} = \mathfrak{G}(c)\,.\,(\mathfrak{f}(b)-\mathfrak{f}(a))$$

(using theorem 47). Hence substituting in (1) and replacing $\mathfrak{H}(x)$ by $\mathfrak{f}(x)-\mathfrak{f}(a)$

$$\int_a^b (\mathfrak{f}(x)-\mathfrak{f}(a))\,\mathfrak{g}(x)\,dx = (\mathfrak{f}(b)-\mathfrak{f}(a))\,\mathfrak{G}(b)-(\mathfrak{f}(b)-\mathfrak{f}(a))\,\mathfrak{G}(c)$$

$$= (\mathfrak{f}(b)-\mathfrak{f}(a))\int_c^b \mathfrak{g},$$

$$\therefore\quad \int_a^b \mathfrak{f}\mathfrak{g} = \mathfrak{f}(a)\int_a^c \mathfrak{g}+\mathfrak{f}(b)\int_c^b \mathfrak{g}.$$

Corollary 2. *If* $\mathfrak{f}\downarrow$, $\mathfrak{f} \geqslant 0$ *and* $D\mathfrak{f}$, $\mathfrak{g} \in \mathsf{R}(a,b)$ *then* $\exists\, c$ *such that* $a \leqslant c \leqslant b$ *and* $\int_a^b \mathfrak{f}\mathfrak{g} = \mathfrak{f}(a)\int_a^c \mathfrak{g}$.

By corollary 1 $\exists\, d$ such that $a \leqslant d \leqslant b$ and

$$(2) \qquad \int_a^b \mathfrak{f}\mathfrak{g} = \mathfrak{f}(a)\int_a^d \mathfrak{g} + \mathfrak{f}(b)\int_d^b \mathfrak{g}.$$

If $\mathfrak{f}(b) = 0$ this is the required form with $c = d$. If $\mathfrak{f}(b) > 0$ then as ξ varies $\int_d^\xi \mathfrak{g}$ takes all values lying between $\int_d^b \mathfrak{g}$ and $\int_d^d \mathfrak{g} = 0$ (since it is a continuous function of ξ theorem 44, p. 145). Also $0 < \mathfrak{f}(b) \leqslant \mathfrak{f}(a)$ and thus $\exists\, c$, $d \leqslant c \leqslant b$ such that

$$\int_d^c \mathfrak{g} = \frac{\mathfrak{f}(b)}{\mathfrak{f}(a)}\int_d^b \mathfrak{g}.$$

Substituting in (2) for $\int_d^b \mathfrak{g}$ we get

$$\int_a^b \mathfrak{f}\mathfrak{g} = \mathfrak{f}(a)\int_a^c \mathfrak{g}.$$

Theorem 49. *Change of variable in an integral. If* $\mathfrak{f} \in \mathsf{R}(a, b)$ *and* $\mathfrak{g}(y)$ *is an increasing differentiable function defined over* (c, d) *such that* $D\mathfrak{g} \in \mathsf{R}(c, d)$ *and* $\mathfrak{g}(c) = a$, $\mathfrak{g}(d) = b$ *then* $\mathfrak{f}\mathfrak{g}(y) . \mathfrak{g}'(y) \in \mathsf{R}(c, d)$ *and*

$$\int_a^b \mathfrak{f} = \int_c^d \mathfrak{f}(\mathfrak{g}(y)) . \mathfrak{g}'(y) . dy.$$

Suppose, first, that $\mathfrak{f}$ is of the form

$$\mathfrak{f}(x) = 0 \quad (x < c_1),$$
$$\mathfrak{f}(x) = k \quad (c_1 \leqslant x \leqslant c_2),$$
$$\mathfrak{f}(x) = 0 \quad (x > c_2),$$

then
$$\int_a^b \mathfrak{f} = k(c_2 - c_1)$$

and if α, β are defined by $\mathfrak{g}(\alpha) = c_1$, $\mathfrak{g}(\beta) = c_2$, then

$$\int_c^d \mathfrak{f}(\mathfrak{g}(y))\, \mathfrak{g}'(y)\, dy = \int_\alpha^\beta k\mathfrak{g}'(y)\, dy$$
$$= k(\mathfrak{g}(\beta) - \mathfrak{g}(\alpha))$$
$$= k(c_2 - c_1)$$

by theorem 47, p. 151, and the theorem is proved in this case.

If $\mathfrak{f}$ is a step-function then it is the sum of a finite number of functions of the above form and the theorem follows by addition.

If $\mathfrak{f} \in \mathsf{R}(a, b)$, then by theorem 41, corollary 3, $\exists$ step-functions $\mathfrak{k}$, $\mathfrak{K}$ such that

$$\mathfrak{k}(x) \leqslant \mathfrak{f}(x) \leqslant \mathfrak{K}(x), \quad \int_a^b (\mathfrak{K} - \mathfrak{k}) < \epsilon.$$

Since $\mathfrak{g}'(y) \geqslant 0$ we have

$$\mathfrak{f}(\mathfrak{g}(y)) \,.\, \mathfrak{g}'(y) \leqslant \mathfrak{K}(\mathfrak{g}(y)) \,.\, \mathfrak{g}'(y)$$

and for any dissection $\mathscr{D}$ of $[c, d]$ the upper Riemann sum of $\mathfrak{f}(\mathfrak{g}(y)) \,.\, \mathfrak{g}'(y)$ is less than or equal to that of $\mathfrak{K}(\mathfrak{g}(y)) \,.\, \mathfrak{g}'(y)$. Hence the upper Riemann integral of $\mathfrak{f}(\mathfrak{g}(y)) \,.\, \mathfrak{g}'(y)$ over $[c, d]$ is less than or equal to

$$\int_c^d \mathfrak{K}(\mathfrak{g}(y)) \,.\, \mathfrak{g}'(y) \,.\, dy = \int_a^b \mathfrak{K},$$

where we have used the fact that the theorem has already been proved for step-functions.

Similarly the lower Riemann integral of $\mathfrak{f}(\mathfrak{g}(y)) \,.\, \mathfrak{g}'(y)$ over $[c, d]$ is not less than

$$\int_a^b \mathfrak{k} > \int_a^b \mathfrak{K} - \epsilon.$$

Thus the upper and lower Riemann integrals of $\mathfrak{f}(\mathfrak{g}(y)) \,.\, \mathfrak{f}'(y)$ over $[c, d]$ differ by at most ϵ and this is true for every $\epsilon > 0$. Thus the upper and lower Riemann integrals are equal and by theorem 41, corollary 1

$$\mathfrak{f}(\mathfrak{g}(y)) \,.\, \mathfrak{g}'(y) \in \mathsf{R}(c, d).$$

Moreover $$\int_a^b \mathfrak{K} \geqslant \int_c^d \mathfrak{f}(\mathfrak{g}(y)) \,.\, \mathfrak{g}'(y)\, dy \geqslant \int_a^b \mathfrak{k}.$$

Also $$\int_a^b \mathfrak{K} \geqslant \int_a^b \mathfrak{f} \geqslant \int_a^b \mathfrak{k}$$

and as $$\int_a^b \mathfrak{K} < \int_a^b \mathfrak{k} + \epsilon$$

it follows that $$\left| \int_a^b \mathfrak{f} - \int_c^d \mathfrak{f}(\mathfrak{g}(y)) \,.\, \mathfrak{g}'(y) \,.\, dy \right| < \epsilon.$$

This is true for every $\epsilon > 0$ and therefore

$$\int_a^b \mathfrak{f} = \int_c^d \mathfrak{f}(\mathfrak{g}(y)) \,.\, \mathfrak{g}'(y) \,.\, dy$$

and the theorem is proved.

In the nth mean-value theorem, theorem 37, we have already had several forms for the remainder. It is possible in certain cases to express this remainder as an integral, as follows.

Suppose that $\mathfrak{f}$ is such that $D^n\mathfrak{f}$ exists for $a \leqslant x \leqslant b$ and $D^n\mathfrak{f} \in \mathsf{R}(a, b)$. Then for $0 \leqslant t \leqslant 1$,

$$\frac{d}{dt}\left(\sum_{r=1}^{n-1} \frac{(b-a)^r}{r!}(1-t)^r D^{(r)}\mathfrak{f}(a+t(b-a))\right)$$
$$= \frac{(b-a)^n}{(n-1)!}(1-t)^{n-1} D^{(n)}\mathfrak{f}(a+t(b-a)) - (b-a)\, D\mathfrak{f}(a+t(b-a)),$$

where D denotes differentiation with respect to t.

The right-hand side of this equation is Riemann integrable with respect to t over $[0, 1]$ and therefore so also is the left-hand side. By theorem 47, integrating over $[0, 1]$ we have

$$-\sum_{r=1}^{n-1} \frac{(b-a)^r}{r!} D^r\mathfrak{f}(a)$$
$$= \int_0^1 \frac{(b-a)^n}{(n-1)!}(1-t)^{n-1} D^{(n)}\mathfrak{f}(a+t(b-a))\, dt - \mathfrak{f}(b) + \mathfrak{f}(a)$$

and on rearranging this becomes

$$\mathfrak{f}(b) = \mathfrak{f}(a) + \sum_{r=1}^{n-1} \frac{(b-a)^r}{r!} D^r\mathfrak{f}(a)$$
$$+ \frac{(b-a)^n}{(n-1)!}\int_0^1 (1-t)^{n-1} D^{(n)}\mathfrak{f}(a+t(b-a))\, dt.$$

This is the nth mean-value theorem with an integral form for the remainder.

Exercises

1. $\mathfrak{f}$ and $\mathfrak{g} \in \mathsf{R}(a, b)$ and $\mathfrak{F}$, $\mathfrak{H}$ are defined by

$$\mathfrak{F}(x) = \int_a^x \mathfrak{f}, \quad \mathfrak{H}(x) = \int_a^x \mathfrak{f}.\mathfrak{g} \quad (a \leqslant x \leqslant b).$$

Moreover, it is given that $D\mathfrak{F} = \mathfrak{f}$ and $\mathfrak{g}$ is continuous. Show that $D\mathfrak{H} = \mathfrak{f}.\mathfrak{g}$, $a \leqslant x \leqslant b$.

2. $\mathfrak{g}$ is continuous over $[a, b]$ and $\mathfrak{f} = D\mathfrak{h}$ where $\mathfrak{h}$ is defined over an open interval containing $[a, b]$. If $\mathfrak{f} \in \mathsf{R}(a, b)$ show that $\mathfrak{f}.\mathfrak{g}$ is the derivative of a continuous function for $a \leqslant x \leqslant b$.

3. Show that if $\mathfrak{f}$ has an integrable derivative for all $x \geqslant 1$ then

$$\sum_{n \leqslant x} \mathfrak{f}(n) = [x].\mathfrak{f}(x) - \int_1^x [t].D\mathfrak{f}(t).dt \quad (x \geqslant 1),$$

where the summation extends over all positive integers not exceeding x and where $[x]$ denotes the integral part of x.

4. $\mathfrak{f}$ is continuous for all x and $\exists X$ such that $\mathfrak{f}(x) = 0$ if $x < X$. Also $\mathfrak{f}(x) \geqslant 0$, $(\mathfrak{f}(x))^2 \geqslant \mathfrak{g}(x)$, where $\mathfrak{g}(x) = \int_X^x \mathfrak{f}$. By considering the differential coefficient of $\mathfrak{g}$, or otherwise, prove that $\mathfrak{g}(x) \geqslant \frac{1}{4}(x-a)$ where a is any number for which $\mathfrak{g}(a) \neq 0$.

5. $\mathfrak{f} \in \mathsf{R}(0,1)$. Prove that

$$\int_0^1 \mathfrak{f}(x) \sin nx\, dx \to 0 \quad \text{as} \quad n \to \infty.$$

6. $\mathfrak{f}$ is such that $D^n\mathfrak{f}$ exists and $D^n\mathfrak{f} \in \mathsf{R}(a,b)$. Prove that

$$\frac{d^{n-1}}{dx^{n-1}}\left(\frac{\mathfrak{f}(b)-\mathfrak{f}(x)}{b-x}\right) = \frac{1}{(b-x)^n}\int_x^b (b-u)^{n-1} D^n\mathfrak{f}(u)\, du.$$

7. $\mathfrak{f}$ and $\mathfrak{g}$ are positive and have continuous positive derivatives for $x \geqslant 0$. If $\mathfrak{f}(0) = 0$ prove that

$$\mathfrak{f}(a)\,\mathfrak{g}(b) \leqslant \int_0^a \mathfrak{g}.D\mathfrak{f} + \int_0^b \mathfrak{f}.D\mathfrak{g} \quad (0 < a \leqslant b).$$

8. $\mathfrak{f}$ has a continuous derivative in $[0,1]$. Show that

$$\tfrac{1}{2}m \leqslant \int_0^1 \mathfrak{f} - \mathfrak{f}(0) \leqslant \tfrac{1}{2}M,$$

where $m = \inf D\mathfrak{f}(x)$, $M = \sup D\mathfrak{f}(x)$, $0 \leqslant x \leqslant 1$. Deduce that

$$\lim_{n\to\infty}\left(n\int_0^1 \mathfrak{f} - \sum_0^{n-1} \mathfrak{f}(r/n)\right) = \tfrac{1}{2}(\mathfrak{f}(1)-\mathfrak{f}(0)).$$

24

THE RIEMANN–STIELTJES INTEGRAL

In this chapter an extension of the Riemann integral is given which is important in itself and by means of which we can both simplify and generalize some of the results concerning the Riemann integral.

Definition 48. *$\mathfrak{f}$ and $\mathfrak{g}$ are two functions defined over $a \leqslant x \leqslant b$. If $\exists$ a number l with the property $\forall \epsilon > 0 \,\exists \delta > 0$ such that for any dissection $\mathscr{D}$ with $\tau(\mathscr{D}) < \delta$,*

$$\left|\sum_{i=0}^{k-1} \mathfrak{f}(\xi_i)\,(\mathfrak{g}(x_{i+1}) - \mathfrak{g}(x_i)) - l\right| < \epsilon,$$

where $\mathscr{D}$ is $x_0, x_1, \ldots, x_k$ and $x_i < \xi_i < x_{i+1}$, then we say that $\mathfrak{f}$ is Riemann–Stieltjes integrable with respect to $\mathfrak{g}$ over $[a, b]$ and that l is its Riemann–Stieltjes integral.

We write

$$\int_a^b \mathfrak{f}(x)\, d\mathfrak{g}(x) \quad \text{or} \quad \int_a^b \mathfrak{f}\, d\mathfrak{g} \quad \text{for} \quad l$$

and extend the scope of this symbol to the case $b \leqslant a$ by writing

$$\int_a^a \mathfrak{f}\, d\mathfrak{g} = 0, \quad \int_a^b \mathfrak{f}\, d\mathfrak{g} = -\int_b^a \mathfrak{f}\, d\mathfrak{g}.$$

$\mathfrak{f}$ and $\mathfrak{g}$ could be complex-valued but we shall only consider real-valued functions. The class of functions Riemann–Stieltjes integrable with respect to $\mathfrak{g}$ over $[a, b]$ will be denoted by $\mathsf{RS}_{\mathfrak{g}}(a, b)$ and we shall use RS-*integrable* as an abbreviation of Riemann–Stieltjes integrable. If $\mathfrak{g}(x)$ is x then $\mathsf{RS}_{\mathfrak{g}}(a, b) = \mathsf{R}(a, b)$ and in fact if $\mathfrak{g}\uparrow$ the theory can be developed in a manner analogous to that used for the Riemann integral.

Exercises. (1) $\mathfrak{g}$ is defined over $[a, b]$. Prove that

$$\int_a^b d\mathfrak{g} = \mathfrak{g}(b) - \mathfrak{g}(a).$$

(2) $\mathfrak{g}$ is defined by $\mathfrak{g}(x) = 0$ if $x < \frac{1}{2}$, $\mathfrak{g}(x) = 1$ if $x \geqslant \frac{1}{2}$. Prove that $\mathfrak{f} \in \mathsf{RS}_{\mathfrak{g}}(a, b)$ if and only if $\mathfrak{f}$ is continuous at $x = \frac{1}{2}$ and that then

$$\int_0^1 \mathfrak{f}\, d\mathfrak{g} = \mathfrak{f}(\tfrac{1}{2}).$$

Elementary properties

(1) If $\mathfrak{f} \in \mathsf{RS}_{\mathfrak{g}}(a, b)$ and $\mathfrak{f} \in \mathsf{RS}_{\mathfrak{h}}(a, b)$ then $\mathfrak{f} \in \mathsf{RS}_{\mathfrak{g}+\mathfrak{h}}(a, b)$ and

$$\int_a^b \mathfrak{f}\, d(\mathfrak{g}+\mathfrak{h}) = \int_a^b \mathfrak{f}\, d\mathfrak{g} + \int_a^b \mathfrak{f}\, d\mathfrak{h}.$$

Also $\mathfrak{f} \in \mathsf{RS}_{-\mathfrak{g}}(a, b)$.

(2) If $\mathfrak{f}_1$ and $\mathfrak{f}_2$ belong to $\mathsf{RS}_{\mathfrak{g}}(a, b)$ then $\mathfrak{f}_1 + \mathfrak{f}_2$ and $k\mathfrak{f}_1$ belong to $\mathsf{RS}_{\mathfrak{g}}(a, b)$. Also

$$\int_a^b k\mathfrak{f}_1\, d\mathfrak{g} = k\int_a^b \mathfrak{f}_1\, d\mathfrak{g}, \quad \int_a^b (\mathfrak{f}_1 + \mathfrak{f}_2)\, d\mathfrak{g} = \int_a^b \mathfrak{f}_1\, d\mathfrak{g} + \int_a^b \mathfrak{f}_2\, d\mathfrak{g}.$$

k is here a constant.

(3) If $\mathfrak{h}$ is defined by $\mathfrak{h}(x) = \mathfrak{g}(a+b-x)$ and $\mathfrak{f} \in \mathsf{RS}_{\mathfrak{g}}(a, b)$, then $\mathfrak{f} \in \mathsf{RS}_{\mathfrak{h}}(a, b)$ and

$$\int_a^b \mathfrak{f}\, d\mathfrak{g} = -\int_a^b \mathfrak{f}\, d\mathfrak{h}.$$

The proofs of these results (1)–(3) are left to the reader.

Theorem 50. *If* $\mathfrak{f}$ *is continuous over* $[a, b]$ *and* $\mathfrak{g}$ *is of bounded variation over* $[a, b]$ *then* $\mathfrak{f} \in \mathsf{RS}_{\mathfrak{g}}(a, b)$.

By theorem 29 there are bounded increasing functions $\mathfrak{h}(x)$, $\mathfrak{k}(x)$ such that

$$\mathfrak{g}(x) = \mathfrak{h}(x) - \mathfrak{k}(x).$$

By property (1) above it is sufficient to show that $\mathfrak{f} \in \mathsf{RS}_{\mathfrak{h}}(a, b)$. The proof of this is omitted. It is analogous to that of theorem 42 (ii) with the simplification that $\mathfrak{f}$ is continuous at all points of $[a, b]$.

Remark. It is not necessarily the case that if $\mathfrak{f}$ and $\mathfrak{g}$ are continuous then

$$\mathfrak{f} \in \mathsf{RS}_{\mathfrak{g}}(a, b).$$

For example, if $\mathfrak{f}(0) = \mathfrak{g}(0) = 0$ and $\mathfrak{f}(x) = x^{\frac{1}{2}} \sin 1/x$, $\mathfrak{g}(x) = x^{\frac{1}{2}} \cos 1/x$, $x > 0$, then

$$\mathfrak{f} \notin \mathsf{RS}_{\mathfrak{g}}(0, 1).$$

Theorem 48a. (*Integration by parts.*) *If* $\mathfrak{f}$ *is of bounded variation and* $\mathfrak{g}$ *is continuous over* $[a, b]$ *then* $\mathfrak{f} \in \mathsf{RS}_{\mathfrak{g}}(a, b)$ *and*

$$\int_a^b \mathfrak{f}\, d\mathfrak{g} = \mathfrak{f}(b)\,\mathfrak{g}(b) - \mathfrak{f}(a)\,\mathfrak{g}(a) - \int_a^b \mathfrak{g}\, d\mathfrak{f}.$$

We consider a particular dissection $\mathscr{D} = x_0, x_1, x_2, \ldots, x_k$ and rewrite the sum

$$\sigma_{\mathscr{D}} = \sum_{i=0}^{k-1} \mathfrak{f}(\xi_i)\,(\mathfrak{g}(x_{i+1}) - \mathfrak{g}(x_i)) \quad (x_i < \xi_i < x_{i+1})$$

and obtain

$$\sigma_{\mathscr{D}} = -\sum_{i=1}^{k-1} (\mathfrak{f}(\xi_i) - \mathfrak{f}(\xi_{i-1}))\,\mathfrak{g}(x_i) - (\mathfrak{f}(\xi_0) - \mathfrak{f}(a))\,\mathfrak{g}(a)$$
$$- (\mathfrak{f}(b) - \mathfrak{f}(\xi_{k-1}))\,\mathfrak{g}(b) + \mathfrak{f}(b)\,\mathfrak{g}(b) - \mathfrak{f}(a)\,\mathfrak{g}(a).$$

Define ξ_{-1} to be a, ξ_k to be b and select y_i such that $a < y_0 < \xi_0$, $\xi_{k-1} < y_k < b$, $y_i = x_i$ for $1 \leqslant i \leqslant k-1$. Then

$$\sigma_{\mathscr{D}} = -\sum_{i=-1}^{k-1} (\mathfrak{f}(\xi_{i+1}) - \mathfrak{f}(\xi_i))\,\mathfrak{g}(y_{i+1}) - (\mathfrak{f}(\xi_0) - \mathfrak{f}(a))\,(\mathfrak{g}(a) - \mathfrak{g}(y_0))$$
$$- (\mathfrak{f}(b) - \mathfrak{f}(\xi_{k-1}))\,(\mathfrak{g}(b) - \mathfrak{g}(y_k)) + \mathfrak{f}(b)\,\mathfrak{g}(b) - \mathfrak{f}(a)\,\mathfrak{g}(a).$$

Now the sum $\sum_{i=-1}^{k-1} (\mathfrak{f}(\xi_{i+1}) - \mathfrak{f}(\xi_i))\,\mathfrak{g}(y_{i+1})$ is one which differs from $\int_a^b \mathfrak{g}\, d\mathfrak{f}$ by an amount which is small when $\tau(\mathscr{D}_1)$ is small, where $\mathscr{D}_1$ is the dissection $\xi_{-1}, \xi_0, \ldots, \xi_{k-1}, \xi_k$. Now $\tau(\mathscr{D}_1) < 2\tau(\mathscr{D})$ and thus $\tau(\mathscr{D}_1) \to 0$ as $\tau(\mathscr{D}) \to 0$. Further as $\tau(\mathscr{D}) \to 0$, $y_0 \to a$, $y_k \to b$ and by the continuity of $\mathfrak{g}$

$$\mathfrak{g}(a) - \mathfrak{g}(y_0) \to 0, \quad \mathfrak{g}(b) - \mathfrak{g}(y_k) \to 0.$$

Hence if $\tau(\mathscr{D})$ is sufficiently small $\sigma_{\mathscr{D}}$ differs from

$$-\int_a^b \mathfrak{g}\, d\mathfrak{f} + \mathfrak{f}(b)\,\mathfrak{g}(b) - \mathfrak{f}(a)\,\mathfrak{g}(a)$$

by not more than an arbitrarily assigned positive amount. Thus $\mathfrak{f} \in \mathsf{RS}_{\mathfrak{g}}(a, b)$ and

$$\int_a^b \mathfrak{f}\, d\mathfrak{g} = \mathfrak{f}(b)\,\mathfrak{g}(b) - \mathfrak{f}(a)\,\mathfrak{g}(a) - \int_a^b \mathfrak{g}\, d\mathfrak{f}.$$

Theorem 51. *If $\mathfrak{f}$ is continuous over $[a, b]$ and $\mathfrak{g}$ is of bounded variation over $[a, b]$ then*

(i) *for $a < c < b$*

$$\int_a^b \mathfrak{f}\, d\mathfrak{g} = \int_a^c \mathfrak{f}\, d\mathfrak{g} + \int_c^b \mathfrak{f}\, d\mathfrak{g},$$

(ii) *if $\mathfrak{V}(c)$ denotes the total variation of $\mathfrak{g}$ in $[a, c]$ then*

$$\left|\int_a^b \mathfrak{f}\, d\mathfrak{g}\right| \leqslant \int_a^b |\mathfrak{f}|\, d\mathfrak{V} \leqslant |\sup \mathfrak{f}(x)| \,.\, \mathfrak{V}(b).$$

(i) The three integrals all exist by theorem 50. The truth of the equality is an immediate consequence of the definition.

(ii) The function $\mathfrak{V}(x)$ is monotonic increasing and bounded. Thus it is of bounded variation and since $|\mathfrak{f}(x)|$ is continuous, $|\mathfrak{f}| \in \mathsf{RS}_{\mathfrak{V}}(a, b)$. For any dissection

$$|\sigma_{\mathscr{D}}| = \left|\sum_{i=0}^{k-1} \mathfrak{f}(\xi_i)\,(\mathfrak{g}(x_{i+1}) - \mathfrak{g}(x_i))\right| \quad (x_i < \xi_i < x_{i+1})$$

$$\leqslant \sum_{i=0}^{k-1} |\mathfrak{f}(\xi_i)| \,.\, |\mathfrak{g}(x_{i+1}) - \mathfrak{g}(x_i)|.$$

Now in the notation of theorem 29, p. 98, using positive and negative variations

$$\mathfrak{V}(x) = P^{\mathfrak{g}}(a, x) - N^{\mathfrak{g}}(a, x),$$

$$\mathfrak{g}(x) - \mathfrak{g}(a) = P^{\mathfrak{g}}(a, x) + N^{\mathfrak{g}}(a, x).$$

Thus

$$|\mathfrak{g}(x_{i+1}) - \mathfrak{g}(x_i)| \leqslant |P^{\mathfrak{g}}(a, x_{i+1}) - P^{\mathfrak{g}}(a, x_i)| + |N^{\mathfrak{g}}(a, x_{i+1}) - N^{\mathfrak{g}}(a, x_i)|$$

$$= \mathfrak{V}(x_{i+1}) - \mathfrak{V}(x_i),$$

and
$$|\sigma_{\mathscr{D}}| \leqslant \sum_{i=0}^{k-1} |\mathfrak{f}(\xi_i)|\, (\mathfrak{V}(x_{i+1}) - \mathfrak{V}(x_i)).$$

It follows by considering a sequence of dissections $\mathscr{D}_n$ for which $\tau(\mathscr{D}_n) \to 0$ that

$$\left|\int_a^b \mathfrak{f}\, d\mathfrak{g}\right| \leqslant \int_a^b |\mathfrak{f}|\, d\mathfrak{V}.$$

Further

$$\sum_{0}^{k-1} |\mathfrak{f}(\xi_i)| \, (\mathfrak{B}(x_{i+1}) - \mathfrak{B}(x_i)) \leqslant \sup |\mathfrak{f}(x)| \, . \sum_{0}^{k-1} (\mathfrak{B}(x_{i+1}) - \mathfrak{B}(x_i))$$
$$= \sup |\mathfrak{f}(x)| \, . \, \mathfrak{B}(b)$$

and this implies that

$$\int_a^b |\mathfrak{f}| \, d\mathfrak{B} \leqslant \sup |\mathfrak{f}(x)| \, . \, \mathfrak{B}(b).$$

The theorem is proved.

Theorem 49a. *(Change of variable.) If $\mathfrak{f} \in \mathsf{RS}_{\mathfrak{g}}(a, b)$ and $\mathfrak{h}(x)$ is a continuous and strictly increasing function defined over $[c, d]$ where $a = \mathfrak{h}(c)$, $b = \mathfrak{h}(d)$ then $\mathfrak{f}(\mathfrak{h}(x)) \in \mathsf{RS}_{\mathfrak{g}(\mathfrak{h})}(c, d)$ and*

$$\int_a^b \mathfrak{f} \, d\mathfrak{g} = \int_c^d \mathfrak{f}(\mathfrak{h}) \, d\mathfrak{g}(\mathfrak{h}).$$

Let $\mathscr{D}_1$ be the dissection $c = t_0 < t_1 < \ldots < t_k = d$ of $[c, d]$ and select η_i $(i = 0, 1, \ldots, k-1)$ such that $t_i < \eta_i < t_{i+1}$. Write $x_i = \mathfrak{h}(t_i)$, $\xi_i = \mathfrak{h}(\eta_i)$, then $x_0, x_1, \ldots, x_k$ is a dissection say $\mathscr{D}$ of $[a, b]$. Also $x_i < \xi_i < x_{i+1}$. Thus

$$\sum_{i=0}^{k-1} \mathfrak{f}(\mathfrak{h}(\eta_i)) \, (\mathfrak{g}(\mathfrak{h}(t_{i+1})) - \mathfrak{g}(\mathfrak{h}(t_i))) = \sum_{i=0}^{k-1} \mathfrak{f}(\xi_i) \, (\mathfrak{g}(x_{i+1}) - \mathfrak{g}(x_i))$$

and this last expression approaches $\int_a^b \mathfrak{f} \, d\mathfrak{g}$ as $\tau(\mathscr{D}) \to 0$. But $\mathfrak{h}$ is uniformly continuous and thus $\tau(\mathscr{D}) \to 0$ if $\tau(\mathscr{D}_1) \to 0$. Thus the expression on the left-hand side above is close to $\int_a^b \mathfrak{f} \, d\mathfrak{g}$ if $\tau(\mathscr{D}_1)$ is sufficiently small. Hence $\mathfrak{f}(\mathfrak{h}) \in \mathsf{RS}_{\mathfrak{g}(\mathfrak{h})}(c, d)$ and

$$\int_c^d \mathfrak{f}(\mathfrak{h}) \, d\mathfrak{g}(\mathfrak{h}) = \int_a^b \mathfrak{f} \, d\mathfrak{g}.$$

Theorem 52. *Reduction of a Riemann–Stieltjes integral to a Riemann integral. If $\mathfrak{f}, \mathfrak{g} \in \mathsf{R}(a, b)$, $\mathfrak{G}(x) = \int_a^x \mathfrak{g}$ and $\mathfrak{f} \in \mathsf{RS}_{\mathfrak{G}}(a, b)$ then*

$$\int_a^b \mathfrak{f} \, d\mathfrak{G} = \int_a^b \mathfrak{f} \, . \, \mathfrak{g}.$$

By theorem 43, $\mathfrak{f}\mathfrak{g} \in \mathsf{R}(a, b)$. Given $\epsilon > 0 \; \exists$ a dissection $\mathscr{D}$ and $\mathfrak{h}$ a step-function on it such that $\int_a^b |\mathfrak{g} - \mathfrak{h}| < \epsilon$ and, writing $\mathfrak{H}(x) = \int_a^x \mathfrak{h}$, the sums

$$\sigma_{\mathscr{D}} = \sum_{i=0}^{k-1} \mathfrak{f}(\xi_i) \, (\mathfrak{G}(x_{i+1}) - \mathfrak{G}(x_i)), \quad \kappa_{\mathscr{D}} = \sum_{i=0}^{k-1} \mathfrak{f}(\xi_i) \, (\mathfrak{H}(x_{i+1}) - \mathfrak{H}(x_i))$$

differ by at most

$$\sum_{i=0}^{k-1} |\mathfrak{f}(\xi_i)| \int_{x_i}^{x_{i+1}} |\mathfrak{g}-\mathfrak{h}| \leqslant \epsilon \sup_{a \leqslant x \leqslant b} |\mathfrak{f}(x)|.$$

Thus

$$\left| \int_a^b \mathfrak{f}\, d\mathfrak{G} - \int_a^b \mathfrak{f}\, d\mathfrak{H} \right|$$

can be made arbitrarily small by an appropriate choice of ϵ. Further we can choose $\mathfrak{h}$ so that also

$$\left| \int_a^b \mathfrak{f}\mathfrak{g} - \int_a^b \mathfrak{f}\mathfrak{h} \right| \leqslant \sup |\mathfrak{f}(x)|\, \epsilon.$$

Thus it is sufficient to establish the theorem when $\mathfrak{g}$ is a step-function. Further we need only consider functions of the form

$$\mathfrak{g}(x) = 0 \ (x < c), \quad \mathfrak{g}(x) = k \ (c \leqslant x \leqslant d), \quad \mathfrak{g}(x) = 0 \ (x > d)$$

since every step-function is the sum of a finite number of functions of the above form and the result for step-functions will then follow by addition. But when $\mathfrak{g}$ is defined as above $\mathfrak{G}(x) = 0$ $(x < c)$, $\mathfrak{G}(x) = k(x-c)$ $(c \leqslant x \leqslant d)$, $\mathfrak{G}(x) = k(d-c)$ $(x > d)$ and

$$\int_a^b \mathfrak{f}\, d\mathfrak{G} = k \int_c^d \mathfrak{f}\, d(x-c) = k \int_c^d \mathfrak{f}, \qquad \int_a^b \mathfrak{f}\mathfrak{g} = k \int_c^d \mathfrak{f}.$$

Thus the integrals are equal in this special case and therefore generally.

If $\mathfrak{G}$ has a R-integrable derivative $D\mathfrak{G}$, then by theorem 47

$$\mathfrak{G}(x) - \mathfrak{G}(a) = \int_a^x D\mathfrak{G}$$

and thus if $\mathfrak{f} \in \mathsf{RS}_{\mathfrak{G}}(a, b)$ and $\mathfrak{f} \in \mathsf{R}(a, b)$ then

$$\int_a^b \mathfrak{f}\, d\mathfrak{G} = \int_a^b \mathfrak{f} . D\mathfrak{G}.$$

Exercises. (1) $\mathfrak{f}$ has a continuous derivative in $a \leqslant x \leqslant b$. Show that

$$\int_a^b \mathfrak{f}\, d\mathfrak{f} = \tfrac{1}{2}(\mathfrak{f}^2(b) - \mathfrak{f}^2(a)).$$

(2) Prove that

$$\int_0^{\pi} \cos x\, d(\sin x) = \tfrac{1}{2}\pi, \quad \int_1^2 x^2\, d(1/x) = -1.$$

(3) Prove that

(i) $$\int_0^1 e^{x^2}\,dx^2 = e-1,$$

(ii) $$\int_0^{\frac{1}{2}\pi} e^{\sin x}\,d(\sin x) = e-1,$$

(iii) $$\int_a^b e^{\mathfrak{f}(x)}\,d\mathfrak{f}(x) = e^{\mathfrak{f}(b)} - e^{\mathfrak{f}(a)} \quad \text{if } \mathfrak{f} \text{ is continuous and } \mathfrak{f}\Uparrow.$$

Second Integral mean-value theorem (*see also theorem* 48, *corollaries* 1, 2, p. 154)

Theorem 53. *If* $\mathfrak{f}$ *is monotonic over* $[a,b]$ *and* $\mathfrak{g} \in \mathsf{R}(a,b)$, *then* $\exists\, c$ *such that* $a \leqslant c \leqslant b$ and

$$\int_a^b \mathfrak{f}\mathfrak{g} = \mathfrak{f}(a)\int_a^c \mathfrak{g} + \mathfrak{f}(b)\int_c^b \mathfrak{g}.$$

Write $\mathfrak{G}(x) = \int_a^x \mathfrak{g}$. Since $\mathfrak{f}$ is monotonic $\mathfrak{f} \in \mathsf{R}(a,b)$ also $\mathfrak{f}$ is of bounded variation and $\mathfrak{G}$ is continuous (theorem 44, corollary 4). Thus by theorem 48a $\mathfrak{f} \in \mathsf{RS}_{\mathfrak{G}}(a,b)$. Hence theorem 52 can be applied and

$$\int_a^b \mathfrak{f}\,d\mathfrak{G} = \int_a^b \mathfrak{f}\mathfrak{g}.$$

By theorem 48a $$\int_a^b \mathfrak{f}\,d\mathfrak{G} = \mathfrak{f}(b)\,\mathfrak{G}(b) - \int_a^b \mathfrak{G}\,d\mathfrak{f}.$$

If $\mathfrak{f}\uparrow$ then from the definition of the RS-integral

$$\inf \mathfrak{G}(x)\int_a^b d\mathfrak{f} \leqslant \int_a^b \mathfrak{G}\,d\mathfrak{f} \leqslant \sup \mathfrak{G}(x)\int_a^b d\mathfrak{f}$$

and if $\mathfrak{f}\downarrow$ similar but reversed inequalities hold. Since $\mathfrak{G}$ is continuous $\exists\, c$ such that

$$a \leqslant c \leqslant b \quad \text{and} \quad \int_a^b \mathfrak{G}\,d\mathfrak{f} = \mathfrak{G}(c)\int_a^b d\mathfrak{f} = \mathfrak{G}(c)\,(\mathfrak{f}(b) - \mathfrak{f}(a)).$$

Hence by substitution we obtain

$$\begin{aligned} \int_a^b \mathfrak{f}\mathfrak{g} &= \mathfrak{f}(b)\,\mathfrak{G}(b) - \mathfrak{G}(c)\,(\mathfrak{f}(b) - \mathfrak{f}(a)) \\ &= \mathfrak{f}(a)\int_a^c \mathfrak{g} + \mathfrak{f}(b)\int_c^b \mathfrak{g}. \end{aligned}$$

Corollary. *If* $\mathfrak{f}\downarrow$, $\mathfrak{f} \geqslant 0$ *and* $\mathfrak{g} \in \mathsf{R}(a,b)$, *then* $\exists\, c$ *such that*

$$a \leqslant c \leqslant b \quad \textit{and} \quad \int_a^b \mathfrak{f}\mathfrak{g} = \mathfrak{f}(a)\int_a^c \mathfrak{g}.$$

Define $\mathfrak{f}_1$ by $\mathfrak{f}_1(x) = \mathfrak{f}(x)$ $(a \leqslant x < b)$, $\mathfrak{f}_1(b) = 0$. Then $\mathfrak{f}_1 \downarrow$ and by the theorem $\exists\, c$ such that $a \leqslant c \leqslant b$ and

$$\int_a^b \mathfrak{f}_1 \mathfrak{g} = \mathfrak{f}_1(a) \int_a^c \mathfrak{g}.$$

Since $$\int_a^b \mathfrak{f}_1 \mathfrak{g} = \int_a^b \mathfrak{f}\mathfrak{g} \quad \text{and} \quad \mathfrak{f}_1(a) = \mathfrak{f}(a)$$

the proof of the corollary is complete.

Worked example

$\mathfrak{f}$ and $\mathfrak{g}$ are defined over $[a, b]$ and for some subinterval $[c, d]$ of $[a, b]$ where $a < c < d < b$, $\mathfrak{f} \notin \mathsf{RS}_{\mathfrak{g}}(c, d)$. Show that $\exists\, x_0$ such that $a < x_0 < b$ and if x_1, x_2 are such that $a < x_1 < x_0 < x_2 < b$ then $\mathfrak{f} \notin \mathsf{RS}_{\mathfrak{g}}(x_1, x_2)$.

Let $I_1 = [c, d]$ and denote by J, K the two half-intervals of I_1, $[c, \frac{1}{2}(c+d)]$ and $[\frac{1}{2}(c+d), d]$. If $\mathfrak{f} \in \mathsf{RS}_{\mathfrak{g}}(J)$ and $\mathfrak{f} \in \mathsf{RS}_{\mathfrak{g}}(K)$, then it is easy to see that $\mathfrak{f} \in \mathsf{RS}_{\mathfrak{g}}(I_1)$.

This is not so thus either $\mathfrak{f} \notin \mathsf{RS}_{\mathfrak{g}}(J)$ or $\mathfrak{f} \notin \mathsf{RS}_{\mathfrak{g}}(K)$. Denote that interval J, K over which $\mathfrak{f}$ is not RS-integrable with respect to $\mathfrak{g}$ by I_2. (If a choice is possible select that half-interval with the same left-hand end-point as I_1.) Generally when $I_1, \ldots, I_n$ have been defined I_{n+1} is a half-interval of I_n selected so that $\mathfrak{f} \notin \mathsf{RS}_{\mathfrak{g}}(I_{n+1})$. By theorem 20, p. 76, $\bigcap_{n=1}^{\infty} I_n$ is a single point say x_0. x_0 clearly has the required property.

Exercises

1. $\mathfrak{f}$ is a step-function. $\mathfrak{g}$ is bounded and continuous at each discontinuity of $\mathfrak{f}$. Prove that $\mathfrak{f} \in \mathsf{RS}_{\mathfrak{g}}(a, b)$.

2. The rational numbers between 0 and 1 are arranged as a sequence $\{r_n\}$ and $\mathfrak{g}$ is defined over $[0, 1]$ by $\mathfrak{g}(x) = \Sigma 1/n^2$, where the sum is over those integers for which $r_n < x$. $\mathfrak{f}$ is continuous. Show that

$$\mathfrak{f} \in \mathsf{RS}_{\mathfrak{g}}(0, 1) \quad \text{and} \quad \int_0^1 \mathfrak{f}\, d\mathfrak{g} = \sum_{n=1}^{\infty} \mathfrak{f}(r_n)/n^2.$$

3. $\mathfrak{f}$ and $\mathfrak{g}$ are convex over $[a, b]$. Show that $\mathfrak{f} \in \mathsf{RS}_{\mathfrak{g}}(a, b)$.

4. $\mathfrak{g} \uparrow$ and $\mathfrak{f}, \mathfrak{h} \in \mathsf{RS}_{\mathfrak{g}}(a, b)$. Show that $\mathfrak{f} . \mathfrak{h} \in \mathsf{RS}_{\mathfrak{g}}(a, b)$.

5. $\mathfrak{g} \uparrow$ and $\mathfrak{f}_1, \mathfrak{f}_2 \in \mathsf{RS}_{\mathfrak{g}}(a, b)$. If $\mathfrak{f}_1 \geqslant \mathfrak{f}_2$ show that

$$\int_a^b \mathfrak{f}_1\, d\mathfrak{g} \geqslant \int_a^b \mathfrak{f}_2\, d\mathfrak{g}.$$

Deduce that if $\mathfrak{f}$ is continuous $\mathfrak{h} \geqslant 0$ and $\mathfrak{f}\mathfrak{h} \in \mathsf{RS}_{\mathfrak{g}}(a, b)$, $\mathfrak{h} \in \mathsf{RS}_{\mathfrak{g}}(a, b)$ then $\exists\, \xi$ such that $a \leqslant \xi \leqslant b$ and

$$\int_a^b \mathfrak{f}\mathfrak{h}\, d\mathfrak{g} = \mathfrak{f}(\xi) \int_a^b \mathfrak{h}\, d\mathfrak{g}.$$

25

IMPROPER INTEGRALS; CONVERGENCE OF INTEGRALS

The Riemann integral in its original form applies only to bounded functions over intervals. It is necessary to extend its scope to functions which are bounded except in the neighbourhood of a finite number of points and to ranges of integration of the form $a \leqslant x$ or $x \leqslant a$ or all real x.

If $\mathfrak{f} \in \mathsf{R}(a, b-\epsilon)$ for every $\epsilon > 0$ and $\int_a^{b-\epsilon} \mathfrak{f} \to l$ as $\epsilon \to 0+$, then we say that $\mathfrak{f}$ has an *improper Riemann integral* over $[a, b]$ and that its value is l. We shall also write $\int_a^b \mathfrak{f}$ in place of l.

For example, if $a = 0$, $b = 1$ and $\mathfrak{f}(x) = (1-x)^{-\frac{1}{2}}$, then

$$\int_0^{1-\epsilon} (1-x)^{-\frac{1}{2}}\,dx = 2(1-\epsilon^{\frac{1}{2}}) \to 2 \quad \text{as} \quad \epsilon \to 0+$$

and $(1-x)^{-\frac{1}{2}}$ is improperly Riemann integrable over $[0, 1]$.

Similarly if $\mathfrak{f} \in \mathsf{R}(a+\epsilon, b)$ for every $\epsilon > 0$ and $\int_{a+\epsilon}^{b} \mathfrak{f} \to l$ as $\epsilon \to 0+$, then again $\mathfrak{f}$ is improperly Riemann integrable over $[a, b]$ and $l = \int_a^b \mathfrak{f}$.

Generally if $\mathfrak{f}$ is defined over $[a, b]$ and there are a finite number of points $c_1, \ldots, c_k$ such that $a < c_1 < \ldots < c_k < b$, and $\mathfrak{f}$ is improperly Riemann integrable over each interval $[a, c_1]$, $[c_1, c_2]$, ..., $[c_{k-1}, c_k]$, $[c_k, b]$, then $\mathfrak{f}$ is said to be improperly Riemann integrable over $[a, b]$ and we shall denote the sum

$$\int_a^{c_1} \mathfrak{f} + \int_{c_1}^{c_2} \mathfrak{f} + \ldots + \int_{c_k}^{b} \mathfrak{f} \quad \text{by} \quad \int_a^b \mathfrak{f}.$$

In this case we write $\mathfrak{f} \in \mathsf{IR}(a, b)$.

Properties of improper Riemann integrals

(*a*) If $\mathfrak{f}$ and $\mathfrak{g}$ belong to $\mathsf{IR}(a, b)$ then so do $\mathfrak{f}+\mathfrak{g}$ and $k\mathfrak{f}$ where k is a constant function. Also

$$\int_a^b (\mathfrak{f}+\mathfrak{g}) = \int_a^b \mathfrak{f} + \int_a^b \mathfrak{g}, \quad \int_a^b k\mathfrak{f} = k\int_a^b \mathfrak{f}.$$

(It is *not* necessarily the case that if $\mathfrak{f} \in \mathsf{IR}(a, b)$ then $\mathfrak{f}^2 \in \mathsf{IR}(a, b)$.)

(b) If $\mathfrak{f} \in \mathrm{IR}(a,b)$ and $[c,d] \subset [a,b]$ then $\mathfrak{f} \in \mathrm{IR}(c,d)$. Also if $a < c < b$ then

$$\int_a^b \mathfrak{f} = \int_a^c \mathfrak{f} + \int_c^b \mathfrak{f}.$$

(c) If $\mathfrak{f}$ and $\mathfrak{g}$ belong to $\mathrm{IR}(a,b)$ and $\mathfrak{f}(x) \geqslant \mathfrak{g}(x)$ for $a \leqslant x \leqslant b$ then

$$\int_a^b \mathfrak{f} \geqslant \int_a^b \mathfrak{g}.$$

(d) If $m \leqslant \mathfrak{f}(x) \leqslant M$, $\mathfrak{g} \geqslant 0$, where $\mathfrak{f}.\mathfrak{g}$ and $\mathfrak{g}$ both belong to $\mathrm{IR}(a,b)$ then

$$m\int_a^b \mathfrak{g} \leqslant \int_a^b \mathfrak{f}\mathfrak{g} \leqslant M\int_a^b \mathfrak{g}.$$

If, in addition, $\mathfrak{f}$ is continuous, then we have the first mean-value theorem: $\exists\, \xi$ such that

$$a \leqslant \xi \leqslant b \quad \text{and} \quad \int_a^b \mathfrak{f}\mathfrak{g} = \mathfrak{f}(\xi)\int_a^b \mathfrak{g}.$$

(e) If $\mathfrak{f} \in \mathrm{IR}(a,b)$, then $\int_a^c \mathfrak{f}$ is a continuous function of c where $a \leqslant c \leqslant b$.

(f) If $\mathfrak{f}$ is such that $D\mathfrak{f}$ exists for $a \leqslant x \leqslant b$ and $D\mathfrak{f} \in \mathrm{IR}(a,b)$ then

$$\int_a^b D\mathfrak{f} = \mathfrak{f}(b) - \mathfrak{f}(a).$$

(g) If $\mathfrak{f}$ and $\mathfrak{g}$ belong to $\mathrm{IR}(a,b)$ and $\mathfrak{F}$, $\mathfrak{G}$ are defined by

$$\mathfrak{F}(x) = \int_a^x \mathfrak{f}, \quad \mathfrak{G}(x) = \int_a^x \mathfrak{g}$$

then

$$\mathfrak{F}(b)\,\mathfrak{G}(b) = \int_a^b \mathfrak{f}\mathfrak{G} + \int_a^b \mathfrak{F}\mathfrak{g}.$$

(h) If $\mathfrak{f} \in \mathrm{IR}(a,b)$ and $\mathfrak{g}(y)\Uparrow$ for $c \leqslant y \leqslant d$, where $\mathfrak{g}(c) = a$, $\mathfrak{g}(d) = b$, $c < d$, $a < b$ and if $D\mathfrak{g} \in \mathrm{R}(c,d)$ then

$$\mathfrak{f}(\mathfrak{g}).D\mathfrak{g} \in \mathrm{IR}(c,d) \quad \text{and} \quad \int_a^b \mathfrak{f} = \int_c^d \mathfrak{f}(\mathfrak{g}).D\mathfrak{g}.$$

The proofs of all these results are left to the reader. They follow from the corresponding results for R-integrable functions by limiting processes.

If $|\mathfrak{f}| \in \mathrm{R}(a,b)$, then $\mathfrak{f}$ is said to be *absolutely improperly Riemann integrable over* $[a,b]$. If $\mathfrak{f} \in \mathrm{IR}(a,b)$ and $|\mathfrak{f}| \notin \mathrm{IR}(a,b)$, then $\mathfrak{f}$ is said to be *conditionally improperly Riemann integrable over* $[a,b]$.

Integrals over infinite ranges

If $\mathfrak{f} \in \mathsf{R}(a, X)$ for every $X > a$ and $\int_a^X \mathfrak{f}$ tends to a finite limit l as $X \to \infty$, then we say that $\mathfrak{f} \in \mathsf{R}(a, \infty)$ and write $\int_a^\infty \mathfrak{f}$ for l. It is customary in these circumstances to say that *the integral* $\int_a^\infty \mathfrak{f}$ *converges*. If $\mathfrak{f} \in \mathsf{R}(a, X)$ for all $X > a$ and $\int_a^X \mathfrak{f}$ does not tend to a finite limit as $X \to \infty$, then we say that *the integral* $\int_a^\infty \mathfrak{f}$ *diverges*.

Similarly we define the set of functions $\mathsf{R}(-\infty, b)$. If $\mathfrak{f} \in \mathsf{R}(a, \infty)$ and $\mathfrak{f} \in \mathsf{R}(-\infty, a)$, then we say that $\mathfrak{f} \in \mathsf{R}(-\infty, \infty)$ and define the integral of $\mathfrak{f}$ over $[-\infty, \infty]$ to be

$$\int_{-\infty}^{\infty} \mathfrak{f} = \int_{-\infty}^{a} \mathfrak{f} + \int_a^\infty \mathfrak{f}.$$

Remark. There is an ambiguity of notation since '$\int_a^\infty \mathfrak{f}$' can be used to mean either the function $\mathfrak{F}(X) = \int_a^X \mathfrak{f}$ defined for $X \geqslant a$, or the limit to which this function converges as $X \to \infty$. To avoid confusion we write 'the integral $\int_a^\infty \mathfrak{f}$' when we mean the function $\mathfrak{F}(X)$ and restrict the use of '$\int_a^\infty \mathfrak{f}$' to the limit of $\mathfrak{F}(X)$ as $X \to \infty$. This ambiguity is similar to that met previously in connexion with the convergence of series, chapter 7.

Properties of the set of functions $\mathsf{R}(a, \infty)$

(*a*) If $\mathfrak{f}$ and $\mathfrak{g}$ belong to $\mathsf{R}(a, \infty)$, then so do $\mathfrak{f} + \mathfrak{g}$, $k\mathfrak{f}$ where k is a constant and

$$\int_a^\infty (\mathfrak{f} + \mathfrak{g}) = \int_a^\infty \mathfrak{f} + \int_a^\infty \mathfrak{g}, \quad \int_a^\infty k\mathfrak{f} = k\int_a^\infty \mathfrak{f}.$$

(It is *not* necessarily the case that if $\mathfrak{f} \in \mathsf{R}(a, \infty)$ then $\mathfrak{f}^2 \in \mathsf{R}(a, \infty)$.)

(*b*) If $\mathfrak{f} \in \mathsf{R}(a, \infty)$, then $\int_X^\infty \mathfrak{f}$ is a continuous function of X for $X > a$. If $\mathfrak{f}$ is continuous at X where $X > a$ then $D\left(\int_X^\infty \mathfrak{f}\right) = -\mathfrak{f}(X)$.

(*c*) If $D\mathfrak{f} \in \mathsf{R}(a, \infty)$ then $\exists\, l$ such that $\mathfrak{f}(x) \to l$ as $x \to \infty$ and

$$\int_a^\infty D\mathfrak{f} = l - \mathfrak{f}(a).$$

(*d*) If $\mathfrak{f}$ and $\mathfrak{g}$ belong to $\mathsf{R}(a, \infty)$,

$$\mathfrak{F}(x) = \int_a^x \mathfrak{f}, \quad \mathfrak{G}(x) = \int_a^x \mathfrak{g},$$

then provided that at least one of $\mathfrak{f}\mathfrak{G}$ or $\mathfrak{F}\mathfrak{g}$ belongs to $\mathsf{R}(a, \infty)$ we have

$$\int_a^\infty \mathfrak{f}\mathfrak{G} + \int_a^\infty \mathfrak{F}\mathfrak{g} = \int_a^\infty \mathfrak{f} . \int_a^\infty \mathfrak{g} \quad \text{(integration by parts).}$$

This result is a consequence of letting $X \to \infty$ in the formula for integration by parts over $[a, X]$.

(*e*) Suppose that $\mathfrak{g}\Uparrow$, $D\mathfrak{g} \in \mathsf{R}(c, d)$, $\mathfrak{g}(c) = a$ and $\mathfrak{g}(y) \to \infty$ as $y \to d-$. Then if $\mathfrak{f} \in \mathsf{R}(a, \infty)$ we have $\mathfrak{f}(\mathfrak{g})\, D\mathfrak{g} \in \mathsf{R}(c, d)$ and

$$\int_a^\infty \mathfrak{f} = \int_c^d \mathfrak{f}(\mathfrak{g})\, D\mathfrak{g}.$$

It is a consequence of (*e*) that we may reduce the consideration of the existence of improper Riemann integrals to the convergence or otherwise of integrals over infinite ranges. Suppose, for example, that $\mathfrak{f} \in \mathsf{R}(a, b-\epsilon)$ for every ϵ satisfying $0 < \epsilon < b-a$. Then putting $y = 1/(b-x)$ we obtain

$$\int_a^{b-\epsilon} \mathfrak{f} = \int_{1/(b-a)}^{1/\epsilon} \mathfrak{f}(b - 1/y)\, dy/y^2 = \int_{1/(b-a)}^{1/\epsilon} \mathfrak{g},$$

where $\mathfrak{g}$ is used in place of $\mathfrak{f}(b-1/y).y^{-2}$. Then $\mathfrak{f} \in \mathsf{IR}(a, b)$ if and only if $\mathfrak{g} \in \mathsf{R}(1/(b-a), \infty)$. Hence we shall not consider improper integrals any further, but the various theorems given below relating to members of $\mathsf{R}(a, \infty)$ can be interpreted as results concerning $\mathsf{IR}(a, b)$.

The general principle of convergence for integrals

Theorem 5e. $\mathfrak{f} \in \mathsf{R}(a, \infty)$ *if and only if* (*a*) $\mathfrak{f} \in \mathsf{R}(a, X)$ *for every* $X > a$, *and* (*b*) $\forall \epsilon > 0\, \exists X$ *such that if* $Y > X$, $Z > X$ *then*

$$\left|\int_Y^Z \mathfrak{f}\right| < \epsilon.$$

This is an immediate consequence of the general principle of convergence, theorem 5*c*, p. 89, applied to the function

$$\mathfrak{F}(x) = \int_a^x \mathfrak{f}.$$

Absolute convergence

We say that *the integral* $\int_a^\infty \mathfrak{f}$ *is absolutely convergent* if and only if $|\mathfrak{f}| \in \mathsf{R}(a, \infty)$.

If $|\mathfrak{f}| \in \mathsf{R}(a, \infty)$ then $\mathfrak{f} \in \mathsf{R}(a, \infty)$ for by the general principle of convergence applied to the integral $\int_a^\infty |\mathfrak{f}|$, $\forall \epsilon > 0 \, \exists X$ such that

$$\left| \int_Y^Z |\mathfrak{f}| \right| < \epsilon \quad \text{if} \quad Y > X, Z > X.$$

But by the modulus inequality for integrals

$$\left| \int_Y^Z \mathfrak{f} \right| \leqslant \left| \int_Y^Z |\mathfrak{f}| \right|$$

and thus by another application of the general principle of convergence $\mathfrak{f} \in \mathsf{R}(a, \infty)$.

Integrals of positive functions

If $\mathfrak{f}(x) \geqslant 0$ for $x \geqslant a$ and $\mathfrak{f} \in \mathsf{R}(a, X)$ for all $X > a$, then the necessary and sufficient condition that $\mathfrak{f} \in \mathsf{R}(a, \infty)$ is that $\int_a^X \mathfrak{f}$ be bounded above for $X > a$. This is because $\int_a^X \mathfrak{f} \uparrow$.

If $0 \leqslant \mathfrak{f}(x) \leqslant \mathfrak{g}(x)$ and $\mathfrak{g} \in \mathsf{R}(a, \infty)$, $\mathfrak{f} \in \mathsf{R}(a, X)$ all $X > a$ then $\mathfrak{f} \in \mathsf{R}(a, \infty)$.

If $0 \leqslant \mathfrak{g}(x) \leqslant \mathfrak{f}(x)$ and $\mathfrak{g} \in \mathsf{R}(a, X)$ for all $X > a$ but $\mathfrak{g} \notin \mathsf{R}(a, \infty)$ then $\mathfrak{f} \notin \mathsf{R}(a, \infty)$.

Exercises. (1) Prove that the functions $1/x^2$, $\sin x/x^2$, $\log x/x^2$ all belong to $\mathsf{R}(1, \infty)$.

(2) Prove that the functions $\sin x$, $1/x$, $|\sin(x^2)|$ do not belong to $\mathsf{R}(1, \infty)$.

Conditionally convergent integrals

If $\mathfrak{f} \in \mathsf{R}(a, \infty)$ and $|\mathfrak{f}| \notin \mathsf{R}(a, \infty)$ then we say that *the integral* $\int_a^\infty \mathfrak{f}$ *is conditionally convergent.*

Analogue of Abel's lemma

If $\mathfrak{f} \downarrow$, $\mathfrak{f}(x) \geqslant 0$ *and* $\mathfrak{g} \in \mathsf{R}(a, b)$ *then*

$$\left| \int_a^b \mathfrak{f}\mathfrak{g} \right| \leqslant \mathfrak{f}(a) \, . \sup_{a \leqslant \xi \leqslant b} \left| \int_a^\xi \mathfrak{g} \right|.$$

This follows from the corollary to theorem 53, p. 164.

Theorem 54. *(a) If* $\mathfrak{g} \in \mathsf{R}(a, \infty)$ *and* $\mathfrak{f}$ *is monotonic and bounded then* $\mathfrak{f}.\mathfrak{g} \in \mathsf{R}(a, \infty)$.

(b) If $\mathfrak{g} \in \mathsf{R}(a, X)$ *for all* $X > a$, $\int_a^X \mathfrak{g}$ *is bounded for* $X > a$, $\mathfrak{f}$ *is monotonic and if* $\mathfrak{f}(x) \to 0$ *as* $x \to \infty$ *then* $\mathfrak{f}\mathfrak{g} \in \mathsf{R}(a, \infty)$.

(*a*) It is clear that $\mathfrak{f}.\mathfrak{g} \in \mathsf{R}(a, X)$ for every $X > a$. We have to show that $\int_a^X \mathfrak{f}\mathfrak{g}$ tends to a finite limit as $X \to \infty$. Suppose that $\mathfrak{f}(x) \to l$ as $x \to \infty$. If $\mathfrak{f}\uparrow$ write $\mathfrak{k}(x) = l - \mathfrak{f}(x)$, if $\mathfrak{f}\downarrow$ write $\mathfrak{k}(x) = \mathfrak{f}(x) - l$. In any case $\mathfrak{k}\downarrow$ and $\mathfrak{k}(x) \geqslant 0$. If $\mathfrak{k}(a) = 0$ then $\mathfrak{k}(x) = 0$ all $x > a$, i.e. $\mathfrak{f}(x) = l$ and the result is trivially true. Suppose then that $\mathfrak{k}(a) > 0$. Since $\mathfrak{g} \in \mathsf{R}(a, \infty)$, $\forall \epsilon > 0 \, \exists X$ such that

$$\left| \int_Y^Z \mathfrak{g} \right| < \epsilon/\mathfrak{k}(a) \quad \text{if} \quad Z \geqslant Y \geqslant X.$$

By the analogue of Abel's lemma it follows that

$$\left| \int_Y^Z \mathfrak{k}\mathfrak{g} \right| < \epsilon \quad (Z \geqslant Y \geqslant X).$$

Thus by the general principle of convergence the integral $\int_a^\infty \mathfrak{k}\mathfrak{g}$ converges, i.e. $\mathfrak{k}\mathfrak{g} \in \mathsf{R}(a, \infty)$. Hence $\mathfrak{f}.\mathfrak{g} \in \mathsf{R}(a, \infty)$.

(*b*) By hypothesis $\exists K$ such that

$$\left| \int_a^X \mathfrak{g} \right| < K \quad \text{for all } X > a.$$

Thus $$\left| \int_Y^Z \mathfrak{g} \right| < 2K \quad \text{for all } Y, Z, \text{ where } Y > a, Z > a.$$

Since $\mathfrak{f}(x) \to 0$ as $x \to \infty$, $\forall \epsilon > 0 \, \exists X$ such that $|\mathfrak{f}(x)| < \epsilon/2K$ if $x \geqslant X > a$. Either $\mathfrak{f}\downarrow$ or $-\mathfrak{f}\downarrow$ and by the analogue of Abel's lemma

$$\left| \int_Y^Z \mathfrak{f}\mathfrak{g} \right| < \epsilon \quad (Y \geqslant X, Z \geqslant X),$$

and thus by the general principle of convergence $\mathfrak{f}\mathfrak{g} \in \mathsf{R}(a, \infty)$.

Exercise. Show that $(\sin x)/x$ and $\sin(x^2)$ belong to $\mathsf{R}(1, \infty)$.

Worked examples

(i) *Prove that*

$$\int_u^\infty e^{-x^2}\, dx = \frac{e^{-u^2}}{2u}\left(1 + O\left(\frac{1}{u^2}\right)\right) \quad (u > 0).$$

Since $0 < e^{-x^2} < x^{-2}$ and $x^{-2} \in \mathsf{R}(u, \infty)$, $u > 0$ it follows that $e^{-x^2} \in \mathsf{R}(u, \infty)$. Making the transformation from x to t where $x = t^{\frac{1}{2}}$ and then integrating by parts we obtain

$$\int_u^X e^{-x^2}\,dx = \frac{1}{2}\int_{u^2}^{X^2} e^{-t}t^{-\frac{1}{2}}\,dt$$

$$= \Big[-\tfrac{1}{2}t^{-\frac{1}{2}}e^{-t}\Big]_{u^2}^{X^2} + \frac{1}{4}\int_{u^2}^{X^2} t^{-\frac{3}{2}}e^{-t}\,dt$$

$$= T_1 + T_2,$$

say. Now $T_1 \to e^{-u^2}/2u$ as $X \to \infty$ and

$$|T_2| \leqslant \frac{1}{4u^3}\int_{u^2}^{\infty} e^{-t}\,dt = O\left(\frac{e^{-u^2}}{u^3}\right).$$

Hence
$$\int_u^{\infty} e^{-x^2}\,dx = \frac{e^{-u^2}}{2u}\left(1 + O\left(\frac{1}{u^2}\right)\right).$$

(ii) *Prove that if $a > 0$, $\mathfrak{f}$ is continuous except at the origin, and*

$$\mathfrak{f} \in \mathsf{IR}(0, a), \quad \mathfrak{g}(x) = \int_x^a \mathfrak{f}(t)/t\,dt,$$

then
$$\mathfrak{g} \in \mathsf{IR}(0, a) \quad \textit{and} \quad \int_0^a \mathfrak{g} = \int_0^a \mathfrak{f}.$$

Let ϵ be such that $a > \epsilon > 0$. Write $\mathfrak{h}(t)$ for $\mathfrak{f}(t)/t$. Then integrating by parts gives

$$\int_\epsilon^a \left(\int_x^a \mathfrak{h}\right)dx = \left[x\int_x^a \mathfrak{h}\right]_\epsilon^a + \int_\epsilon^a \mathfrak{f}$$

$$= -\epsilon\int_\epsilon^a \mathfrak{h} + \int_\epsilon^a \mathfrak{f}.$$

Thus the required result will follow if we can show that

$$\epsilon\int_\epsilon^a \mathfrak{h} \to 0 \quad \text{as} \quad \epsilon \to 0+.$$

Choose ϵ_1 so that $\epsilon < \epsilon_1 < a$ and $\epsilon_1 \to 0$, $\epsilon/\epsilon_1 \to 0$ as $\epsilon \to 0+$. For example, we could suppose that $\epsilon < 1$ and take $\epsilon_1 = \epsilon^{\frac{1}{2}}$. Then

$$\epsilon\int_\epsilon^a \mathfrak{h} = \epsilon\int_\epsilon^{\epsilon_1} \mathfrak{h} + \epsilon\int_{\epsilon_1}^a \mathfrak{h}$$

and applying the corollary to theorem 53 to each integral separately we have

$$\epsilon\int_\epsilon^a \mathfrak{h} = \int_\epsilon^d \mathfrak{f} + \frac{\epsilon}{\epsilon_1}\int_{\epsilon_1}^c \mathfrak{f},$$

where $\epsilon \leqslant d \leqslant \epsilon_1$, $\epsilon_1 \leqslant c \leqslant a$. Now as $\epsilon \to 0+$ we have $\epsilon_1 \to 0$ and thus $d \to 0$. Since $\lim\limits_{\epsilon\to 0+} \int_\epsilon^a \mathfrak{f}$ exists it follows that $\int_\epsilon^d \mathfrak{f} \to 0$. Also $\epsilon/\epsilon_1 \to 0$ as $\epsilon \to 0+$ and $\int_{\epsilon_1}^c \mathfrak{f}$ is bounded. Thus

$$\epsilon \int_\epsilon^a \mathfrak{h} \to 0 \quad \text{as} \quad \epsilon \to 0+$$

and this, as we have seen, implies the required result.

(iii) *Establish the existence of the finite limit* $\lim\limits_{x\to 0+} \left(\int_x^{\frac{1}{2}\pi} \log(\sin t)\,dt\right)$ *and evaluate this limit by means of the identity*

$$\tfrac{1}{2}\sin\frac{\pi}{2n}\,.\,\sin\frac{2\pi}{2n}\,.\,\sin\frac{3\pi}{2n}\,.\,\ldots\,.\,\sin\frac{n\pi}{2n} = \frac{n^{\frac{1}{2}}}{2^n}.$$

Since $\quad 0 < \sin t \leqslant 1, \quad \log(\sin t) \leqslant 0 \quad$ for $\quad 0 < t \leqslant \frac{1}{2}\pi.$

Thus $\int_x^{\frac{1}{2}\pi} \log(\sin t)\,dt \uparrow$ function of x and to establish the existence of the limit it is sufficient to show that $\int_x^{\frac{1}{2}\pi} \log(\sin t)\,dt$ tends to a finite limit as $x \to 0+$ through any sequence of values. In particular we shall do this for the sequence $\{\pi/2n\}$ and evaluate the limit at the same time. Now $\log(\sin t) \uparrow$ for $0 < t \leqslant \frac{1}{2}\pi$ and thus

$$\frac{\pi}{2n}\sum_1^{n-1}\log\sin\left(\frac{r\pi}{2n}\right) \leqslant \sum_1^{n-1}\int_{r\pi/2n}^{(r+1)\pi/2n}\log(\sin t)\,dt = \int_{\pi/2n}^{\frac{1}{2}\pi}\log(\sin t)\,dt$$

$$\leqslant \frac{\pi}{2n}\sum_2^{n}\log\sin\left(\frac{r\pi}{2n}\right).$$

But $$\sum_1^{n-1}\log\left(\sin\frac{r\pi}{2n}\right) = \log\left(\prod_1^{n-1}\sin\frac{r\pi}{2n}\right) = \log\left(\frac{n^{\frac{1}{2}}}{2^{(n-1)}}\right),$$

$$\sum_2^{n}\log\sin\left(\frac{r\pi}{2n}\right) = \log\left(\frac{n^{\frac{1}{2}}}{2^{n-1}}\right) - \log\left(\sin\frac{\pi}{2n}\right).$$

Thus

$$\int_{\pi/2n}^{\frac{1}{2}\pi}\log(\sin t)\,dt = \frac{\pi}{2n}\log\frac{n^{\frac{1}{2}}}{2^n} + o(1) = -\tfrac{1}{2}\pi\log 2 + o(1) \quad \text{as} \quad n \to \infty.$$

By the remarks made above this implies both that

$$\lim_{x\to 0+}\int_x^{\frac{1}{2}\pi}\log(\sin t)\,dt$$

exists and its value is $-\frac{1}{2}\pi\log 2$.

Exercises

1. $\mathfrak{f}$ is defined for $x \geqslant 0$ by

$$\mathfrak{f}(0) = 0, \quad \mathfrak{f}(x) = D(x^k \sin \pi/x) \quad (x > 0).$$

Find for what values of k (*a*) $\mathfrak{f} \in \mathsf{R}(0,1)$, (*b*) $\mathfrak{f} \in \mathsf{IR}(0,1)$.

2. Prove that $(\log x)^n (1+x)^{-1} \in \mathsf{IR}(0,1)$ if $n > 0$.

3. By considering the integral

$$\int_0^1 x^a \frac{1-x^n}{1-x} dx$$

prove that if $a > -1$ then

$$\frac{1}{a+1} + \frac{1}{a+2} + \ldots + \frac{1}{a+n} = \sum_{r=1}^{n} \frac{(-1)^{r-1}(r-1)!\binom{n}{r}}{(a+1)(a+2)\ldots(a+r)}.$$

Prove also that if $a > -1$ then

$$\sum_{r=1}^{n} \frac{(r-1)!}{(a+1)(a+2)\ldots(a+r)} = \sum_{r=1}^{n} \frac{(-1)^{r-1}\binom{n}{r}}{a+r}.$$

4. Determine whether or not the following functions of ϵ tend to finite limits as $\epsilon \to 0+$.

$$(a) \int_\epsilon^1 \left| \frac{1}{x} \sin \frac{\pi}{x} \right| dx, \quad (b) \int_\epsilon^1 \frac{1}{x(\log x)^{(1+p)}} dx, \quad \text{where } p > 0.$$

5. Prove that

$$\int_{k-\frac{1}{2}}^{k+\frac{1}{2}} \frac{1}{x} dx = \frac{1}{k} + O\left(\frac{1}{k^3}\right) \quad \text{as} \quad k \to \infty.$$

Find the sum $\sum_{n+1}^{2n} \frac{1}{n}$ with an error $O\left(\frac{1}{n^2}\right)$.

6. Discuss the convergence of the integrals

$$(a) \int_1^\infty \frac{x^{1-\alpha}}{1+x} dx, \quad (b) \int_1^\infty \left| \frac{\sin x}{x^\alpha} \right| dx, \quad (c) \int_e^\infty \left| \frac{\sin x}{\log x} \right| dx,$$

$$(d) \int_0^1 \frac{dx}{x^\alpha (1 + \log(1/x))},$$

where α is a positive number.

7. Discuss the convergence of the integrals

$$(a) \int_1^\infty \frac{x^k}{(1+x)^l (1+(\log x))^2} dx, \quad (b) \int_0^\infty \left| \frac{\sin x(1-\cos x)}{x^k} \right| dx$$

for all real values of k and l.

8. $\mathfrak{f}$ is a positive decreasing function and $X > Y > 0$. Show that

$$\lim_{n\to\infty} \int_Y^X \frac{\mathfrak{f}(x) \sin nx}{x} dx = 0.$$

9. Prove that if

$$u_n = \int_0^{\frac{1}{2}\pi} \sin 2nx \cot x \, dx, \quad v_n = \int_0^{\frac{1}{2}\pi} \frac{\sin 2nx}{x} \, dx,$$

then

$$u_n = \tfrac{1}{2}\pi \quad \text{and} \quad \lim_{n\to\infty} v_n = \int_0^{\infty} \frac{\sin x}{x} \, dx.$$

Prove also (by integration by parts or otherwise) that $\lim_{n\to\infty} (u_n - v_n) = 0$ and deduce that

$$\int_0^{\infty} \frac{\sin x}{x} \, dx = \tfrac{1}{2}\pi.$$

10. Prove that if $a \geqslant 0$, $b \geqslant 0$ then

$$\int_0^X \left(\frac{\cos ax - \cos bx}{x}\right) dx = \log(b/a) + O(1/X^2) \quad \text{as} \quad X \to \infty.$$

11. Prove that as $X \to \infty$

$$\int_0^X \left(|\sin x| - \frac{2}{\pi}\right) dx = O(1), \quad \int_0^X \frac{|\sin x|}{(x+X)^2} \, dx = \frac{1}{\pi X} + O(1/X^2).$$

12. $\mathfrak{f}(x) \geqslant 0$ for $x \geqslant 0$ and $\mathfrak{f} \in \mathsf{R}(0, X)$ for every $X > 0$. If

$$\int_0^X \mathfrak{f} \to \infty \quad \text{as} \quad X \to \infty \quad \text{prove that} \quad \int_0^X \left(\frac{\mathfrak{f}(x)}{\int_0^x \mathfrak{f}}\right) dx \to \infty \quad \text{as} \quad X \to \infty.$$

The series Σa_n of positive summands diverges. Find a sequence $\{\epsilon_n\}$ such that $\epsilon_n \to 0$ as $n \to \infty$ and $\Sigma a_n \epsilon_n$ diverges.

13. Show that the integral

$$\int_a^{\infty} x^{-n} e^{\sin x} \sin 2x \, dx$$

converges if $n > 0$, $a > 0$.

14. Establish the existence and convergence of the integral

$$\int_0^{\infty} x^{-\alpha} \sin(x^{1-\beta}) \, dx,$$

where $1 > \alpha > \beta > 0$.

26

FURTHER TESTS FOR THE CONVERGENCE OF SERIES

In certain circumstances there is a close connexion between the behaviour of the integral $\int_1^{\infty} \mathfrak{f}$ and that of the series $\Sigma \mathfrak{f}(n)$. In particular if $\mathfrak{f}$ is positive and monotonic they both converge or both diverge.

***Theorem* 55.** *If $\mathfrak{f}$ is defined for $x > 0$, $\mathfrak{f}\downarrow$ and $\mathfrak{f} \geqslant 0$, then*

$$\int_0^N \mathfrak{f} - \sum_0^N \mathfrak{f}(n)$$

tends to a finite limit as $N \to \infty$.

Write

$$k_N = \int_0^N \mathfrak{f} - \sum_0^N \mathfrak{f}(n).$$

Then since $\mathfrak{f}\downarrow$ $\quad k_{N+1} - k_N = \int_N^{N+1} \mathfrak{f} - \mathfrak{f}(N+1) \geqslant 0.$

Thus $\{k_N\}\uparrow$.

Write

$$l_N = \int_0^N \mathfrak{f} - \sum_0^{N-1} \mathfrak{f}(n)$$

then by a similar argument $\{l_N\}\downarrow$. Also $l_N - k_N = \mathfrak{f}(N) \geqslant 0$. Thus $k_N \leqslant l_N \leqslant l_1$, that is to say k_N is bounded above. Hence k_N converges to a finite limit.

***Corollary*.** *The Cauchy–Maclaurin integral test. If $\mathfrak{f}$ is defined for $x > 0$, $\mathfrak{f}\downarrow$ and $\mathfrak{f} \geqslant 0$, then $\Sigma\mathfrak{f}(n)$ converges if and only if $\int_0^\infty \mathfrak{f}$ converges.*

If $\int_0^\infty \mathfrak{f}$ converges, then $\int_0^N \mathfrak{f}$ tends to a finite limit as $N \to \infty$ and therefore so also does

$$\sum_0^N \mathfrak{f}(n) = \int_0^N \mathfrak{f} - k_N.$$

On the other hand if $\sum_0^N \mathfrak{f}(n)$ converges to a finite limit as $N \to \infty$, then $\mathfrak{f}(n) \to 0$ as $n \to \infty$ and

$$\int_0^N \mathfrak{f} = \sum_0^N \mathfrak{f}(n) + k_N$$

also tends to a finite limit as $N \to \infty$. Now if $N < X < N+1$, then

$$\left| \int_0^X \mathfrak{f} - \int_0^N \mathfrak{f} \right| \leqslant \mathfrak{f}(N) \to 0 \quad \text{as} \quad N \to \infty,$$

and thus in this case $\int_0^X \mathfrak{f}$ converges to a finite limit as $X \to \infty$.

The logarithmic scale of comparison series

In theorem 55 take $\mathfrak{f}(x) = 1/(x+1)$. Then

$$1 + 1/2 + 1/3 + \ldots + 1/n - \log n$$

tends to a finite limit as $n \to \infty$. This limit is called Euler's constant and is denoted by γ: $1 > \gamma > 0$.

Next take
$$\mathfrak{f}(x) = [x . \log x . \dots . \log_{r-1}(x) . (\log_r(x))^{\alpha}]^{-1},$$
where $\quad \log_s(x) = \log(\log_{s-1}(x)), \dots, \log_2(x) = \log(\log x)$.

Take a so laıge that $\mathfrak{f}$ is defined for $x > a$ (say $a > e^{e^{\cdot^{\cdot^{e}}}}$, where there are $r-1$ e's). Then
$$\int_a^X \mathfrak{f} = \frac{1}{1-\alpha}((\log_r X)^{1-\alpha} - (\log_r(a))^{1-\alpha}) \quad (\alpha \neq 1)$$
$$= \log_{r+1}(X) - \log_{r+1}(a) \quad (\alpha = 1).$$

Hence the series $\Sigma \mathfrak{f}(n)$ converges if $\alpha > 1$ and diverges if $\alpha \leqslant 1$. These series form a useful class of comparison series. In fact the Maclaurin integral test has little utility beyond establishing the convergence conditions of this class of series.

***Theorem* 56.** *Gauss's test for convergence. The positive series Σa_n is such that*
$$\frac{a_n}{a_{n+1}} = 1 + \frac{\mu}{n} + o\left(\frac{1}{n \log n}\right) \quad as \quad n \to \infty.$$
Then Σa_n converges if $\mu > 1$ and diverges if $\mu \leqslant 1$.

If $\mu > 1$ write $\mu = 1 + 2k$, $k > 0$ and $b_n = n^{-1-k}$. Then Σb_n converges and
$$\frac{b_n}{b_{n+1}} = \left(\frac{n+1}{n}\right)^{1+k} = 1 + \frac{1+k}{n} + O\left(\frac{1}{n^2}\right).$$

Thus $\exists N$ such that $b_n/b_{n+1} < a_n/a_{n+1}$ for $n > N$. Hence Σa_n converges.

If $\mu < 1$ the proof is similar to that above writing $\mu = 1 - 2k$ and using the divergence of Σn^{-1+k}.

If $\mu = 1$ write
$$b_n = \frac{1}{n \log n}.$$
Then
$$\frac{b_n}{b_{n+1}} = \left(1 + \frac{1}{n}\right)\left(1 + \frac{\log\left(1 + \frac{1}{n}\right)}{\log n}\right) = 1 + \frac{1}{n} + \frac{1}{n \log n} + O\left(\frac{1}{n^2}\right)$$
(by the expansion $\log\left(1 + \frac{1}{n}\right) = \frac{1}{n} - \frac{1}{2n^2} + \dots$ and the corollary to theorem 15).

Thus $\exists\, N$ such that if $n > N$ then

$$\frac{b_n}{b_{n+1}} > \frac{a_n}{a_{n+1}}.$$

Since Σb_n diverges this implies that Σa_n diverges.

Worked examples

(i) *Prove that* $$\lim_{\alpha \to 0+} \lim_{n \to \infty} \left(\alpha \sum_1^n r^{-1-\alpha}\right) = 1.$$

For $\alpha > 0$ $$\sum_1^n r^{-1-\alpha} > \int_1^{n+1} x^{-1-\alpha}\,dx > \sum_2^n r^{-1-\alpha}$$

and thus
$$\sum_1^n r^{-1-\alpha} = \int_1^{n+1} x^{-1-\alpha}\,dx + \chi_n$$
$$= \alpha^{-1} - \alpha^{-1}(n+1)^{-\alpha} + \chi_n,$$

where $0 < \chi_n < 1$. But $\Sigma r^{-1-\alpha}$ converges and thus $\chi_n \to l$ as $n \to \infty$ and $0 \leqslant l \leqslant 1$. l varies with α but $\lim_{\alpha \to 0+} \alpha l = 0$ and thus

$$\lim_{\alpha \to 0+} \lim_{n \to \infty} \left(\alpha \sum_1^n r^{-1-\alpha}\right) = 1.$$

(ii) *Σu_n is a divergent series of positive summands and* $U_n = \sum_1^n u_r$. *By considering* $\int_{U_n}^{U_{n+1}} x^{-\alpha}\,dx$ *prove that* $\Sigma u_{n+1} U_n^{-\alpha}$ *diverges if* $0 < \alpha \leqslant 1$ *and that* $\Sigma u_n U_n^{-\alpha}$ *converges if* $\alpha > 1$. *Prove further that* $\Sigma u_n U_n^{-\alpha}$ *diverges if* $0 < \alpha \leqslant 1$.

If $0 < \alpha \leqslant 1$ the integral $\int_1^\infty x^{-\alpha}\,dx$ diverges. Also

$$\int_{U_n}^{U_{n+1}} x^{-\alpha}\,dx \leqslant (U_{n+1} - U_n)/U_n^{\alpha} = u_{n+1} U_n^{-\alpha}$$

and thus $\Sigma u_{n+1} U_n^{-\alpha}$ diverges.

If $\alpha > 1$ the integral $\int_0^\infty x^{-\alpha}\,dx$ converges. Also

$$\int_{U_n}^{U_{n+1}} x^{-\alpha}\,dx \geqslant \frac{U_{n+1} - U_n}{U_{n+1}^{\alpha}} = u_{n+1} U_{n+1}^{-\alpha}.$$

Therefore $\Sigma u_n U_n^{-\alpha}$ converges if $\alpha > 1$.

Finally, either $u_n > \frac{1}{2}U_n$ for infinitely many integers n or $u_n \leqslant \frac{1}{2}U_n$ for all n sufficiently large, say $n > N$. In the first case

$$u_n U_n^{-\alpha} > \tfrac{1}{2}U_n^{1-\alpha}.$$

But $U_n \to \infty$ and $\alpha \leqslant 1$, thus the summands of $\Sigma u_n U_n^{-\alpha}$ do not tend to zero as $n \to \infty$ and $\Sigma u_n U_n^{-\alpha}$ diverges. In the second case $U_{n+1} \leqslant \frac{3}{2}U_n$ for $n > N$ and thus $u_{n+1}U_{n+1}^{-\alpha} \geqslant (\frac{2}{3})^\alpha u_{n+1} U_n^{-\alpha}$, since $\alpha > 0$. But $\Sigma u_{n+1} U_n^{-\alpha}$ diverges and thus $\Sigma u_n U_n^{-\alpha}$ diverges.

Exercises

1. If $0 < r_1 < r_2 < \ldots$ and $\mathfrak{n}(u)$ is a function of u defined by $\mathfrak{n}(u) = n$ for $r_n \leqslant u < r_{n+1}$ show that

$$\rho^{-1}\sum_1^N r_n^{-\rho} = \int_{r_1}^{r_N} \frac{\mathfrak{n}(u)}{u^{\rho+1}}\,du + (N-1)\rho^{-1}r_N^{-\rho} \quad (\rho > 0).$$

By taking for r_n the value of u for which $u^p = n$ $(n = 1, 2, \ldots)$ show that if $\mathfrak{n}(u)$ is equal to the integral part of u^p $(0 < p < \rho)$, then $\Sigma r_n^{-\rho}$ converges.

2. $\mathfrak{f}$ is a positive decreasing function defined for $x > 0$ and h is a positive number. Show that

$$\int_h^{(m+1)h} \mathfrak{f} \leqslant h \sum_{n=1}^{m} \mathfrak{f}(nh) \leqslant \int_0^{mh} \mathfrak{f}.$$

By taking $\mathfrak{f}(x) = e^{-x^2}$ prove that

$$\lim_{t\to 1-} (1-t)^{\frac{1}{2}}(t + t^4 + t^9 + t^{16} + \ldots) = \int_0^\infty e^{-x^2}\,dx.$$

3. $\mathfrak{f}$ is defined for $x \geqslant 1$ and is such that $D^2\mathfrak{f} = O(x^{-\alpha})$, $\alpha > 1$. Prove that

$$\int_n^{n+1} (\mathfrak{f}(x) - \mathfrak{f}(n+\tfrac{1}{2}))\,dx = O(n^{-\alpha}) \quad \text{as} \quad n \to \infty$$

and

$$\sum_1^n \mathfrak{f}(m+\tfrac{1}{2}) = \int_1^{n+1} \mathfrak{f} + C + O(n^{1-\alpha}),$$

where C is a constant. Examine the special case $\mathfrak{f}(x) = \log(x - \frac{1}{2})$.

4. a, b are positive constants and $0 < s < 1$. Show that a positive constant A can be found such that if

$$\mathfrak{f}(n) = (a+b)^{-s} + (a+2b)^{-s} + \ldots + (a+nb)^{-s} - \frac{(a+nb)^{1-s}}{b(1-s)} - A$$

then $\mathfrak{f}(n) \to 0$ as $n \to \infty$.

Prove further that $\mathfrak{f}(n) - \mathfrak{f}(n+1) = O(n^{-1-s})$ and hence that $\mathfrak{f}(n) = O(n^{-s})$ as $n \to \infty$.

5. Show that

$$0 < \frac{1}{n}\left(\frac{n}{1} + \frac{n-1}{2} + \frac{n-2}{3} + \ldots + \frac{1}{n} - \log(n!)\right) < 1$$

and that the expression in the middle of this inequality tends to γ, Euler's constant as $n \to \infty$.

6. Prove that if $s > 1$ then

$$\sum_{r=1}^{n}\left(\frac{s}{r}-\frac{1}{r^s}\right)-s\log n$$

tends to a finite limit as n tends to infinity and that if this limit is $\mathfrak{f}(s)$ then

$$0 \leqslant \mathfrak{f}(s)+\frac{1}{s-1} \leqslant s-1.$$

7. Does the series whose nth term is

$$\frac{1.3.5.....2n-1}{2.4.6.....2n}$$

converge or diverge?

8. Let p be a given number greater than unity. For any positive integer n the sequence of integers $n^{(0)}, n^{(1)}, n^{(2)}, \ldots$ is defined as follows

$$n^{(0)} = n, \quad \frac{n^{(i)}}{p} \leqslant n^{(i+1)} < \frac{n^{(i)}}{p}+1 \quad (i = 0, 1, 2, \ldots).$$

Show that $$\frac{n}{p^s} \leqslant n^{(s)} < \frac{n}{p^s}+\frac{p}{p-1} \quad (s = 1, 2, \ldots).$$

Deduce that the positive series Σa_n converges if $\lim a_n/a_{n(1)} < 1/p$.

9. $\{a_n\}$ and $\{b_n\}$ are positive sequences such that if

$$\frac{a_{n+1}}{a_n}\frac{b_n}{b_{n+1}} = 1+c_n,$$

then Σc_n converges absolutely. Show that

$$\Sigma\left(\log\left(\frac{a_{n+1}}{b_{n+1}}\right)-\log\left(\frac{a_n}{b_n}\right)\right)$$

converges and deduce that a_n/b_n tends to a finite non-zero limit as $n \to \infty$. If $a_{n+1}/a_n = 1-\lambda/n+d_n$, where Σd_n converges absolutely, show that Σa_n converges or diverges according as $\lambda > 1$ or $\lambda \leqslant 1$.

10. $I_\lambda(x)$ is defined by $I_\lambda(x) = \int_x^\pi \tfrac{1}{2}\sin\lambda t.\operatorname{cosec}\tfrac{1}{2}t.dt$ $(0 < x < 2\pi)$. Prove that if x is fixed $I_\lambda(x) \to 0$ as $\lambda \to \infty$. By considering $I_{n+\frac{1}{2}}(x)-I_{n-\frac{1}{2}}(x)$ show that the series $\sum_1^\infty \frac{\sin nx}{n}$ converges for $0 < x < 2\pi$ and find its sum.

11. The series $$1-\tfrac{1}{2}+\tfrac{1}{3}-\tfrac{1}{4}+\ldots+\frac{(-1)^{n-1}}{n}+\ldots$$

is rearranged so that the order of the positive terms amonst themselves is unchanged and the order of the negative terms amongst themselves is unchanged, whilst the first n terms of the rearranged series contain $\mathfrak{f}(n)$ positive and $\mathfrak{g}(n)$ negative terms. If

$$0 < \lambda = \varliminf\frac{\mathfrak{f}(n)}{\mathfrak{g}(n)} < \varlimsup\frac{\mathfrak{f}(n)}{\mathfrak{g}(n)} = \mu < \infty$$

show that the rearranged series oscillates and its limits of oscillation are $\frac{1}{2}\log 4\lambda$ and $\frac{1}{2}\log 4\mu$.

If the rearranged series consists of p_1 positive terms followed by p_2 negative terms followed by p_3 positive terms, etc., where $p_n = 2^{n-1}$ show that the rearranged series oscillates finitely and determine its limits of oscillation.

12. Prove that if $\mathfrak{f}$ has a continuous second derivative for $n-\frac{1}{2} \leqslant x \leqslant n+\frac{1}{2}$ then

$$\mathfrak{f}(n) = \int_{n-\frac{1}{2}}^{n+\frac{1}{2}} \mathfrak{f} - \frac{1}{8}\int_{-\frac{1}{2}}^{+\frac{1}{2}} (1-2|t|)^2 . D^2\mathfrak{f}(n+t) . dt.$$

Prove that if θ is a real non-zero constant the sequence of partial sums of $\Sigma e^{i\theta\sqrt{n}} n^{-\beta}$ is bounded when $\beta = \frac{1}{2}$ and that in this case the series diverges. Hence or otherwise find all real values of β for which the series converges.

13. For every $\epsilon > 0$, $e^{-\epsilon x}\mathfrak{f}(x) \in \mathsf{R}(0,\infty)$ and $\lim_{t\to\infty} \mathfrak{f}(t)$ exists. Prove that

$$\lim_{\epsilon\to 0+} \epsilon \int_0^\infty e^{-\epsilon t}\mathfrak{f}(t)\,dt = \lim_{t\to\infty} \mathfrak{f}(t).$$

27

UNIFORM CONVERGENCE

We have defined 'Σa_n converges' to mean that the sequence S_N, where $S_N = \sum_1^N a_n$ tends to a limit say l. If often happens that the series in which we are interested have summands that are functions of a real variable say $\Sigma\mathfrak{f}_n(x)$ where $x \in E$. The sum of such a series is also a function of x say $\mathfrak{f}(x)$, and it may be important to be able to assert that $\int_a^b \mathfrak{f} = \Sigma \int_a^b \mathfrak{f}_n$ or that $D\mathfrak{f} = \Sigma D\mathfrak{f}_n$. These equations are not always true even when all the expressions actually exist. One criterion that when appropriately applied will ensure their validity is *uniform convergence*.

Definition 49. *The sequence of functions $\{\mathfrak{s}_n(x)\}$ defined for $x \in E$ is said to converge uniformly over E to $\mathfrak{s}(x)$ if $\forall \epsilon > 0 \,\exists$ an integer N such that $|\mathfrak{s}_n(x) - \mathfrak{s}(x)| < \epsilon$ for $x \in E$ and $n \geqslant N$, where N depends on ϵ but is independent of x.*

Theorem 57. *The general principle of uniform convergence. The necessary and sufficient condition for the uniform convergence of $\{\mathfrak{s}_n(x)\}$ over E is that $\forall \epsilon > 0 \,\exists$ an integer N which is independent of x and is such that*

$$|\mathfrak{s}_n(x) - \mathfrak{s}_m(x)| < \epsilon \quad (x \in E,\ m \geqslant N,\ n \geqslant N).$$

Necessity. If $\mathfrak{s}_n(x)$ converges uniformly to $\mathfrak{s}(x)$ then $\forall \epsilon > 0 \exists$ an integer N independent of x such that

$$|\mathfrak{s}_n(x) - \mathfrak{s}(x)| < \tfrac{1}{2}\epsilon \quad (n \geqslant N,\ x \in E).$$

Thus

$$|\mathfrak{s}_n(x) - \mathfrak{s}_m(x)| \leqslant |\mathfrak{s}_n(x) - \mathfrak{s}(x)| + |\mathfrak{s}(x) - \mathfrak{s}_m(x)| < \epsilon$$
$$(n \geqslant N,\ m \geqslant N,\ x \in E)$$

and the condition is satisfied.

Sufficiency. By the general principle of convergence applied to the sequence $\{\mathfrak{s}_n(x)\}$ for each fixed x it follows that $\mathfrak{s}_n(x)$ converges to a limit which since it varies with x, we denote by $\mathfrak{s}(x)$.

Now $\forall \epsilon > 0 \exists N$ independent of x such that

$$|\mathfrak{s}_n(x) - \mathfrak{s}_m(x)| < \tfrac{1}{2}\epsilon \quad (x \in E,\ m \geqslant N,\ n \geqslant N).$$

Let $m \to \infty$ in this inequality and we deduce that

$$|\mathfrak{s}_n(x) - \mathfrak{s}(x)| \leqslant \tfrac{1}{2}\epsilon < \epsilon \quad (x \in E,\ n \geqslant N).$$

Since N is independent of x this means that $\mathfrak{s}_n(x) \to \mathfrak{s}(x)$ uniformly.

Remark. As an example consider the convergence of the two sequences

$$\mathfrak{s}_n(x) = \frac{1}{nx} \quad \text{and} \quad \mathfrak{t}_n(x) = \frac{x}{n}.$$

If $E = (0, 1)$ then $\mathfrak{s}_n(x) \to 0$, $\mathfrak{t}_n(x) \to 0$ as $n \to \infty$ for $x \in E$. But $\mathfrak{t}_n(x) \to 0$ uniformly over E, whereas $\mathfrak{s}_n(x)$ does not tend to zero uniformly over E. If $E = \{x | x > 1\}$ then $\mathfrak{s}_n(x) \to 0$ uniformly and $\mathfrak{t}_n(x) \to 0$ but not uniformly over E.

The series $\Sigma \mathfrak{f}_n(x)$ is said to be uniformly convergent to the sum $\mathfrak{f}(x)$ over E if and only if the corresponding sequence of partial sums $\left\{\sum_1^N \mathfrak{f}_n(x)\right\}$ is uniformly convergent to $\mathfrak{f}(x)$ for $x \in E$, i.e. if and only if $\forall \epsilon > 0 \exists$ an integer N independent of x such that

$$\left|\mathfrak{f}(x) - \sum_1^n \mathfrak{f}_r(x)\right| < \epsilon \quad \text{for} \quad x \in E \quad \text{and} \quad n \geqslant N.$$

Theorem 57a. *The necessary and sufficient condition that $\Sigma \mathfrak{f}_n(x)$ be uniformly convergent over E is that $\forall \epsilon > 0 \exists$ an integer N independent of x such that $\left|\sum_l^m \mathfrak{f}_n(x)\right| < \epsilon$ for $m > l \geqslant N$, $x \in E$.*

The proof is omitted. The theorem is merely a restatement of theorem 57.

Exercises. (1) Show that Σx^n is uniformly convergent over $[0, \frac{1}{2}]$ and not uniformly convergent over $[0, 1]$.

(2) $\Sigma \mathfrak{f}_n(x)$ is uniformly convergent over E and $\mathfrak{f}_n(x) \geqslant 0$. If

$$0 \leqslant \mathfrak{g}_n(x) \leqslant \mathfrak{f}_n(x), \quad x \in E \quad (n = 1, 2, 3, \ldots)$$

show that $\Sigma \mathfrak{g}_n(x)$ is uniformly convergent over E.

(3) $\Sigma |\mathfrak{f}_n(x)|$ is uniformly convergent over E. Show that $\Sigma \mathfrak{f}_n(x)$ is uniformly convergent over E.

Properties of uniformly convergent series

(i) *If $\mathfrak{f}_n(x)$ is continuous at x in E and $\Sigma \mathfrak{f}_n(x)$ converges uniformly to $\mathfrak{f}(x)$ over E then $\mathfrak{f}(x)$ is continuous at x in E.*

By the uniformity of convergence given $\forall \epsilon > 0 \, \exists N$ such that

$$\left| \mathfrak{f}(x) - \sum_1^N \mathfrak{f}_n(x) \right| < \tfrac{1}{3}\epsilon \quad (x \in E),$$

where N is independent of x. By the continuity of $\mathfrak{f}_n(x)$, given $c \in E$ $\exists \, \delta_n > 0$ such that

$$|\mathfrak{f}_n(x) - \mathfrak{f}_n(c)| < \frac{1}{3N}\epsilon \quad \text{if} \quad |x - c| < \delta_n \quad (x \in E, \, 1 \leqslant n \leqslant N).$$

Thus writing $\delta = \min\limits_{1 \leqslant n \leqslant N} \delta_n$,

$$|\mathfrak{f}(x) - \mathfrak{f}(c)| \leqslant \left| \mathfrak{f}(x) - \sum_1^N \mathfrak{f}_n(x) \right| + \left| \sum_1^N \mathfrak{f}_n(x) - \sum_1^N \mathfrak{f}_n(c) \right|$$
$$+ \left| \sum_1^N \mathfrak{f}_n(c) - \mathfrak{f}(c) \right| < \epsilon \quad \text{if} \quad |x - c| < \delta \quad \text{and} \quad x \in E.$$

Hence $\mathfrak{f}$ is continuous in E at c.

Remark. We often use the converse of this result, the convergence of a sequence of continuous functions to a non-continuous limit must be non-uniform.

Exercise. Show that $\Sigma(x^{n+1} - x^n)$ converges non-uniformly over $0 \leqslant x \leqslant 1$.

(ii) *If $\mathfrak{f}_n(x)$ is continuous for $a \leqslant x \leqslant b$ and $\Sigma \mathfrak{f}_n(x)$ converges uniformly to $\mathfrak{f}(x)$ over $[a, b]$ then*

$$\int_a^b \mathfrak{f} = \sum_1^\infty \int_a^b \mathfrak{f}_n.$$

By (i) $\mathfrak{f}$ is continuous and thus $\mathfrak{f} \in \mathsf{R}(a, b)$. By the uniformity of convergence $\forall \epsilon > 0 \, \exists N$ such that

$$\left| \mathfrak{f}(x) - \sum_1^n \mathfrak{f}_r(x) \right| < \tfrac{1}{3}\epsilon \quad \text{if} \quad a \leqslant x \leqslant b \quad \text{and} \quad n \geqslant N.$$

Hence $$\left|\int_a^b \mathfrak{f} - \sum_1^n \int_a^b \mathfrak{f}_r\right| \leqslant \int_a^b \left|\mathfrak{f} - \sum_1^n \mathfrak{f}_r\right| < \tfrac{1}{3}\epsilon(b-a),$$

provided that $n \geqslant N$. Since ϵ is any positive number it follows that $\Sigma \int_a^b \mathfrak{f}_n$ converges to $\int_a^b \mathfrak{f}$.

(iii) *If $\Sigma \mathfrak{f}_n(x)$ converges to $\mathfrak{f}(x)$ for $a \leqslant x \leqslant b$ and if each function $\mathfrak{f}_n(x)$ has a continuous derivative $D\mathfrak{f}_n(x)$ for $a \leqslant x \leqslant b$ such that $\Sigma D\mathfrak{f}_n(x)$ converges uniformly to $\mathfrak{g}(x)$ over $[a, b]$, then for $a < x < b$ $\mathfrak{f}$ is differentiable and $D\mathfrak{f} = \mathfrak{g}$.*

By (ii) and the uniform convergence of $\Sigma D\mathfrak{f}_n(x)$ over the sub-interval $[a, y]$ of $[a, b]$ we have (by theorem 47)

$$\int_a^y \mathfrak{g} = \sum_1^\infty \int_a^y D\mathfrak{f}_n = \sum_1^\infty (\mathfrak{f}_n(y) - \mathfrak{f}_n(a)) = \mathfrak{f}(y) - \mathfrak{f}(a)$$

and differentiating (by theorem 45)

$$\mathfrak{g}(y) = D\mathfrak{f}(y).$$

Tests for uniform convergence of series

Theorem 58 (*Weierstrass M test or comparison principle*). *If the sequence of functions $\{\mathfrak{f}_n(x)\}$ is such that $|\mathfrak{f}_n(x)| \leqslant M_n$ for $x \in E$ and if ΣM_n converges then $\Sigma \mathfrak{f}_n(x)$ converges uniformly over E.*

Since ΣM_n converges $\forall \epsilon > 0 \, \exists N$ such that

$$\left|\sum_l^m M_n\right| < \epsilon \quad \text{if} \quad m > l \geqslant N.$$

But $|\mathfrak{f}_n(x)| \leqslant M_n$ and this means (since $M_n \geqslant 0$) that

$$\left|\sum_l^m \mathfrak{f}_n(x)\right| \leqslant \sum_l^m M_n = \left|\sum_l^m M_n\right| < \epsilon \quad (m > l \geqslant N).$$

Now N has been defined in terms of the properties of a series that does not involve x. Thus N is independent of x and it follows from the general principle of convergence, theorem 57a, that $\Sigma \mathfrak{f}_n(x)$ converges uniformly over E.

Theorem 59. *If $\Sigma \mathfrak{f}_n(x)$ converges uniformly over E and $\{v_n\}$ is a bounded real monotonic sequence then $\Sigma v_n . \mathfrak{f}_n(x)$ converges uniformly over E.*

The proof is omitted. See proof of theorem 8 (a).

If the power series $\Sigma a_{n-1} x^{n-1}$, x real, has a radius of convergence R then for $R_1 < R$ the series is uniformly convergent over $[-R_1, R_1]$ (as may be seen from the Weierstrass M test). Further the series

$\Sigma n a_n x^{n-1}$ has radius of convergence R and we conclude from properties (i), (ii) and (iii) of uniform convergence that

(i) $\mathfrak{f}(x) = \sum_0^\infty a_n x^n$ is a continuous function of x if $|x| \leqslant R_1$,

(ii) $\int_0^x \mathfrak{f} = \sum_0^\infty a_n x^{n+1}/(n+1) \quad (|x| \leqslant R_1)$,

(iii) $D\mathfrak{f} = \sum_0^\infty n a_n x^{n-1} \qquad (|x| \leqslant R_1)$.

Since these properties hold for $|x| \leqslant R_1 < R$ and any R_1 they hold for $|x| < R$. (Note that the consequences of uniform convergence hold for $|x| < R$ but it is *not* the case that $\Sigma a_n x^n$ must be uniformly convergent for $|x| < R$. For example, $\sum_0^\infty x^n$ is not uniformly convergent over $|x| < 1$.)

Theorem 60. *Abel's limit theorem. If Σa_n converges then $\lim_{x\to 1-} \sum_0^\infty a_n x^n$ exists and is equal to $\sum_0^\infty a_n$.*

Since Σa_n converges it follows that $a_n \to 0$ and thus the radius of convergence of $\Sigma a_n x^n$ is at least unity. Hence $\sum_0^\infty a_n x^n$ is defined for $0 < x < 1$. Further, $\forall \epsilon > 0 \;\exists$ an integer N such that

$$\left|\sum_L^M a_n\right| < \epsilon \quad \text{if} \quad M > L \geqslant N.$$

By Abel's lemma, chapter 8, p. 47,

$$\left|\sum_N^M a_n x^n\right| < x^N \epsilon < \epsilon \quad \text{for} \quad M > N \quad \text{and} \quad 0 < x < 1.$$

Since this is true for all $M > N$ we have also

$$\left|\sum_N^\infty a_n x^n\right| \leqslant x^N \epsilon < \epsilon.$$

Thus
$$\left|\sum_0^\infty a_n - \sum_0^\infty a_n x^n\right| \leqslant \left|\sum_0^N a_n(1-x^n)\right| + \epsilon + \epsilon$$

and
$$\overline{\lim_{x\to 1-}} \left|\sum_0^\infty a_n - \sum_0^\infty a_n x^n\right| \leqslant 2\epsilon.$$

But ϵ is any positive number. Hence

$$\sum_0^\infty a_n x^n \to \sum_0^\infty a_n \quad \text{as} \quad x \to 1-.$$

Uniform convergence of integrals

We say that the integral $\int_a^\infty \mathfrak{f}(x, y)\,dx$ is uniformly convergent for $y \in E$ if and only if $\forall \epsilon > 0 \, \exists \, Y$ such that

$$\left| \int_a^X \mathfrak{f}(x, y)\,dx - \int_a^\infty \mathfrak{f}(x, y)\,dx \right| = \left| \int_X^\infty \mathfrak{f}(x, y)\,dx \right| < \epsilon$$

for $X > Y$, $y \in E$, where Y *can be chosen independent of* y.

There is a general principle of uniform convergence of integrals and tests for convergence analogous to those for series.

Worked examples

(i) $\Sigma \mathfrak{v}_n(x)$ *converges uniformly for* $x \in E$ *where* E *is a set of real numbers;* $y \in \bar{E}$ *and* $\lim_{x \to y} \mathfrak{v}_n(x) = w_n$. *Show that* Σw_n *converges and that*

$$\lim_{x \to y} \sum_1^\infty \mathfrak{v}_n(x) = \sum_1^\infty w_n.$$

$\forall \epsilon > 0 \, \exists \, N$ such that

$$\left| \sum_L^M \mathfrak{v}_n(x) \right| < \epsilon \quad \text{if} \quad M, L \geqslant N \quad \text{and} \quad x \in E$$

(by the uniformity of convergence of $\Sigma \mathfrak{v}_n(x)$). Therefore letting $x \to y$ it follows that

$$\left| \sum_L^M w_n \right| \leqslant \epsilon$$

and by the general principle of convergence for series theorem 5*a*, p. 41, it follows that Σw_n converges. Also

$$\left| \sum_1^\infty w_n - \sum_1^\infty \mathfrak{v}_n(x) \right| \leqslant \left| \sum_1^N w_n - \sum_1^N \mathfrak{v}_n(x) \right| + \left| \sum_{N+1}^\infty w_n \right| + \left| \sum_{N+1}^\infty \mathfrak{v}_n(x) \right|$$

$$\leqslant \left| \sum_1^N (w_n - \mathfrak{v}_n(x)) \right| + 2\epsilon$$

$$\leqslant 3\epsilon \quad \text{if} \quad |x - y| < \delta,$$

where δ is some appropriate positive number. Since this is true for every $\epsilon > 0$, we have

$$\sum_1^\infty w_n = \lim_{x \to y} \sum_1^\infty \mathfrak{v}_n(x).$$

(ii) *Show that both* $\Sigma x^n(1 - x)/n$ *and* $\Sigma x^n(1 - x^n)/n$ *converge uniformly over* $[0, 1 - \delta]$ *where* $0 < \delta < 1$ *but that only one of these series converges uniformly over* $[0, 1]$.

Since $0 \leqslant 1-x \leqslant 1-x^n$ we expect the first series to be uniformly convergent over $[0, 1]$ and the second series not to be uniformly convergent over $[0, 1]$.

Consider the first series. We have

$$\sum_N^M \frac{x^n(1-x)}{n} = \sum_N^M \left(\frac{x^n}{n} - \frac{x^{n+1}}{n}\right) = \sum_{N+1}^M x^n\left(\frac{1}{n} - \frac{1}{n-1}\right) + \frac{x^N}{N} - \frac{x^{M+1}}{M}.$$

Thus if $|x| \leqslant 1$

$$\left|\sum_N^M \frac{x^n(1-x)}{n}\right| \leqslant \sum_{N+1}^M \left(\frac{1}{n-1} - \frac{1}{n}\right) + \frac{1}{N} + \frac{1}{M} \leqslant \frac{2}{N}.$$

Let L be an integer greater than $2/\epsilon$. Then if $M \geqslant N \geqslant L$ it follows that

$$\left|\sum_N^M \frac{x^n(1-x)}{n}\right| < \epsilon.$$

But L is an integer independent of x. By theorem 57 *a* $\Sigma x^n(1-x)/n$ converges uniformly over $[0, 1]$. This implies that it converges uniformly over $[0, 1-\delta]$ where $0 < \delta < 1$.

Consider next $\Sigma x^n(1-x^n)/n$. If $0 \leqslant x \leqslant 1-\delta$ then

$$|x|^n\,|1-x^n|/n \leqslant (1-\delta)^n$$

and since $\Sigma(1-\delta)^n$ converges if $\delta > 0$ it follows by the Weierstrass M test that the series $\Sigma x^n(1-x^n)/n$ converges uniformly over $[0, 1-\delta]$ if $\delta > 0$. Finally, consider the convergence of $\Sigma x^n(1-x^n)/n$ over $[0, 1]$. When $x = 1-1/N$ and $N \leqslant n \leqslant 2N$ we have

$$\frac{x^n(1-x^n)}{n} \geqslant \frac{(1-1/N)^{2N}(1-(1-1/N)^N)}{2N} \sim \frac{e^{-2}(1-e^{-1})}{2N}.$$

Thus
$$\left|\sum_N^{2N} \frac{x^n(1-x^n)}{n}\right| > \frac{e^{-2}(1-e^{-1})}{3}$$

for N sufficiently large and this implies that $\Sigma x^n(1-x^n)/n$ does not converge uniformly over $[0, 1]$. Alternatively, we may argue as follows. For $0 \leqslant x < 1$

$$\sum_1^\infty \frac{x^n(1-x^n)}{n} = -\log(1-x) + \log(1-x^2) = \log(1+x)$$

and for $x = 1$
$$\sum_1^\infty \frac{x^n(1-x^n)}{n} = 0.$$

There is thus a discontinuity in the sum of the series at $x = 1$ and the convergence cannot be uniform over $0 \leqslant x \leqslant 1$.

(iii) *Show that the function* $\mathfrak{f}(x) = \sum_1^\infty x^{\frac{3}{2}}/(1+n^2x^2)$ *is a continuous function of* x *for* $x \geq 0$.

It is sufficient to show that the series converges uniformly over $[0, K]$ for every $K > 0$. If $x > 0$

$$\sum_{N+1}^{M} \frac{1}{1+n^2x^2} < \int_N^M \frac{dy}{1+x^2y^2} = \int_{xN}^{xM} \frac{dv}{x(1+v^2)}$$

$$= \frac{1}{x}(\tan^{-1} xM - \tan^{-1} xN) < \frac{\pi}{2x}$$

and thus if $0 \leq x \leq \delta$

$$\sum_{N+1}^{M} \frac{x^{\frac{3}{2}}}{1+n^2x^2} \leq \frac{\pi}{2} x^{\frac{1}{2}} \leq \frac{\pi\delta^{\frac{1}{2}}}{2}.$$

$\forall \epsilon > 0$ choose first $\delta > 0$ such that $\frac{1}{2}\pi\delta^{\frac{1}{2}} < \epsilon$ and then L such that if

$$M \geq N \geq L, \quad \frac{K^{\frac{3}{2}}}{\delta^2} \sum_{N+1}^{M} \frac{1}{n^2} < \epsilon.$$

If $\delta > 0$ is fixed and $0 < \delta \leq x \leq K$ then

$$\frac{x^{\frac{3}{2}}}{1+n^2x^2} \leq \frac{K^{\frac{3}{2}}}{\delta^2} \frac{1}{n^2}.$$

Thus for $0 \leq x \leq K$ $\qquad \sum_{N+1}^{M} \frac{x^{\frac{3}{2}}}{1+n^2x^2} < \epsilon.$

The series converges uniformly over $[0, K]$ and this implies the required result.

(iv) *Evaluate the integral*

$$\int_0^1 \frac{\log(1-x)}{x} dx.$$

The function $(\log(1-x))/x$ is defined and continuous for all x satisfying $0 < x < 1$. It is not defined at $x = 0$ nor at $x = 1$ and we must therefore investigate its existence as an improper Riemann integral at both these limits.

For $0 < x < 1$ we have $\log(1-x) = -\sum_{n=1}^{\infty} x^n/n$. Since $-\Sigma x^{n-1}/n$ converges uniformly over $[\epsilon, a]$, to $(\log(1-x))/x$ if $0 < \epsilon < a < 1$ it follows that

$$\int_\epsilon^a \frac{\log(1-x)}{x} dx = -\int_\epsilon^a \left(\sum_1^\infty \frac{x^{n-1}}{n}\right) dx = -\sum_1^\infty \int_\epsilon^a \frac{x^{n-1}}{n} dx$$

$$= -\sum_1^\infty \frac{a^n - \epsilon^n}{n^2}.$$

Now $\sum_1^\infty \frac{x^n}{n^2}$ is uniformly convergent for $0 \leqslant x \leqslant 1$. Thus as $\epsilon \to 0+$, $\sum_1^\infty \epsilon^n/n^2 \to 0$, and the improper Riemann integral

$$\int_0^a (\log(1-x))/x$$

exists and is equal to $-\sum_1^\infty a^n/n^2$. But further by uniform convergence again $\sum_1^\infty a^n/n^2 \to \sum_1^\infty 1/n^2$ as $a \to 1-$. Thus the improper Riemann integral of $(\log(1-x))/x$ exists over $[0, 1]$ and is equal to

$$-\sum_1^\infty 1/n^2 = -\tfrac{1}{6}\pi.$$

Exercises

1. Discuss the uniform convergence of

$$(a)\ \Sigma \frac{x}{(1+x)^n}, \quad (b)\ \Sigma \frac{(-1)^n}{n^x \log(1+n)},$$

$$(c)\ \Sigma \frac{x}{(nx+1)((n+2)x+1)}, \quad \text{for} \quad x \geqslant 0.$$

2. Show that for every $\delta > 0$ the *sequence* $\{x^n/(x^{2n}+1)\}$ converges uniformly over $[0, 1-\delta]$ and over $[1+\delta, 2]$ but does not converge uniformly over $[0, 2]$. Show that the sequence $\{(1-x)x^n/(x^{2n}+1)\}$ is uniformly convergent over $[0, 2]$.

3. Find for what values of x, α, β satisfying $x \geqslant 0$, $\alpha \geqslant 0$ and $\beta \geqslant 1$, $\Sigma n^\alpha x^\beta/(1+n^2x^2)$ converges. Show that it converges uniformly over $[0, 1]$ if and only if $0 \leqslant \alpha < 1$, $\alpha + 1 < \beta$ and that it converges uniformly over $x \geqslant 1$ if and only if $0 \leqslant \alpha < 1$, $\beta \leqslant 2$.

4. By actual summation or otherwise prove that the sum of the infinite series

$$\frac{|x|}{1+|x|} + \frac{|x|}{(1+|x|)^2} + \frac{|x|}{(1+|x|)^3} + \dots$$

exists for all real x but that it has a discontinuity. State the bearing of this on the uniformity of convergence of the series.

5. $\mathfrak{f}(x)$ is continuous in $[0, 1]$. Prove that the *sequence* $\{x^n \mathfrak{f}(x)\}$ converges uniformly over $[0, 1]$ if and only if $\mathfrak{f}(1) = 0$.

6. For $x \geqslant 0$, $\mathfrak{s}_n(x)$ is defined by

$$\mathfrak{s}_n(x) = \sin\frac{1}{n}\sin\frac{2}{n}\left(\sin^2\frac{1}{n} + x\cos^2\frac{1}{n}\right)^{-1} \quad (n = 1, 2, \dots).$$

Discuss the convergence and uniform convergence of the sequence $\{\mathfrak{s}_n(x)\}$ over $x \geqslant 0$.

7. The sequence $\{\mathfrak{s}_n(x)\}$ converges to $\mathfrak{s}(x)$ for $x \in E$ and

$$\sup_{x \in E} |\mathfrak{s}_n(x) - \mathfrak{s}(x)| = M_n.$$

Prove that the condition $M_n \to 0$ is necessary and sufficient for the convergence to be uniform.

If $\mathfrak{s}_n(x) = \sum_1^n \frac{\sin^3 mx}{m}$ prove that $\mathfrak{s}_n\left(\frac{\pi}{2n}\right) > \frac{1}{3}$.

If k is a real constant discuss the convergence and uniform convergence of $\sum \frac{\sin^3 mx}{m^k}$ over $[0, \frac{1}{2}\pi]$.

8. The functions $\mathfrak{f}_n(x)$ are such that

$$\begin{aligned} \sum_1^N \mathfrak{f}_n(x) &= x^{2+(1/(2N-1))} \sin(1/x) \quad (x \neq 0) \\ &= 0 \quad (x = 0). \end{aligned}$$

Show that $\Sigma D\mathfrak{f}_n(x)$ does not converge uniformly in any interval containing $x = 0$ as an interior point but that nevertheless

$$D\left(\sum_1^\infty \mathfrak{f}_n(x)\right) = \sum_1^\infty D\mathfrak{f}_n(x).$$

9. Show that if $\alpha > 1$ and $0 < x < 1$ then

$$\sum_{n=0}^\infty \frac{(-x)^n}{n+\alpha} = x^{-\alpha} \int_0^x \frac{t^{\alpha-1}}{1+t} dt.$$

Does this result hold when $x = 1$?

10. Prove that $\Sigma(1/n) \sin(\pi x^n)$ converges uniformly over $[0, 1-\delta]$ where $0 < \delta < 1$. If $\mathfrak{f}(x)$ denotes its sum prove that

$$\sum_1^\infty \frac{1}{n} \int_0^1 \sin(\pi x^n)\, dx = \int_0^1 \mathfrak{f}(x)\, dx.$$

11. Suppose that $\{\mathfrak{f}_n(x)\}$ is any sequence of functions such that

$$\mathfrak{f}_{n+1}(x) = D\mathfrak{f}_n(x) \quad (n = 0, \pm 1, \pm 2, \ldots)$$

for all real x. Show that if the series $\sum_0^\infty \mathfrak{f}_n(x)$ and $\sum_0^\infty \mathfrak{f}_{-n}(x)$ converge uniformly over every finite interval then $\sum_{-\infty}^\infty \mathfrak{f}_n(x) = e^x \sum_{-\infty}^\infty \mathfrak{f}_n(0)$.

12. Let

$$\mathfrak{s}_n(x) = \frac{nx}{1+n^2x^p} \quad (p > 0).$$

Show that $\mathfrak{s}_n(x) \to 0$ as $n \to \infty$ for $0 \leqslant x \leqslant 1$. Find for what values of p $\{\mathfrak{s}_n(x)\}$ converges uniformly over $[0, 1]$ and show that the relation

$$\lim_{n \to \infty} \int_0^1 \mathfrak{s}_n(x)\, dx = \int_0^1 \lim_{n \to \infty} \mathfrak{s}_n(x)\, dx$$

is true for $p = 2$ but not for $p = 4$.

13. $\mathfrak{f}$ is defined for all real x by $\mathfrak{f}(x) = (-1)^n$, $n < x < n+1$, $\mathfrak{f}(n) = 0$, n an integer. Prove that the sum of $\Sigma 2^{-n}\mathfrak{f}(2^{n-1}x)$ is continuous for all values of x other than the even integers.

Further examples

14. $\mathfrak{f}_n(x)$ is differentiable over $[0, 1]$ $\Sigma\mathfrak{f}_n(x)$ converges and for a certain fixed K independent of N,

$$0 \leqslant \sum_1^N D\mathfrak{f}_n(x) \leqslant K \quad (0 \leqslant x \leqslant 1).$$

Show that $\Sigma\mathfrak{f}_n(x)$ converges uniformly over $[0, 1]$. Discuss the convergence and uniform convergence over $[0, 1]$ of $\Sigma((-1)^n)/n \tanh nx$.

15. $\Sigma\mathfrak{u}_n(x)$ converges at every point of $[a, b]$ to the function $\mathfrak{s}(x)$, and each $\mathfrak{u}_n(x)$ is continuous throughout $[a, b]$. Corresponding to any assigned pair ϵ, p of positive numbers and any point x of the interval $[a, b]$, $\mathfrak{N} = \mathfrak{N}(\epsilon, p, x)$ is defined as the least integer greater than p for which

$$\left| \mathfrak{s}(x) - \sum_{n=1}^{\mathfrak{N}} \mathfrak{u}_n(x) \right| < \epsilon.$$

Prove that $\mathfrak{s}(x)$ is continuous if $\mathfrak{N}(\epsilon, p, x)$ is bounded above over $[a, b]$ for each pair ϵ, p even if the convergence is not uniform.

16. Σu_n converges absolutely. The functions $\mathfrak{v}_n(x)$ are such that $|\mathfrak{v}_n(x)| \leqslant V$ for all x in $[a, b]$ and all positive integers n where V is a constant. Prove that $\Sigma u_n \mathfrak{v}_n(x)$ converges uniformly over $[a, b]$.

If Σu_n is given to be conditionally convergent prove that the result is still true if, in addition, $\mathfrak{v}_n(x)$ is a monotonic sequence in n for each fixed x of $[a, b]$.

17. Show that $\Sigma(\sin nx)/n^p$: (i) converges uniformly for all real x if $p > 1$, and (ii) converges uniformly over $[a, b]$ where $0 < a < b < 2\pi$ if $p > 0$. Writing

$$\mathfrak{s}_N(x) = \sum_1^N \frac{\sin nx}{n^p}$$

show that if $p \leqslant 1$ $\qquad \mathfrak{s}_{2N}\left(\frac{\pi}{4N}\right) - \mathfrak{s}_N\left(\frac{\pi}{4N}\right) \geqslant \frac{1}{2\sqrt{2}}$

and deduce that in this case the series does not converge uniformly over any interval containing $x = 0$.

18. Prove that if

$$\mathfrak{f}_n(x) = \frac{\sin x + \sin 2x + \ldots + \sin nx}{n},$$

then $\mathfrak{f}_n(x) \to 0$ as $n \to \infty$.

Determine whether $\mathfrak{f}_n(x)$ converges to its limit uniformly (i) over $[\frac{1}{2}\pi, \pi]$, (ii) over $[0, \frac{1}{2}\pi]$.

19. If $\mathfrak{f}_n(x) = a(b-c)\,x^{na-1} + b(c-a)\,x^{nb-1} + c(a-b)\,x^{nc-1}$, where a, b, c are fixed positive integers, show that

$$\int_0^X \sum_{n=1}^{\infty} \mathfrak{f}_n(x)\,dx \quad \text{and} \quad \sum_{n=1}^{\infty} \left(\int_0^X \mathfrak{f}_n(x)\,dx \right)$$

exist when $0 < X \leqslant 1$ and evaluate them: (i) when $0 < X < 1$, (ii) when $X = 1$.

20. Σa_n, Σb_n and Σc_n all converge where $c_n = a_1 b_n + a_2 b_{n-1} + \ldots + a_n b_1$. Deduce from Abel's limit theorem that

$$\left(\sum_1^\infty a_n\right)\left(\sum_1^\infty b_n\right) = \left(\sum_1^\infty c_n\right).$$

28

FUNCTIONS OF TWO REAL VARIABLES. CONTINUITY AND DIFFERENTIABILITY

This chapter contains an introduction to the theory of functions defined over ordered pairs of real numbers.

The function $\mathfrak{f}(x_1, x_2)$ is *continuous* at (y_1, y_2) if and only if it is defined in a neighbourhood of (y_1, y_2) and given $\epsilon > 0$, $\exists\, \delta > 0$ such that

$$|\mathfrak{f}(x_1, x_2) - \mathfrak{f}(y_1, y_2)| < \epsilon \quad \text{if} \quad [(x_1 - y_1)^2 + (x_2 - y_2)^2]^{\frac{1}{2}} < \delta.$$

This definition is the same as if we regarded $\mathfrak{f}$ as a function of the complex variable $x_1 + ix_2$.

We shall write $\mathbf{x}$ in place of (x_1, x_2) and $\mathfrak{d}(\mathbf{x}, \mathbf{y})$ in place of $[(x_1 - y_1)^2 + (x_2 - y_2)^2]^{\frac{1}{2}}$ and sometimes refer to $\mathbf{x}$, $\mathbf{y}$ as *vectors*. If $\mathbf{y}$ is fixed and $\mathbf{x}$ variable then we use $\mathbf{x} \to \mathbf{y}$ in place of $\mathfrak{d}(\mathbf{x}, \mathbf{y}) \to 0$.

Definition 50. *$\mathfrak{f}(\mathbf{x})$ is differentiable at $\mathbf{y}$ if and only if it is defined in a neighbourhood of $\mathbf{y}$ such that*

$$\mathfrak{f}(\mathbf{x}) - \mathfrak{f}(\mathbf{y}) = X_1(x_1 - y_1) + X_2(x_2 - y_2) + o(\mathfrak{d}(\mathbf{x}, \mathbf{y})) \quad \textit{as} \quad \mathbf{x} \to \mathbf{y},$$

where X_1, X_2 are numbers independent of $\mathbf{x}$.

X_1, X_2 will usually depend on $\mathbf{y}$.

If $\mathfrak{f}(\mathbf{x})$ is differentiable at $\mathbf{y}$ it is also continuous at $\mathbf{y}$. If $\mathfrak{f}(\mathbf{x})$ and $\mathfrak{g}(\mathbf{x})$ are differentiable at $\mathbf{y}$ then so are $a\mathfrak{f}(\mathbf{x}) + b\mathfrak{g}(\mathbf{x})$, a, b constants, $\mathfrak{f}(\mathbf{x}).\mathfrak{g}(\mathbf{x})$, $1/\mathfrak{f}(\mathbf{x})$ provided $\mathfrak{f}(\mathbf{y}) \neq 0$.

If $\mathfrak{f}(\mathbf{x})$ is differentiable at $\mathbf{y}$ and $x_1 = \mathfrak{g}_1(\mathbf{u})$, $x_2 = \mathfrak{g}_2(\mathbf{u})$ are differentiable functions of $\mathbf{u}$ at $\mathbf{v}$, where $y_1 = \mathfrak{g}_1(\mathbf{v})$, $y_2 = \mathfrak{g}_2(\mathbf{v})$, then $\mathfrak{h}(\mathbf{u}) = \mathfrak{f}(\mathfrak{g}_1(\mathbf{u}), \mathfrak{g}_2(\mathbf{u}))$ is differentiable at $\mathbf{v}$. For

$$\begin{aligned}\mathfrak{h}(\mathbf{u}) - \mathfrak{h}(\mathbf{v}) &= \mathfrak{f}(\mathfrak{g}_1(\mathbf{u}), \mathfrak{g}_2(\mathbf{u})) - \mathfrak{f}(\mathfrak{g}_1(\mathbf{v}), \mathfrak{g}_2(\mathbf{v})) \\ &= X_1(\mathfrak{g}_1(\mathbf{u}) - \mathfrak{g}_1(\mathbf{v})) + X_2(\mathfrak{g}_2(\mathbf{u}) - \mathfrak{g}_2(\mathbf{v})) + o(\mathfrak{d}(\mathbf{x}, \mathbf{y})),\end{aligned}$$

where $\mathbf{x} \to \mathbf{y}$ by the differentiability of $\mathfrak{f}$. Thus using the differentiability of $\mathfrak{g}_1(\mathbf{u})$, $\mathfrak{g}_2(\mathbf{u})$ we have

$$(1)\quad \mathfrak{h}(\mathbf{u})-\mathfrak{h}(\mathbf{v}) = X_1[L_1(u_1-v_1)+M_1(u_2-v_2)+o(\mathfrak{d}(\mathbf{u},\mathbf{v}))] \\ +X_2[L_2(u_1-v_1)+M_2(u_2-v_2)+o(\mathfrak{d}(\mathbf{u},\mathbf{v}))]+o(\mathfrak{d}(\mathbf{x},\mathbf{y})),$$

where L_1, M_1, L_2, M_2 are appropriate numbers independent of $\mathbf{u}$. Now

$$\begin{aligned}\mathfrak{d}(\mathbf{x},\mathbf{y}) &= [(x_1-y_1)^2+(x_2-y_2)^2]^{\frac{1}{2}} \\ &= [(\mathfrak{g}_1(\mathbf{u})-\mathfrak{g}_1(\mathbf{v}))^2+(\mathfrak{g}_2(\mathbf{u})-\mathfrak{g}_2(\mathbf{v}))^2]^{\frac{1}{2}} \\ &= [(L_1(u_1-v_1)+M_1(u_2-v_2)+o(\mathfrak{d}(\mathbf{u},\mathbf{v})))^2 \\ &\quad +(L_2(u_1-v_1)+M_2(u_2-v_2)+o(\mathfrak{d}(\mathbf{u},\mathbf{v})))^2]^{\frac{1}{2}}.\end{aligned}$$

Now $|u_1-v_1| \leqslant \mathfrak{d}(\mathbf{u},\mathbf{v})$, $|u_2-v_2| \leqslant \mathfrak{d}(\mathbf{u},\mathbf{v})$, thus

$$\mathfrak{d}(\mathbf{x},\mathbf{y}) \leqslant \mathfrak{d}(\mathbf{u},\mathbf{v})\,[(|L_1|+|M_1|+o(1))^2+(|L_2|+|M_2|+o(1))^2]^{\frac{1}{2}}.$$

i.e. $$\mathfrak{d}(\mathbf{x},\mathbf{y}) = O(\mathfrak{d}(\mathbf{u},\mathbf{v})) \quad \text{as} \quad \mathfrak{d}(\mathbf{u},\mathbf{v}) \to 0.$$

Hence substituting in (1) above and rearranging

$$\mathfrak{h}(\mathbf{u})-\mathfrak{h}(\mathbf{v}) = (X_1L_1+X_2L_2)(u_1-v_1) \\ +(X_1M_1+X_2M_2)(u_2-v_2)+o(\mathfrak{d}(\mathbf{u},\mathbf{v})),$$

as $\mathfrak{d}(\mathbf{u},\mathbf{v}) \to 0$. This implies the differentiability of $\mathfrak{h}(\mathbf{u})$ at $\mathbf{v}$ since $X_1L_1+X_2L_2$ and $X_1M_1+X_2M_2$ are numbers independent of $\mathbf{u}$.

Remarks. (1) If $\mathfrak{f}(\mathbf{x})$ is differentiable at $\mathbf{y}$ and

$$\mathfrak{f}(\mathbf{x})-\mathfrak{f}(\mathbf{y}) = X_1(x_1-y_1)+X_2(x_2-y_2)+o(\mathfrak{d}(\mathbf{x},\mathbf{y})),$$

then the function $X_1(x_1-y_1)+X_2(x_2-y_2)$ is called the *differential* at $\mathbf{y}$ of $\mathfrak{f}$ and is denoted by $d\mathfrak{f}$. If we write dx_1 for (x_1-y_1) and dx_2 for (x_2-y_2) then

$$d\mathfrak{f} = X_1dx_1+X_2dx_2.$$

This is sometimes a convenient and compact notation but we shall not employ it in this book.

(2) We could have treated the differentiation of a function of a single variable in a similar manner by saying that $\mathfrak{f}(x)$ is differentiable at y if and only if

$$\mathfrak{f}(x)-\mathfrak{f}(y) = X(x-y)+o(|x-y|)$$

as $x \to y$, where X is a number independent of x.

(3) The concept of differentiability can be extended to functions of more than two variables. For example, $\mathfrak{f}(x_1,x_2,x_3)$ is differentiable at (y_1,y_2,y_3) if it is defined in a neighbourhood of (y_1,y_2,y_3) and

$$\mathfrak{f}(x_1,x_2,x_3)-\mathfrak{f}(y_1,y_2,y_3) = X_1(x_1-y_1)+X_2(x_2-y_2)+X_3(x_3-y_3)+o(\eta),$$

as $(x_1,x_2,x_3) \to (y_1,y_2,y_3)$ where $\eta = ((x_1-y_1)^2+(x_2-y_2)^2+(x_3-y_3)^2)^{\frac{1}{2}}$ and X_1, X_2, X_3 are numbers independent of x_1, x_2, x_3.

Exercises. (1) Show that the function $|x|+|y|$ is not differentiable at $(0, 0)$.

(2) Show that if $\mathfrak{f}(x)$ and $\mathfrak{g}(x)$ are differentiable functions of the single variable x at x_1 and x_2 respectively then $\mathfrak{f}(x).\mathfrak{g}(y)$ is a differentiable function of (x, y) at (x_1, x_2).

Partial differentiation

Definition 51. *If* $\lim_{h\to 0} \dfrac{\mathfrak{f}(y_1+h, y_2)-\mathfrak{f}(y_1, y_2)}{h}$ *exists we call it the partial derivative of* $\mathfrak{f}$ *with respect to* x_1 *at* $\mathbf{y}$ *and denote it by* $\partial\mathfrak{f}/\partial x_1$ *or* $(\partial\mathfrak{f}/\partial x_1)\mathbf{y}$ *or* $\mathfrak{f}_1(\mathbf{y})$ *or* $\left(\dfrac{\partial\mathfrak{f}}{\partial x_1}\right)_{(y_1, y_2)}$. *Similarly* $\partial\mathfrak{f}/\partial x_2$ *or* $\mathfrak{f}_2(\mathbf{y})$ *is defined to be*

$$\lim_{h\to 0} \frac{\mathfrak{f}(y_1, y_2+h)-\mathfrak{f}(y_1, y_2)}{h}$$

if this limit exists.

If $\mathfrak{f}(\mathbf{x})$ is differentiable at $\mathbf{y}$ then the partial derivatives of $\mathfrak{f}(\mathbf{x})$ exist at $\mathbf{y}$ and

$$\mathfrak{f}(\mathbf{x})-\mathfrak{f}(\mathbf{y}) = \mathfrak{f}_1(\mathbf{y})(x_1-y_1)+\mathfrak{f}_2(\mathbf{y})(x_2-y_2)+o(\mathfrak{d}(\mathbf{x}, \mathbf{y})).$$

But the existence of the partial derivatives does not imply the differentiability of $\mathfrak{f}(\mathbf{x})$. For example, if $\mathfrak{f} = |x.y|^{\frac{1}{2}}$, then $\partial\mathfrak{f}/\partial x$ and $\partial\mathfrak{f}/\partial y$ both exist and are zero at $(0, 0)$. Since, for example, $|\mathfrak{f}(x, x)-\mathfrak{f}(0, 0)| \neq o(|x|)$ as $x\to 0$ it follows that $\mathfrak{f}$ is not differentiable at $(0, 0)$.

It is convenient, in order to avoid the use of suffixes, to change the notation. In what follows we use (x, y) as the variable instead of (x_1, x_2).

Theorem 61. *If* $\partial\mathfrak{f}/\partial x$ *is a continuous function of* (x, y) *at* (a, b) *and if* $\partial\mathfrak{f}/\partial y$ *exists at* (a, b), *then* $\mathfrak{f}(x, y)$ *is differentiable at* (a, b).

By the mean-value theorem

$$\mathfrak{f}(a+h, b+k)-\mathfrak{f}(a, b+k) = h\mathfrak{f}_1(a+\theta h, b+k)$$

where $0<\theta<1$ and $|h|$, $|k|$ are sufficiently small. But $\mathfrak{f}_1(x, y)$ is continuous at (a, b), hence

$$\mathfrak{f}_1(a+\theta h, b+k) = \mathfrak{f}_1(a, b)+o(1) \quad \text{as} \quad h^2+k^2\to 0.$$

Further $\partial\mathfrak{f}/\partial y$ exists at (a, b), that is to say,

$$\mathfrak{f}(a, b+k)-\mathfrak{f}(a, b) = k(\mathfrak{f}_2(a, b)+o(1)) \quad \text{as} \quad k\to 0.$$

Combining all these relations we have

$$\mathfrak{f}(a+h, b+k)-\mathfrak{f}(a, b) = h\mathfrak{f}_1(a, b)+k\mathfrak{f}_2(a, b)+o(\eta),$$

where $\eta = (h^2+k^2)^{\frac{1}{2}}$, as $\eta\to 0$.

Thus $\mathfrak{f}(x, y)$ is differentiable at (a, b).

It is not necessarily the case that if the second-order mixed partial derivatives $\partial/\partial x\,(\partial\mathfrak{f}/\partial y)$ and $\partial/\partial y\,(\partial\mathfrak{f}/\partial x)$ exist at (a, b) then they are equal, for example if $\mathfrak{f}(x, y)$ is defined by

$$\mathfrak{f}(0,0) = 0, \quad \mathfrak{f}(x,y) = xy(x^2-y^2)/(x^2+y^2), \quad (x,y) \neq (0,0).$$

There are, however, various criteria that ensure that this inversion of the orders of differentiation is permissible. The simplest is the following.

Theorem 62. *If $\partial/\partial x\,(\partial\mathfrak{f}/\partial y)$ and $\partial/\partial y\,(\partial\mathfrak{f}/\partial x)$ exist and are continuous at (a, b) then they are equal at (a, b).*

Consider $\Delta = \mathfrak{f}(a+h, b+k) - \mathfrak{f}(a+h, b) - \mathfrak{f}(a, b+k) + \mathfrak{f}(a, b)$. We may write this as $\mathfrak{g}(a+h) - \mathfrak{g}(a)$, where $\mathfrak{g}(x) = \mathfrak{f}(x, b+k) - \mathfrak{f}(x, b)$. Then by the mean-value theorem

$$\Delta = h\mathfrak{g}'(a+\theta h) = h(\mathfrak{f}_1(a+\theta h, b+k) - \mathfrak{f}_1(a+\theta h, b)) \quad (0 < \theta < 1)$$

$$= hk\frac{\partial}{\partial y}\left(\frac{\partial f}{\partial x}\right)_{(a+\theta h,\, b+\theta_1 k)} \qquad (0 < \theta_1 < 1).$$

By the continuity of $\partial/\partial y\,(\partial\mathfrak{f}/\partial x)$,

$$\Delta = hk\left(\frac{\partial}{\partial y}\left(\frac{\partial\mathfrak{f}}{\partial x}\right)_{(a,b)} + o(1)\right) \quad \text{as} \quad h^2+k^2 \to 0.$$

By a similar argument interchanging the variables

$$\Delta = hk\left(\frac{\partial}{\partial x}\left(\frac{\partial\mathfrak{f}}{\partial y}\right)_{(a,b)} + o(1)\right) \quad \text{as} \quad h^2+k^2 \to 0.$$

Equating these two expressions for Δ, dividing by hk and letting both $h \to 0$ and $k \to 0$, we conclude that

$$\frac{\partial}{\partial y}\left(\frac{\partial\mathfrak{f}}{\partial x}\right)_{(a,b)} = \frac{\partial}{\partial x}\left(\frac{\partial\mathfrak{f}}{\partial y}\right)_{(a,b)}.$$

The nth mean-value theorem

We next prove the nth mean-value theorem. We shall refer to the set of points

$$\{(x,y)\,|\,x = \lambda x_0 + (1-\lambda)\,x_1, \quad y = \lambda y_0 + (1-\lambda)\,y_1,\ 0 \leqslant \lambda \leqslant 1\}$$

as the *segment* joining (x_0, y_0) and (x_1, y_1). The points (x_0, y_0) and (x_1, y_1) are called the *end-points* of the segment.

Theorem 63. *$\mathfrak{f}(x, y)$ is defined in a set which includes the segment joining the points (x_0, y_0) and (x_1, y_1) and is differentiable, together with all its partial derivatives of order less than or equal to $n-2$ at* *all points of the segment (including the end-points). The partial derivatives of order $n-1$ are differentiable, except possibly at (x_0, y_0) and (x_1, y_1) and are continuous along the whole segment (including the end-points). Under these circumstances there exists a number θ, $0 < \theta < 1$ such that*

$$\mathfrak{f}(x_1, y_1) = \mathfrak{f}(x_0, y_0) + \sum_{1}^{n-1} \frac{1}{r!} (H^{(r)}(\mathfrak{f}))_{(x_0, y_0)} + \frac{1}{n!} (H^{(n)}(\mathfrak{f}))_{(x_0+\theta(x_1-x_0)), (y_0+\theta(y_1-y_0))},$$

where $H^{(1)}(\mathfrak{f})$ denotes the function

$$(x_1 - x_0)\frac{\partial \mathfrak{f}}{\partial x} + (y_1 - y_0)\frac{\partial \mathfrak{f}}{\partial y}$$

and $H^{(r)}(\mathfrak{f})$ the function $H^{(r)}(\mathfrak{f}) = H^{(1)}H^{(r-1)}(\mathfrak{f})$.

Write

$$\mathfrak{g}(t) = \mathfrak{f}(x_0 + (x_1 - x_0)t,\ y_0 + (y_1 - y_0)t) \quad (0 \leqslant t \leqslant 1).$$

Then $\mathfrak{g}(t)$ has derivatives of orders up to n at all points of the interval $[0, 1]$ except possibly 0 and 1 and further

$$\frac{d^r \mathfrak{g}}{dt^r} = H^{(r)}(\mathfrak{f})$$

is a combination of rth order partial derivatives of $\mathfrak{f}$. Thus $d^r\mathfrak{g}/dt^r$ is continuous for $0 \leqslant t \leqslant 1$ and $1 \leqslant r \leqslant n-1$. The mean-value theorem for functions of one variable gives

$$\mathfrak{g}(1) = \mathfrak{g}(0) + \sum_{1}^{n-1} \frac{1}{r!}\left(\frac{d^r\mathfrak{g}}{dt^r}\right)_{t=0} + \frac{1}{n!}\left(\frac{d^n\mathfrak{g}}{dt^n}\right)_{t=\theta},$$

and this is the statement of the theorem.

Corollary. If the conditions of the theorem hold for every integer $n > 0$ and if the remainder tends to zero as $n \to \infty$ then

$$\mathfrak{f}(x_1, y_1) = \mathfrak{f}(x_0, y_0) + \sum_{1}^{\infty} \frac{1}{r!} (H^{(r)}(\mathfrak{f}))_{(x_0, y_0)}.$$

Implicit functions

The functions that we have considered so far have been explicit functions such as $y = \mathfrak{f}(x_1)$, $y = \mathfrak{f}(x_1, x_2)$ and so on. Here given x_1 or (x_1, x_2) as the case may be, belonging to suitable sets, we know that there is a certain rule by which we can calculate y. It often occurs in practice that the connexion between say y and x or between y and (x_1, x_2) is not given in this form, but is given by some condition on the pair (x, y), or the triad (y, x_1, x_2) usually of the form

$$\mathfrak{g}(y, x_1) = 0 \quad \text{or} \quad \mathfrak{h}(y, x_1, x_2) = 0.$$

The next theorem gives conditions under which this relation can be expressed explicitly as $y = \mathfrak{f}(x)$ or $y = \mathfrak{f}(x_1, x_2)$. We shall only deal in detail with the case of two variables, three or more variables may be handled by similar methods.

Theorem 64. *$\mathfrak{f}(x_1, x_2)$ is defined for $|x_1| \leqslant A$, $|x_2| \leqslant B$ and $\mathfrak{f}(0,0) = 0$. If*

(i) $\mathfrak{f}(x_1, x_2)$ is $\Uparrow$ *as a function of x_2 with x_1 kept fixed;*

(ii) *$\mathfrak{f}(x_1, x_2)$ is continuous as a function of x_2 with x_1 kept fixed;*

(iii) *$\mathfrak{f}(x_1, x_2)$ is continuous as a function of x_1 with x_2 kept fixed;*

then for some $\alpha > 0$, there is a function $\mathfrak{g}(x_1)$ defined for $|x_1| \leqslant \alpha$ and such that $\mathfrak{f}(x_1, \mathfrak{g}(x_1)) = 0$, $|x_1| \leqslant \alpha$. Moreover, the function $\mathfrak{g}(x_1)$ is continuous. Further if $\mathfrak{f}(x_1, x_2)$ is differentiable at $(0,0)$ and if $\mathfrak{f}_2(0,0) \neq 0$ then $\mathfrak{g}(x_1)$ is differentiable at $x_1 = 0$.

From (i) and the fact that $\mathfrak{f}(0,0) = 0$, it follows that

$$\mathfrak{f}(0, B) > 0 > \mathfrak{f}(0, -B).$$

From (iii) $\exists\, \alpha > 0$ such that if $|x_1| \leqslant \alpha$ then

$$\mathfrak{f}(x_1, B) > 0 > \mathfrak{f}(x_1, -B).$$

From (i) and (ii) it follows that corresponding to each x_1 satisfying $|x_1| \leqslant \alpha$ there is a unique value of x_2 say $x_2 = \mathfrak{g}(x_1)$ such that

$$\mathfrak{f}(x_1, \mathfrak{g}(x_1)) = 0.$$

We show next that $\mathfrak{g}(x_1)$ is continuous at say y_1. Now by (i) and the definition of $\mathfrak{g}(y_1)$, for any $\epsilon > 0$ sufficiently small, (so that $|\mathfrak{g}(y_1)| + \epsilon \leqslant B$), we have

$$\mathfrak{f}(y_1, \mathfrak{g}(y_1) + \epsilon) > 0 > \mathfrak{f}(y_1, \mathfrak{g}(y_1) - \epsilon).$$

By (iii) $\exists\, \delta > 0$ such that

$$\mathfrak{f}(x_1, \mathfrak{g}(y_1) + \epsilon) > 0 > \mathfrak{f}(x_1, \mathfrak{g}(y_1) - \epsilon) \quad \text{provided} \quad |x_1 - y_1| < \delta.$$

By (i) and the definition of $\mathfrak{g}(x_1)$,

$$\mathfrak{g}(y_1) + \epsilon > \mathfrak{g}(x_1) > \mathfrak{g}(y_1) - \epsilon \quad \text{if} \quad |x_1 - y_1| < \delta.$$

This means that $\mathfrak{g}(x_1)$ is continuous at y_1.

Finally, the differentiability of $\mathfrak{f}(x_1, x_2)$ at $(0,0)$ implies that

$$\mathfrak{f}(x_1, x_2) = X_1 x_1 + X_2 x_2 + o(\mathfrak{d}) \quad \text{as} \quad \mathfrak{d} \to 0,$$

where $\mathfrak{d} = \sqrt{(x_1^2 + x_2^2)}$ and X_1, X_2 are numbers. Further

$$X_2 = \frac{\partial \mathfrak{f}}{\partial x_2} \neq 0.$$

Thus $\qquad (*) \quad 0 = X_1 x_1 + X_2 \mathfrak{g}(x_1) + o(x_1^2 + \mathfrak{g}(x_1)^2)^{\frac{1}{2}}.$

Now $X_2 \neq 0$ and we may choose $\epsilon > 0$, $K > 0$ where ϵ is small and K large so that

$$(**) \quad |X_2| > \epsilon\left(1+\frac{1}{K^2}\right)^{\frac{1}{2}} + \frac{|X_1|}{K}.$$

But then from (*) for x_1 sufficiently near to 0

$$|X_2 \mathfrak{g}(x_1)| < |X_1 x_1| + \epsilon(x_1^2 + \mathfrak{g}(x_1)^2)^{\frac{1}{2}}$$

and if $|\mathfrak{g}(x_1)| > K|x_1|$ we conclude that

$$|X_2| < \frac{|X_1|}{K} + \epsilon\left(\frac{1}{K^2}+1\right)^{\frac{1}{2}}.$$

Since this contradicts (**) we conclude that for all such x_1, $|\mathfrak{g}(x_1)| < K|x_1|$, i.e. substituting in (*)

$$X_1 x_1 + X_2 \mathfrak{g}(x_1) = o\,|x_1|,$$

i.e. $\mathfrak{g}(x_1)$ is differentiable at 0 and its differential coefficient is $-X_1/X_2$.

Results on inversion of orders of integration and differentiation

Partial differentiation and integration with respect to one variable.

(i) *If $\mathfrak{f}(x,y)$ is a continuous function of (x,y) in $a \leqslant x \leqslant b$, $c \leqslant y \leqslant d$ then $\mathfrak{F}(y) = \int_a^b \mathfrak{f}(x,y)\,dx$ is a continuous function of y in $c \leqslant y \leqslant d$.*

By theorem 26, corollary, p. 91, $\mathfrak{f}(x,y)$ is uniformly continuous for $a \leqslant x \leqslant b$, $c \leqslant y \leqslant d$. Thus given $\epsilon > 0\, \exists\, \delta > 0$ such that if $|y-y'| < \delta$ then

$$|\mathfrak{f}(x,y') - \mathfrak{f}(x,y)| < \frac{\epsilon}{b-a},$$

$$\therefore \quad |\mathfrak{F}(y) - \mathfrak{F}(y')| \leqslant \int_a^b |\mathfrak{f}(x,y) - \mathfrak{f}(x,y')|\,dx < \epsilon \quad \text{if} \quad |y-y'| < \delta,$$

i.e. $\mathfrak{F}(y)$ is continuous.

(ii) *If $\partial\mathfrak{f}/\partial y$ is a continuous function of (x,y) in $a \leqslant x \leqslant b$, $c \leqslant y \leqslant d$, then $\mathfrak{F}(y)$ is differentiable and*

$$\frac{d\mathfrak{F}(y)}{dy} = \int_a^b \mathfrak{f}_2(x,y)\,dx,$$

$$\frac{\mathfrak{F}(y) - \mathfrak{F}(y_0)}{y-y_0} = \int_a^b \frac{\mathfrak{f}(x,y) - \mathfrak{f}(x,y_0)}{y-y_0}\,dx$$

$$= \int_a^b \left(\frac{\partial\mathfrak{f}}{\partial y}\right)_{(x,\,g(x))} dx \quad \text{(mean value theorem)},$$

where $y < \mathfrak{g}(x) < y_0$ or $y_0 < \mathfrak{g}(x) < y$. By continuity of $\dfrac{\partial \mathfrak{f}}{\partial y}$ this is equal to

$$\int_a^b \left(\frac{\partial \mathfrak{f}}{\partial y}\right)_{(x,\,y_0)} dx + o(1) \quad \text{as} \quad y \to y_0.$$

Thus $\dfrac{d\mathfrak{J}(y)}{dy}$ exists and is equal to $\displaystyle\int_a^b \frac{\partial \mathfrak{f}}{\partial y}\,dx$.

These results can be extended to integrals over infinite ranges as follows.

(iii) *If* $\mathfrak{f}(x,y)$ *is a continuous function of* (x,y) *in* $a \leqslant x \leqslant \infty$, $c \leqslant y \leqslant d$ *and if* $\displaystyle\int_a^\infty \mathfrak{f}(x,y)\,dx$ *is uniformly convergent in* $c \leqslant y \leqslant d$, *then* $\mathfrak{J}(y) = \displaystyle\int_a^\infty \mathfrak{f}(x,y)\,dx$ *is a continuous function of* y.

Given $\epsilon > 0 \;\exists\; X_0$ such that

$$\left|\int_{X_0}^\infty \mathfrak{f}(x,y)\,dx\right| < \tfrac{1}{3}\epsilon \quad (c \leqslant y \leqslant d),$$

$$\mathfrak{J}(y) - \mathfrak{J}(y_0) = \int_a^{X_0} \{\mathfrak{f}(x,y) - \mathfrak{f}(x,y_0)\}\,dx + \int_{X_0}^\infty \mathfrak{f}(x,y)\,dx - \int_{X_0}^\infty \mathfrak{f}(x,y_0)\,dx.$$

Since $\mathfrak{f}(x,y)$ is continuous (and therefore uniformly continuous over any bounded closed set, theorem 26, p. 91) $\exists\; \delta > 0$ such that if $X_0 > a$

$$|\mathfrak{f}(x,y) - \mathfrak{f}(x,y_0)| < \epsilon/3(X_0 - a) \quad (a \leqslant x \leqslant X_0,\; |y - y_0| < \delta).$$

Thus
$$\left|\int_a^{X_0} (\mathfrak{f}(x,y) - \mathfrak{f}(x,y_0))\,dx\right| < \tfrac{1}{3}\epsilon \quad (|y - y_0| < \delta).$$

Hence
$$|\mathfrak{J}(y) - \mathfrak{J}(y_0)| < \epsilon \quad \text{if} \quad |y - y_0| < \delta$$

and (i) is proved.

Remark. This is the analogue for integrals of the result proved in chapter 27, that the sum of a uniformly convergent series of continuous functions is continuous.

(iv) *If* $\mathfrak{f}_2(x,y)$ *is a continuous function of* (x,y) *in* $a \leqslant x \leqslant \infty$, $c \leqslant y \leqslant d$ *and* $\displaystyle\int_a^\infty \mathfrak{f}_2(x,y)\,dx$ *is uniformly convergent, then* $\mathfrak{J}(y)$ *is differentiable for* $c < y < d$ *and*

$$\frac{d\mathfrak{J}(y)}{dy} = \int_a^\infty \mathfrak{f}_2(x,y)\,dx.$$

Given $\epsilon > 0 \;\exists\; X_0$ such that

$$\left|\int_{X_1}^{X_2} \mathfrak{f}_2(x,\eta)\,dx\right| < \epsilon \quad (c \leqslant \eta \leqslant d).$$

Write

$$\mathfrak{g}(y) = \int_{X_1}^{X_2} \mathfrak{f}(x, y)\, dx,$$

then

$$\left|\int_{X_1}^{X_2} \frac{\mathfrak{f}(x,y)-\mathfrak{f}(x,y_0)}{y-y_0}\, dx\right| = \left|\frac{\mathfrak{g}(y)-\mathfrak{g}(y_0)}{y-y_0}\right|$$

$$= |\mathfrak{g}'(\eta)| \quad \text{by the mean-value theorem}$$

$$= \left|\int_{X_1}^{X_2} \mathfrak{f}_2(x,\eta)\, dx\right|.$$

Now

$$\frac{\mathfrak{J}(y)-\mathfrak{J}(y_0)}{y-y_0} = \int_a^{X_0} \frac{\mathfrak{f}(x,y)-\mathfrak{f}(x,y_0)}{y-y_0}\, dx + \lim_{X\to\infty} \int_{X_0}^{X} \frac{\mathfrak{f}(x,y)-\mathfrak{f}(x,y_0)}{y-y_0}\, dx.$$

Thus

$$\overline{\lim_{y\to y_0}} \frac{\mathfrak{J}(y)-\mathfrak{J}(y_0)}{y-y_0} \leqslant \int_a^{X_0} \mathfrak{f}_2(x,y_0)\, dx + \epsilon \leqslant \int_a^{\infty} \mathfrak{f}_2(x,y_0)\, dx + 2\epsilon.$$

Similarly $$\underline{\lim_{y\to y_0}} \frac{\mathfrak{J}(y)-\mathfrak{J}(y_0)}{y-y_0} \geqslant \int_a^{\infty} \mathfrak{f}_2(x,y_0)\, dx - 2\epsilon$$

and since ϵ is any positive number,

$$\lim_{y\to y_0} \frac{\mathfrak{J}(y)-\mathfrak{J}(y_0)}{y-y_0} \quad \text{exists and is equal to} \quad \int_a^{\infty} \mathfrak{f}_2(x,y_0)\, dx.$$

(v) *If* $\mathfrak{f}(x,y)$ *is continuous in* $a \leqslant x \leqslant b$, $c \leqslant y \leqslant d$, *then*

$$\int_a^b \left\{\int_c^y \mathfrak{f}(x,t)\, dt\right\} dx = \int_c^y \left\{\int_a^b \mathfrak{f}(x,t)\, dx\right\} dt.$$

By (ii) applied to a finite interval $[a, b]$

$$\frac{d}{dy}\int_a^b \left\{\int_c^y \mathfrak{f}(x,t)\, dt\right\} dx = \int_a^b \mathfrak{f}(x,y)\, dx.$$

The right-hand side of this equation is continuous as a function of y. Thus it is Riemann integrable and, using theorem 47,

$$\int_a^b \left\{\int_c^y \mathfrak{f}(x,t)\, dt\right\} dx = \int_c^y \left\{\int_a^b \mathfrak{f}(x,t)\, dx\right\} dt,$$

and this is the required result.

(vi) *If* $\mathfrak{f}(x,y)$ *is a continuous function of* (x,y) *in* $a \leqslant x \leqslant \infty$, $c \leqslant y \leqslant d$ *and* $\int_a^{\infty} \mathfrak{f}(x,y)\, dx$ *is uniformly convergent in* $c \leqslant y \leqslant d$ *then*

$$\int_a^{\infty} \left\{\int_c^y \mathfrak{f}(x,t)\, dt\right\} dx = \int_c^y \left\{\int_a^{\infty} \mathfrak{f}(x,t)\, dx\right\} dt.$$

Given $\epsilon > 0$ there exists X_0 such that

$$\left|\int_{X_1}^{X_2} \mathfrak{f}(x,t)\,dx\right| < \epsilon \quad (X_2, X_1 > X_0,\ c \leqslant t \leqslant d).$$

Thus for some ψ with $|\psi| \leqslant \epsilon(d-c)$ and every $X_1 > X_0$ (ψ depending on X_1),

$$\int_c^y \left\{\int_a^\infty \mathfrak{f}(x,t)\,dx\right\} dt = \int_c^y \left\{\int_a^{X_1} \mathfrak{f}(x,t)\,dx\right\} dt + \psi$$

$$= \int_a^{X_1} \left\{\int_c^y \mathfrak{f}(x,t)\,dt\right\} dx + \psi,$$

by (iii).

Thus
$$\int_a^\infty \left\{\int_c^y \mathfrak{f}(x,t)\,dt\right\} dx = \lim_{X_1\to\infty} \int_a^{X_1} \left\{\int_c^y \mathfrak{f}(x,t)\,dt\right\} dx$$

$$= \int_c^y \left\{\int_a^\infty \mathfrak{f}(x,t)\,dt\right\} dx.$$

Remark. The existence of the integral on the right-hand side of this equation implies the existence of that on the left.

Worked examples

(i) $\mathfrak{f}(x,y,z)$ *is expressed as a function* $\mathfrak{g}(u,v)$, *where* $x = a(u+v)$, $y = b(u-v)$, $z = uv$, $a \neq 0$, $b \neq 0$. *Show that if* $\mathfrak{f}(x,y,z)$ *has continuous partial derivatives of the first and second order then*

$$2\left(\frac{\partial^2 \mathfrak{g}}{\partial u^2} + \frac{\partial^2 \mathfrak{g}}{\partial v^2}\right) = \left(\frac{x^2}{a^2} + \frac{y^2}{b^2}\right)\frac{\partial^2 \mathfrak{f}}{\partial z^2}$$

$$+4\left(a^2\frac{\partial^2 \mathfrak{f}}{\partial x^2} + b^2\frac{\partial^2 \mathfrak{f}}{\partial y^2} + x\frac{\partial^2 \mathfrak{f}}{\partial x\,\partial z} - y\frac{\partial^2 \mathfrak{f}}{\partial y\,\partial z}\right),$$

$$\begin{aligned}
&\mathfrak{g}(u,v_0) - \mathfrak{g}(u_0,v_0)\\
&= \mathfrak{f}(a(u+v_0), b(u-v_0), uv_0) - \mathfrak{f}(a(u_0+v_0), b(u_0-v_0), u_0v_0)\\
&= \mathfrak{f}(a(u+v_0), b(u-v_0), uv_0) - \mathfrak{f}(a(u_0+v_0), b(u-v_0), uv_0)\\
&\quad + \mathfrak{f}(a(u_0+v_0), b(u-v_0), uv_0) - \mathfrak{f}(a(u_0+v_0), b(u_0-v_0), uv_0)\\
&\quad + \mathfrak{f}(a(u_0+v_0), b(u_0-v_0), uv_0) - \mathfrak{f}(a(u_0+v_0), b(u_0-v_0), u_0v_0)\\
&= a(u-u_0)(\mathfrak{f}_x + o(1)) + b(u-u_0)(\mathfrak{f}_y + o(1)) + v_0(u-u_0)(\mathfrak{f}_z + o(1)),
\end{aligned}$$

where we have used both the existence and continuity of the first-order partial derivatives

$$\frac{\partial \mathfrak{f}}{\partial x} = \mathfrak{f}_x, \quad \frac{\partial \mathfrak{f}}{\partial y} = \mathfrak{f}_y, \quad \frac{\partial \mathfrak{f}}{\partial z} = \mathfrak{f}_z.$$

Hence $\partial\mathfrak{g}/\partial u = \mathfrak{g}_u$ exists and

$$\mathfrak{g}_u = a\mathfrak{f}_x + b\mathfrak{f}_y + v\mathfrak{f}_z.$$

Similarly
$$g_v = a\mathfrak{f}_x - b\mathfrak{f}_y + u\mathfrak{f}_z.$$

Now although $\mathfrak{g}_u$ is not expressed above entirely as a function of x, y, z (as it should be if we are to calculate $(\mathfrak{g}_u)_u$ simply by repeating the formula), yet it involves v and not u and in calculating $\partial/\partial u$, v is a constant. Hence

$$\begin{aligned}\mathfrak{g}_{uu} &= a[a\mathfrak{f}_{xx} + b\mathfrak{f}_{xy} + v\mathfrak{f}_{xz}] + b[a\mathfrak{f}_{yx} + b\mathfrak{f}_{yy} + v\mathfrak{f}_{yz}] + v[a\mathfrak{f}_{zx} + b\mathfrak{f}_{zy} + v\mathfrak{f}_{zz}]\\ &= a^2\mathfrak{f}_{xx} + 2ab\mathfrak{f}_{xy} + 2av\mathfrak{f}_{xz} + b^2\mathfrak{f}_{yy} + v^2\mathfrak{f}_{zz} + 2bv\mathfrak{f}_{yz}\end{aligned}$$

since by theorem 62 $\mathfrak{f}_{xy} = \mathfrak{f}_{yx}$, etc.

Similarly

$$\mathfrak{g}_{vv} = a^2\mathfrak{f}_{xx} + b^2\mathfrak{f}_{yy} + u^2\mathfrak{f}_{zz} - 2ab\mathfrak{f}_{xy} + 2au\mathfrak{f}_{xz} - 2bu\mathfrak{f}_{yz}$$

and hence

$$2(\mathfrak{g}_{uu} + \mathfrak{g}_{vv}) = 4a^2\mathfrak{f}_{xx} + 4b^2\mathfrak{f}_{yy} + \left(\frac{x^2}{a^2} + \frac{y^2}{b^2}\right)\mathfrak{f}_{zz} + 4x\mathfrak{f}_{xz} - 4y\mathfrak{f}_{yz}.$$

(ii) *A real function $\mathfrak{u} = \mathfrak{u}(x, y)$ has all four partial derivatives of the second order continuous in a set D and satisfies the inequalities*

$$\mathfrak{u} > 0, \quad \frac{\partial^2\mathfrak{u}}{\partial x^2} > 0, \quad \frac{\partial^2\mathfrak{u}}{\partial x^2}\frac{\partial^2\mathfrak{u}}{\partial y^2} - \left(\frac{\partial^2\mathfrak{u}}{\partial x\,\partial y}\right)^2 > 0$$

throughout D. Prove that if p is a constant greater than 1 the function $\mathfrak{v} = \mathfrak{u}^p$ (where the real positive value is to be taken) has the same properties.

We must first show that the partial derivatives of the second order of $\mathfrak{v} = \mathfrak{u}^p$ exist and are continuous in D. We have

$$\frac{\partial\mathfrak{v}}{\partial x} = p\,.\,\mathfrak{u}^{p-1}\frac{\partial\mathfrak{u}}{\partial x}, \quad \frac{\partial\mathfrak{v}}{\partial y} = p\mathfrak{u}^{p-1}\frac{\partial\mathfrak{u}}{\partial y}$$

and since $\mathfrak{u} > 0$, $\mathfrak{u}^{p-1}$ is a continuous function of (x, y) in D. Thus $\partial\mathfrak{v}/\partial x$ and $\partial\mathfrak{v}/\partial y$ are continuous functions of (x, y) in D. Next we have

$$\frac{\partial^2\mathfrak{v}}{\partial x^2} = p(p-1)\,\mathfrak{u}^{p-2}\left(\frac{\partial\mathfrak{u}}{\partial x}\right)^2 + p\mathfrak{u}^{p-1}\frac{\partial^2\mathfrak{u}}{\partial x^2},$$

$$\frac{\partial^2\mathfrak{v}}{\partial x\,\partial y} = p(p-1)\,\mathfrak{u}^{p-2}\left(\frac{\partial\mathfrak{u}}{\partial x}\right)\left(\frac{\partial\mathfrak{u}}{\partial y}\right) + p\mathfrak{u}^{p-1}\frac{\partial^2\mathfrak{u}}{\partial x\,\partial y},$$

$$\frac{\partial^2\mathfrak{v}}{\partial y^2} = p(p-1)\,\mathfrak{u}^{p-2}\left(\frac{\partial\mathfrak{u}}{\partial y}\right)^2 + p\mathfrak{u}^{p-1}\frac{\partial^2\mathfrak{u}}{\partial y^2}$$

and all three functions $\mathfrak{v}_{xx}$, $\mathfrak{v}_{xy}$, $\mathfrak{v}_{yy}$ are continuous in D.

Next we have $\mathfrak{v} > 0$ in D and, since $p > 1$, all the terms in the expression for $\mathfrak{v}_{xx}$ are positive. Hence $\mathfrak{v}_{xx} > 0$ in D and similarly $\mathfrak{v}_{yy} > 0$ in D. Finally

$$\frac{\partial^2\mathfrak{v}}{\partial x^2}\frac{\partial^2\mathfrak{v}}{\partial y^2}-\left(\frac{\partial^2\mathfrak{v}}{\partial x\,\partial y}\right)^2 = p^2\mathfrak{u}^{2p-2}\left\{\frac{\partial^2\mathfrak{u}}{\partial x^2}\frac{\partial^2\mathfrak{u}}{\partial y^2}-\left(\frac{\partial^2\mathfrak{u}}{\partial x\,\partial y}\right)^2\right\}+p^2(p-1)\,\mathfrak{u}^{2p-3}$$
$$\times\left\{\frac{\partial^2\mathfrak{u}}{\partial x^2}\left(\frac{\partial\mathfrak{u}}{\partial y}\right)^2+\frac{\partial^2\mathfrak{u}}{\partial y^2}\left(\frac{\partial\mathfrak{u}}{\partial x}\right)^2-2\frac{\partial\mathfrak{u}}{\partial x}\frac{\partial\mathfrak{u}}{\partial y}\left(\frac{\partial^2\mathfrak{u}}{\partial x\,\partial y}\right)\right\}.$$

Of the two terms in brackets on the right-hand side the first is positive by the given hypotheses. For the second we write $\mathfrak{u}_{xx}$ for $\partial^2\mathfrak{u}/\partial x^2$, etc., ... and then E for the expression.

$$\begin{aligned} E &= \mathfrak{u}_{xx}\mathfrak{u}_y^2+\mathfrak{u}_{yy}\mathfrak{u}_x^2-2\mathfrak{u}_x\mathfrak{u}_y\mathfrak{u}_{xy} \\ &= (\mathfrak{u}_y\mathfrak{u}_{xx}^{\frac12}+\mathfrak{u}_x\mathfrak{u}_{yy}^{\frac12})^2-2\mathfrak{u}_x\mathfrak{u}_y(\mathfrak{u}_{xx}^{\frac12}\mathfrak{u}_{yy}^{\frac12}+\mathfrak{u}_{xy}) \\ &= (\mathfrak{u}_y\mathfrak{u}_{xx}^{\frac12}-\mathfrak{u}_x\mathfrak{u}_{yy}^{\frac12})^2+2\mathfrak{u}_x\mathfrak{u}_y(\mathfrak{u}_{xx}^{\frac12}\mathfrak{u}_{yy}^{\frac12}-\mathfrak{u}_{xy}). \end{aligned}$$

Since $\mathfrak{u}_{xx}\mathfrak{u}_{yy}-\mathfrak{u}_{xy}^2 > 0$ we have $\mathfrak{u}_{xx}^{\frac12}\mathfrak{u}_{yy}^{\frac12}\pm\mathfrak{u}_{xy} > 0$ (for $\mathfrak{u}_{yy} > 0$ and thus also $\mathfrak{u}_{xx} > 0$). If $\mathfrak{u}_x\mathfrak{u}_y < 0$ then the first expression for E shows that $E > 0$. If $\mathfrak{u}_x\mathfrak{u}_y > 0$ then the second expression shows that $E > 0$.

Thus in any case $E > 0$ and this implies that

$$\frac{\partial^2\mathfrak{v}}{\partial x^2}\frac{\partial^2\mathfrak{v}}{\partial y^2}-\left(\frac{\partial^2\mathfrak{v}}{\partial x\,\partial y}\right)^2 > 0 \quad \text{in} \quad D.$$

Exercises

1. Determine which of the following functions $\mathfrak{f}(x,y)$ are: (i) continuous, and (ii) differentiable at $(0,0)$.

(*a*) $\mathfrak{f}(x,y) = \arg(x+iy)$, where

$$0 \leqslant \arg(x+iy) < 2\pi \quad (x,y) \neq (0,0), \quad \mathfrak{f}(0,0) = 0.$$

(*b*) $|x+iy|$, $\mathfrak{f}(0,0) = 0$; (*c*) $\mathfrak{f}(x,y) = |x+iy-1|$.

2. $\mathfrak{f}(x,y)$ is defined as follows

$$\begin{aligned} \mathfrak{f}(x,y) &= y \quad \text{if} \quad x = y^2, \\ &= 0 \quad \text{otherwise.} \end{aligned}$$

Prove that (i) $\mathfrak{f}(x,y)$ is continuous at $(0,0)$; (ii) for fixed l, m $\mathfrak{f}(lt,mt)$ is a differentiable function of t at $t = 0$; (iii) $\mathfrak{f}(x,y)$ is not differentiable at $(0,0)$.

3. Three real functions of two real variables (x,y) are zero at $(0,0)$ and are defined elsewhere for positive p as follows:

$$\text{(i)}\ (|x|^p+|y|^p)^{1/p}, \quad \text{(ii)}\ |xy|^p, \quad \text{(iii)}\ \sum_1^\infty a_n|xy|^{p/n},$$

where Σa_n is a convergent series of real positive summands.

Find for what values of p they are (*a*) continuous at $(0,0)$, (*b*) differentiable at $(0,0)$.

4. If at all points of the plane $\mathfrak{f}(x,y)$ is continuous with respect to x and has a bounded partial derivative $\mathfrak{f}_y(x,y)$, prove that $\mathfrak{f}(x,y)$ is continuous with respect to (x,y).

5. $\mathfrak{f}(x,y)$ is defined by $\mathfrak{f}(0,0) = 0$,

$$\mathfrak{f}(x,y) = \frac{(x+y)^2(x-y)}{x^2+y^2}, \quad (x,y) \neq (0,0).$$

Find $(\partial\mathfrak{f}/\partial y)$ at $(0,0)$ and explain why $\mathfrak{f}(x,y) = 0$ does not determine y as a single-valued function of x near the origin.

6. Examine whether the equation $x+y+ax^2+bxy+cy^2 = 0$, defines y as a unique function vanishing at $x = 0$ and continuous in a neighbourhood of $x = 0$.

7. In the square $0 \leqslant x \leqslant 1$, $0 \leqslant y \leqslant 1$, $\mathfrak{F}(x,y)$ is an increasing function in each of the variables x, y provided the other is kept fixed. By considering $\mathfrak{F}(x,x)$ or otherwise prove that there are points (x,y) at which $\mathfrak{F}(x,y)$ is continuous as a function of (x,y).

8. Show that if

$$\mathfrak{f}(x,y) = \frac{xy(x^2-y^2)}{x^2+y^2}, \quad (x,y) \neq (0,0), \quad \mathfrak{f}(0,0) = 0,$$

then

$$\frac{\partial}{\partial x}\left(\frac{\partial\mathfrak{f}}{\partial y}\right) \neq \frac{\partial}{\partial y}\left(\frac{\partial\mathfrak{f}}{\partial x}\right) \quad \text{at} \quad (0,0).$$

9. $\mathfrak{f}(x,y) = 0$ if $x = 0$ or $y = 0$ and for all other (x,y)

$$\mathfrak{f}(x,y) = x^2\tan^{-1}\frac{y}{x} - y^2\tan^{-1}\frac{x}{y}, \quad \text{where} \quad -\tfrac{1}{2}\pi \leqslant \tan^{-1}\frac{x}{y} \leqslant \tfrac{1}{2}\pi.$$

Show that

$$\frac{\partial}{\partial y}\left(\frac{\partial\mathfrak{f}}{\partial x}\right) \neq \frac{\partial}{\partial x}\left(\frac{\partial\mathfrak{f}}{\partial y}\right) \quad \text{at} \quad (0,0).$$

10. The relation $\mathfrak{f}(x,y) = 0$ defines y uniquely as a function of x; and $\mathfrak{f}_x$, $\mathfrak{f}_y$ are differentiable and non-zero in $a < x < b$, $c < x < d$. Find d^2y/dx^2 in terms of the partial derivatives of $\mathfrak{f}(x,y)$.

11. $\mathfrak{f}(x,y,z)$ is such that $\mathfrak{f}_x(x,y,z)$ exists and is a continuous function of (x,y,z) in a neighbourhood of (a,b,c), $\mathfrak{f}_y(a,y,z)$ exists and is a continuous function of (y,z) in a neighbourhood of (b,c), $\mathfrak{f}_z(a,b,c)$ exists. Show that $\mathfrak{f}(x,y,z)$ is differentiable at (a,b,c).

12. $\mathfrak{f}(x,y)$ is defined by $\mathfrak{f}(x,y) = x^p(1-x)^q y^p(1-y)^q$, $0 < x < 1$, $0 < y < 1$, $\mathfrak{f}(x,y) = 0$, all other values of (x,y). Find for what values of p and q the function is differentiable for all values of (x,y).

13. The double series Σa_{mn} is absolutely convergent and the function $\mathfrak{f}(x,y)$ is defined for $-1 \leqslant x \leqslant 1$, $-1 \leqslant y \leqslant 1$, by $\mathfrak{f}(x,y) = \sum_{m,n} a_{mn}x^m y^n$. Show that $\mathfrak{f}(x,y)$ and all its partial derivatives are differentiable at $(0,0)$.

14. Show that if $\mathfrak{f}(x, y)$ can be expanded as a Taylor series in a neighbourhood of the point (x, y) the function has a maximum at the point if

$$\frac{\partial \mathfrak{f}}{\partial x} = \frac{\partial \mathfrak{f}}{\partial y} = 0, \quad \frac{\partial^2 \mathfrak{f}}{\partial x^2} < 0, \quad \frac{\partial^2 \mathfrak{f}}{\partial x^2}\frac{\partial^2 \mathfrak{f}}{\partial y^2} > \left(\frac{\partial^2 \mathfrak{f}}{\partial x\, \partial y}\right)^2.$$

Show that if $b^2 > 4ac$, the function $e^x(ax^2 + bxy^2 + cy^4)$ has neither maximum nor minimum at the point $(0, 0)$. If in addition $bc > 0$, the function is a minimum or a maximum at $(-2, \pm(b/c)^{\frac{1}{2}})$ according as $c \gtrless 0$; if $ab > 0$ the function is a maximum or a minimum at $(-2, 0)$ according as $a \gtrless 0$.

15. Prove that

$$\int_0^\infty \frac{\cos ax - \cos bx}{x^2}\, dx = (b-a)\int_0^\infty \frac{\sin x}{x}\, dx, \quad \text{where} \quad b > a > 0.$$

16. By differentiating with respect to y and integrating by parts, prove that

$$\int_0^\infty e^{-x^2} \cos 2xy\, dx = \tfrac{1}{2}\pi^{\frac{1}{2}} e^{-y^2}.$$

(It may be assumed that $\int_0^\infty e^{-x^2}\, dx = \tfrac{1}{2}\pi^{\frac{1}{2}}$.)

17. Given that $$\int_0^\infty e^{-x^2}\, dx = \tfrac{1}{2}\pi^{\frac{1}{2}},$$

show that $$\left(\frac{2}{\pi}\right)^{\frac{1}{2}} \int_0^\infty e^{-\frac{1}{2}tx^2}\, dx = \frac{1}{t^{\frac{1}{2}}} \quad (t > 0)$$

and that $$\left(\frac{2}{\pi}\right)^{\frac{1}{2}} \int_0^\infty x^{2n} e^{-\frac{1}{2}x^2}\, dx = 1.3.5.....2n-1.$$

HINTS ON THE SOLUTION OF EXERCISES AND ANSWERS TO EXERCISES

Chapter 1

(1) Let X be an enumerable set and Y a subset of X. If $Y = \phi$ then Y is enumerable by convention. Otherwise there is a member y of Y. Since X is enumerable there is a function $\mathfrak{f}(n)$ defined over the positive integers and whose range is X. Define a new function $\mathfrak{g}$ by

$$\mathfrak{g}(n) = \mathfrak{f}(n) \quad \text{if} \quad \mathfrak{f}(n) \in Y,$$
$$\mathfrak{g}(n) = y \quad \text{if} \quad \mathfrak{f}(n) \notin Y.$$

Then the function $\mathfrak{g}$ is defined over the positive integers and its range is Y. Thus Y is enumerable.

(2) The positive rational numbers are in 1-1 correspondence with the ordered pairs of positive integers (i, j), where i and j have no common factors. This is a subset of all ordered pairs of positive integers and this last set is enumerable; for example, define $\mathfrak{f}$ by $\mathfrak{f}(n) = (i, j)$, where $i-1$ is the power of 2 dividing n and $j-1$ is the power of 3 dividing n. The positive rationals are in 1-1 correspondence with the negative rationals, thus this last set is also enumerable.

(3) For a non-negative integer i define $\mathfrak{g}$ by $\mathfrak{g}(i) = -\frac{1}{2}i$ (i even); $\mathfrak{g}(i) = \frac{1}{2}(i+1)$ (i odd). For each positive integer n define $\mathfrak{f}$ by $\mathfrak{f}(n) = (\mathfrak{g}(a_1), \ldots, \mathfrak{g}(a_k))$, where $n = 2^{a_1}.3^{a_2}.5^{a_3} \ldots p_k^{a_k}$, p_k being the kth prime*. Then $\mathbf{r}(\mathfrak{f})$ is the set of all ordered finite sets of integers, which is therefore enumerable.

(4) X can be enumerated as $x_1, x_2, \ldots, x_n, \ldots$. The finite subsets of X are in 1-1 correspondence with the finite subsets of the positive integers formed by their indices. Use questions 1 and 3. (In question 3 the finite sets are ordered, here they are not ordered but this does not matter.)

(5) Define a new member of V, say u, by $n \in u$ if and only if $n \notin v_n$.

(6) $\exists\ \mathfrak{f}$ with $\mathbf{d}(\mathfrak{f}) = \mathfrak{P}$, $\mathbf{r}(\mathfrak{f}) = X$. Define $\mathfrak{g}$ by $\mathfrak{g}(n)$ is the least integer for which $\mathfrak{f}(\mathfrak{g}(n)) = \mathfrak{f}(n)$. Let $\mathbf{r}(g) = T$. T is an infinite set of integers in 1-1 correspondence with X by $\mathfrak{f}$. Arrange T in order of increasing magnitude (any set of positive integers has a least) $t_1, t_2, t_3, \ldots, t_p, \ldots$. Define a new function $\mathfrak{h}$ by $\mathfrak{h}(p) = \mathfrak{f}(t_p)$. Then $\mathfrak{h}$ sets up a 1-1 correspondence between $\boldsymbol{P}$ and X.

(7) By question 6 if X were infinite there would be a 1-1 correspondence between X and $\boldsymbol{P}$. But then both the set corresponding to the even integers and that corresponding to the odd integers are infinite: in contradiction with the given hypotheses.

(9) Since $\mathfrak{g}(n)$ is an increasing sequence of positive integers $\mathfrak{g}(n) \geqslant n$. Either $\mathfrak{g}(n) = n$ all n or there is a least integer m for which $\mathfrak{g}(m) > m$. If m is even $\mathfrak{g}(m) = 2\mathfrak{g}(\frac{1}{2}m) = m$ (since $\frac{1}{2}m < m$). Contradiction. If m is odd $\mathfrak{g}(m+1) = 2\mathfrak{g}(\frac{1}{2}(m+1)) = m+1$ (since $\frac{1}{2}(m+1) < m$, for $\mathfrak{g}(1) = 1$ given). Thus when m is odd,

$$\mathfrak{g}(m) < \mathfrak{g}(m+1) = m+1$$

therefore $\mathfrak{g}(m) = m$. Contradiction, therefore $\mathfrak{g}(n) = n$ for all n.

* A prime integer is a positive integer whose only integral factors are itself and unity. p_k is the largest prime factor of n.

(10) Let $\mathfrak{h}$ be a subsequence of $\mathfrak{g}$ then $\mathfrak{h} = \mathfrak{g}(\mathfrak{k})$, where $\mathfrak{k}$ is a strictly increasing set of integers. But $\mathfrak{g}$ is a subsequence of $\mathfrak{f}$ thus $\mathfrak{g} = \mathfrak{f}(\mathfrak{j})$, where $\mathfrak{j}$ is a strictly increasing sequence of integers. Thus $\mathfrak{h} = \mathfrak{f}(\mathfrak{j}(\mathfrak{k}))$ and as $\mathfrak{j}(\mathfrak{k})$ is a strictly increasing sequence of integers, $\mathfrak{h}$ is a subsequence of $\mathfrak{f}$.

Chapter 2

(1) (i) Unbounded below, bounded above, $\sup X = 1$.
(ii) Bounded above and below, $\sup X = 1$, $\inf X = 0$.
(iii) Bounded above and below, $\inf X = -\sqrt{2}$, $\sup X = \sqrt{2}$.
(iv) Bounded below, unbounded above, $\inf X = 1$.
(v) Bounded above and below, $\inf X = \sup X = 0$.

(2) See the characterization of $\sup A$ in the text.

(5) $A \cup B \supset A$. Thus $\sup (A \cup B)$ is an upper bound of A. Hence $\sup (A \cup B) \geqslant \sup A$ and similarly $\sup (A \cup B) \geqslant \sup B$. If $a \in A \cup B$ then $a \in A$ or $a \in B$. Thus $a \leqslant \sup A$ or $a \leqslant \sup B$ and in either case $a \leqslant \max (\sup A, \sup B)$. Hence $\sup (A \cup B) = \max (\sup A, \sup B)$.

Chapter 3

(1) If $a_0 + \ldots + a_n x^n$ is bounded over $x \geqslant 0$ by say M, then

$$\left| x^n \left(a_n + \frac{a_{n-1}}{x} + \ldots + \frac{a_0}{x^n} \right) \right| < M.$$

If
$$x > \max \left(1, \frac{|a_0| + \ldots + |a_{n-1}|}{2\,|a_n|} \right)$$

then $$\left| a_n + \frac{a_{n-1}}{x} + \ldots + \frac{a_0}{x^n} \right| \geqslant |a_n| - \left| \frac{a_{n-1}}{x} \right| - \ldots - \left| \frac{a_0}{x^n} \right| > \tfrac{1}{2}\,|a_n|$$

and thus $|x^n| < 2M/|a_n|$. But this is impossible since we can take x as large as we please.

(2) $|\mathfrak{fg}| < M\,|\mathfrak{g}|$; $|\mathfrak{g}| < |\mathfrak{fg}|/\delta$.

(4) For each integer n $\exists$ a point say a_n' of A such that

$$a_n' > \sup A - 1/n$$

and another point a_n'' such that $a_n'' < \inf A + 1/n$. Take B to be the union of all the points a_n' and a_n''.

(5) $n + n^2 \geqslant 2$ for $n \geqslant 1$.

(6)
$$a_n = 1/(n+1) \quad (n \text{ odd}),$$
$$= 1 - 1/n \quad (n \text{ even}).$$

(7) $$a_n = 0 \quad \text{if} \quad n = 3k \quad (k \text{ an integer})$$
$$= 1 \quad \text{if} \quad n = 3k+1$$
$$= -1 \quad \text{if} \quad n = 3k+2,$$
$$b_n = a_{n+1}.$$

(8) $a_n \leqslant \sup a_n$, $b_n \leqslant \sup b_n$, therefore $a_n b_n \leqslant \sup a_n . \sup b_n$. Thus $\sup a_n . \sup b_n$ is an upper bound of $a_n b_n$ and thus greater than or equal to the least upper bound of $\{a_n b_n\}$, $\sup a_n b_n$.

(10) $a_n \geqslant \inf a_n$ thus $1/a_n \leqslant 1/\inf a_n$. Hence $\sup (1/a_n) \leqslant 1/\inf a_n$. Similarly $1/a_n \leqslant \sup (1/a_n)$ thus

$$a_n \geqslant 1/(\sup (1/a_n)) \quad \text{and} \quad \inf a_n \geqslant 1/(\sup (1/a_n)).$$

(11) $a_n \leqslant \sup a_n$. Thus $(a_1 + \ldots + a_n)/n \leqslant \sup a_n$.

(14) $a_n = 2$, $b_n = 1$, then $a_n = O(b_n)$, but $a_1 \ldots a_n \neq O(b_1 \ldots b_n)$. $a_1 = 1, a_{2n} = 1/n, a_{2n+1} = n, b_n = 1$, then $a_1 \ldots a_n = O(b_1 \ldots b_n)$ but $a_n \neq O(b_n)$.

Chapter 4

(1) $0 < (n+1)/(n^2-n+1) < 4/n$,

$$1 > (n^4-n)/(n^4+n^2)$$
$$= (1-1/n^3)(1+1/n^2)^{-1} > (1-1/n^3)(1-1/n^2),$$
$$(n+1)^{\frac{1}{2}} - n^{\frac{1}{2}} = 1/((n+1)^{\frac{1}{2}} + n^{\frac{1}{2}}).$$

(2) $$|a_n - l| \leqslant |\mathscr{R}(a_n) - \mathscr{R}(l)| + |\mathscr{I}(a_n) - \mathscr{I}(l)|,$$
$$|\mathscr{R}(a_n) - \mathscr{R}(l)| \leqslant |a_n - l|.$$

(5) $\forall \epsilon > 0 \, \exists N$ such that $|1/a_n| < \epsilon$ for all $n > N$. But $a_n > 0$. Thus $0 < 1/a_n < \epsilon$ for all $n > N$, i.e. $a_n > 1/\epsilon$ for all $n > N$.

(6) $\forall \epsilon > 0 \, \exists N$ such that $|a_n - b_n| < \epsilon$ for all $n > N$. But $a_n - b_n > 0$. Thus $a_n < b_n + \epsilon < \epsilon$ for all $n > N$ (since $b_n < 0$). Hence since $a_n > 0$, $|a_n| < \epsilon$ for $n > N$, i.e. $a_n \to 0$ as $n \to \infty$.

(8) If the statement in the question is false then $\exists$ an integer M such that if n is any integer $\geqslant M$ then $\exists$ an integer $\mathfrak{k}(n)$ such that $\mathfrak{k}(n) > n$ and $a_{\mathfrak{k}(n)} > a_n$. Let $n_1 = \mathfrak{k}(M)$ and $n_i = \mathfrak{k}(n_{i-1})$ $(i = 2, 3, \ldots)$. Then $a_{n_i} \geqslant a_M > 0$. Thus $a_{n_i} \nrightarrow 0$. But $\{a_n\}$ is a null sequence and

all its subsequences must tend to zero. Thus $a_{n_i} \to 0$. This contradiction establishes the required result.

(9) Write $b_1 = a_1$, $b_n = a_n - a_{n-1}$ $(n = 2, \ldots)$.

(15) $a_n = 1/n$, $b_n = 1/n^2$ gives $a_n - b_n = o(1)$ but $a_n \sim b_n$ is false.

(16) $a_n = n$, $b_n = n+1$.

(20) $$\frac{1}{N}\sum_1^N (a_{n+2} - 2a_{n+1} + a_n) = (a_{N+2} - a_{N+1} - a_2 + a_1)/N.$$

By worked example (iii)

$$\frac{a_{N+2} - a_{N+1}}{N} \to k.$$

Also $$\frac{N}{N+2} \to 1.$$

Thus $$\frac{a_{N+2} - a_{N+1}}{N+2} \to k \quad \text{as} \quad N \to \infty.$$

Hence $$\frac{a_{N+2}}{N+2} - \frac{a_{N+1}}{N+1} + a_{N+1}\left(\frac{1}{N+1} - \frac{1}{N+2}\right) \to k$$

and if $a_N = o(N^2)$ this means that $a_N/N^2 \to k$ by question 9 or worked example (iii) again. But $a_N = o(N^2)$, therefore $k = 0$.

Chapter 5

(1) If $a_{n+1} \geqslant a_n$, we have to show that

$$\frac{a_1 + a_2 + \ldots + a_{n+1}}{n+1} \geqslant \frac{a_1 + a_2 + \ldots + a_n}{n},$$

i.e. $na_{n+1} \geqslant a_1 + a_2 + \ldots + a_n$, and this follows from the fact that $a_{n+1} \geqslant a_n \geqslant a_{n-1} \geqslant \ldots \geqslant a_1$.

(2) Assume that $a_n > 0$, $b_n > 0$ then $a_{n+1} > 0$, $b_{n+1} > 0$. Thus by induction $a_n > 0$, $b_n > 0$ for all n.

$$\begin{aligned} \text{(i)}\quad a_{n+1} - b_{n+1} &= \tfrac{1}{2}(a_n^{\frac{1}{2}} - b_n^{\frac{1}{2}})^2 > 0, \quad \therefore\ a_n > b_n,\\ a_{n+1} - a_n &= \tfrac{1}{2}(b_n - a_n) < 0, \quad \therefore\ a_n \downarrow,\\ b_{n+1} - b_n &= b_n^{\frac{1}{2}}(a_n^{\frac{1}{2}} - b_n^{\frac{1}{2}}) > 0, \quad \therefore\ b_n \uparrow, \end{aligned}$$

therefore $a_n \to l$, $b_n \to m$. But from the second relation above

$$b_n - a_n = 2(a_{n+1} - a_n) \to 0, \quad \therefore\ l = m.$$

(3) $a_n - a_{n-1} \leqslant a_{n+1} - a_n$, therefore $\{a_n - a_{n-1}\}\uparrow$. But bounded, therefore $a_n - a_{n-1} \to k$ as $n \to \infty$. If $k \neq 0$ suppose $k > 0$ (other case similar), then $a_n - a_{n-1} > \frac{1}{2}k$ for $n \geqslant N$ where N is an appropriate large integer. Thus $a_M - a_{N-1} = \sum_N^M (a_n - a_{n-1}) > \frac{1}{2}(M-N)k$. But this contradicts the boundedness of $\{a_n\}$. Thus $k \leqslant 0$. Similarly $k \geqslant 0$. Hence $k = 0$.

(4) Sum as a geometric progression.

(5) Approximate to α and β by rational sequences.

(6) $a_n^p - l^p = (a_n - l)(a_n^{p-1} + a_n^{p-2}l + \ldots + l^{p-1})$. As $a_n \to l$, $\{a_n\}$ is bounded and thus $\exists K$ such that $|a_n^p - l^p| < K|a_n - l|$. Hence $a_n^p \to l^p$ as $n \to \infty$.

$$a_n - l = (a_n^{1/p} - l^{1/p})(a_n^{(p-1)/p} + a_n^{(p-2)/p}l^{1/p} + \ldots + l^{(p-1)/p}) \quad (a_n > 0),$$

it follows that

$$a_n^{(p-1)/p} + a_n^{(p-2)/p}l^{1/p} + \ldots + l^{(p-1)/p} > l^{(p-1)/p}$$

and thus $$|a_n^{1/p} - l^{1/p}| < |a_n - l|/l^{(p-1)/p}.$$

(10) (i) $\{(1+n^{-2})^{n^2}\}$ is a subsequence of $\{(1+n^{-1})^n\}$.

(ii) $(1+n^{-1})^n \to e > 2$. Thus $(1+n^{-1})^{n^2} > 2^n$ for large n.

(iii) $(1+n^{-2})^{n^2} < 3$ for large n. Thus $1 \leqslant (1+n^{-2})^n < 3^{1/n}$ for large n.

(iv) Write $a_n = (1+\theta_n)/n$. Then $\theta_n \to 0$ as $n \to \infty$ and $1 + (a_n/n)$ lies between $1 + (1/n)$ and $(1+(1/n))(1+(\theta_n/n))$. By an argument similar to that in worked example (iv)

$$\left|1 - \left(1 + \frac{\theta_n}{n}\right)^n\right| < e|\theta_n| \to 0 \text{ as } n \to \infty.$$

(11) Write $v_n = a_n - a_{n-1}$; $u_n = b_n - b_{n-1}$; $v_1 = a_1$; $u_1 = b_1$. Then $v_n/u_n \to l$ as $n \to \infty$, $u_n > 0$, $u_1 + u_2 + \ldots + u_n = b_n \to \infty$ as $n \to \infty$ and we have to show that $(v_1 + v_2 + \ldots + v_n)/(u_1 + u_2 + \ldots + u_n) \to l$. Now $\forall \epsilon > 0 \, \exists N$ such that $|(v_n/u_n) - l| < \epsilon$ for all $n > N$. Thus

$$\left|\frac{v_1 + \ldots + v_n}{u_1 + \ldots + u_n} - l\right| \leqslant \left|\frac{v_1 + \ldots + v_N}{u_1 + \ldots + u_n}\right| + \left|\frac{(v_{N+1} - lu_{N+1}) + \ldots + (v_n - lu_n)}{u_1 + \ldots + u_n}\right| + \left|\frac{l(u_1 + \ldots + u_N)}{u_1 + \ldots + u_n}\right|.$$

The first and last terms tend to zero as $n \to \infty$ and the middle one is less than

$$\epsilon\left(\frac{u_{N+1}+\ldots+u_n}{u_1+\ldots+u_n}\right) < \epsilon.$$

(12) Apply (11) with $a_n = 1^p+\ldots+n^p$, $b_n = n^{p+1}$. Then

$$\frac{a_{n+1}-a_n}{b_{n+1}-b_n} = \frac{(n+1)^p}{(n+1)^{p+1}-n^{p+1}} = \frac{(n+1)^p}{(p+1)\,n^p+O(n^{p-1})} \to \frac{1}{p+1}$$

as $n \to \infty$.

Chapter 6

(1)
$$\begin{aligned}\underline{\lim}\, a_n + \underline{\lim}\, b_n &= \lim\,(\inf_{p\geqslant 1} a_{n+p}) + \lim\,(\inf_{p\geqslant 1} b_{n+p})\\ &= \lim\,(\inf_{p\geqslant 1} a_{n+p} + \inf_{p\geqslant 1} b_{n+p})\\ &\leqslant \lim\,(\inf_{p\geqslant 1}(a_{n+p}+b_{n+p}))\\ &= \underline{\lim}\,(a_n+b_n)\end{aligned}$$

because $\inf\limits_{p\geqslant 1} a_{n+p} + \inf\limits_{p\geqslant 1} b_{n+p}$ is less than or equal to $a_{n+p}+b_{n+p}$ for every $p \geqslant 1$. Thus it is a lower bound of $a_{n+p}+b_{n+p}$ $(p \geqslant 1)$ and must be less than or equal to the greatest of these lower bounds, i.e. $\inf\limits_{p\geqslant 1}(a_{n+p}+b_{n+p})$. Next $\forall\, \epsilon > 0\, \exists\, q$ such that $a_{n+q} \leqslant \inf\limits_{p\geqslant 1} a_{n+p} + \epsilon$ and thus

$$\begin{aligned}\inf_{p\geqslant 1}(a_{n+p}+b_{n+p}) &\leqslant a_{n+q}+b_{n+q}\\ &\leqslant \inf_{p\geqslant 1} a_{n+p} + \epsilon + \sup_{p\geqslant 1} b_{n+p}.\end{aligned}$$

This is true for every $\epsilon > 0$. Hence

$$\inf_{p\geqslant 1}(a_{n+p}+b_{n+p}) \leqslant \inf_{p\geqslant 1} a_{n+p} + \sup_{p\geqslant 1} b_{n+p}.$$

and this by an argument similar to the above implies that

$$\underline{\lim}\,(a_n+b_n) \leqslant \underline{\lim}\, a_n + \overline{\lim}\, b_n.$$

The other inequalities are proved similarly. To prove that $\underline{\lim}\,(a_n - a_{n+1}) \leqslant 0$ apply the inequality

$$\underline{\lim}\,(a_n+b_n) \leqslant \underline{\lim}\, a_n + \overline{\lim}\, b_n$$

with b_n replaced by $-a_{n+1}$ and use the fact that

$$\overline{\lim}\,(-a_{n+1}) = \overline{\lim}\,(-a_n) = -\underline{\lim}\, a_n.$$

(2) Take

$$a_{3n} = 1, \quad a_{3n-1} = -1, \quad a_{3n-2} = 0 \quad (n = 1, 2, \ldots),$$
$$b_{3n} = -1, \quad b_{3n-1} = 0, \quad b_{3n-2} = 1 \quad (n = 1, 2, \ldots).$$

(3) Use $$(\inf_{p\geqslant 1} a_{n+p})(\inf_{p\geqslant 1} b_{n+p}) \leqslant \inf_{p\geqslant 1}(a_{n+p}b_{n+p})$$
$$\leqslant \inf_{p\geqslant 1}(a_{n+p})(\sup_{p\geqslant 1} b_{n+p}).$$

(4) Take

$$a_{3n} = 1, \quad a_{3n-1} = 2, \quad a_{3n-2} = -2 \quad (n = 1, 2, \ldots),$$
$$b_{3n} = 2, \quad b_{3n-1} = -2, \quad b_{3n-2} = 1 \quad (n = 1, 2, \ldots).$$

(5) (*a*) Take

$$a_{2n} = 1, \quad a_{2n-1} = 2, \quad (n = 1, 2, \ldots).$$

(*b*) Take

$$a_{3n} = 1, \quad a_{3n-1} = 2, \quad a_{3n-2} = 2 \quad (n = 1, 2, \ldots).$$

If (*a*) and (*b*) are both satisfied then

$$a_n . a_{n+1} \to l, \quad a_n . a_{n+1} . a_{n+2} \to m.$$

Also $$l \geqslant (\underline{\lim}\, a_n)^2 > 0, \quad \therefore\; a_{n+2} = \frac{a_n a_{n+1} a_{n+2}}{a_n a_{n+1}} \to \frac{m}{l}.$$

Thus $\overline{\lim}\, a_n = \underline{\lim}\, a_n$ in contradiction with the given hypotheses.

(6) If $|a_n| \to 0$ then $a_n \to 0$ and thus $\overline{\lim}\, a_n = \underline{\lim}\, a_n$. Hence $l \neq 0$. Show that $\overline{\lim}\, a_n = |l|$ and $\underline{\lim}\, a_n = -|l|$.

(7) Since $$|a_n - z| \leqslant |z - w| + |a_n - w|,$$
$$\overline{\lim}\, |a_n - z| \leqslant \overline{\lim}\, (|z - w| + |a_n - w|)$$
$$= |z - w| + \overline{\lim}\, |a_n - w|.$$

Similarly, since $$|a_n - z| \geqslant |z - w| - |z_n - w|,$$
$$\underline{\lim}\, |a_n - z| \geqslant \underline{\lim}\, (|z - w| - |a_n - w|)$$
$$= |z - w| + \underline{\lim}\, (-|a_n - w|)$$
$$= |z - w| - \overline{\lim}\, |a_n - w|.$$

(8) $|z_n| - |a_n| \leqslant |a_n + z_n| \leqslant |z_n| + |a_n|$. As $n \to \infty$ $|z_n| \to M$ and $|a_n| \to 0$. Thus both $|z_n| + |a_n| \to M$ and $|z_n| - |a_n| \to M$. Hence $|a_n + z_n| \to M$ as $n \to \infty$.

(9) By (3) with a_n and b_n interchanged,

$$\underline{\lim}\, b_n . \overline{\lim}\, a_n \leqslant \overline{\lim}\, a_n b_n \leqslant \overline{\lim}\, b_n . \overline{\lim}\, a_n.$$

Since $\{b_n\}$ converges $\lim b_n = \overline{\lim}\, b_n = \underline{\lim}\, b_n$.

(10) Use (1).

(11) If $\overline{\lim}\, b_n - \underline{\lim}\, b_n = \chi > 0$ there is an infinite set of integers J such that $b_n < \overline{\lim}\, b_n - \frac{1}{2}\chi$ if $n \in J$. Define a_n by $a_n = \frac{1}{2}\chi$ if $n \in J$, $a_n = 0$ if $n \notin J$. Then $\overline{\lim}\, (a_n + b_n) = \overline{\lim}\, b_n$ and $\overline{\lim}\, a_n = \frac{1}{2}\chi > 0$. The conditions stated in the question are not fulfilled. Hence

$$\overline{\lim}\, b_n - \underline{\lim}\, b_n \leqslant 0.$$

But in all cases $\overline{\lim}\, b_n - \underline{\lim}\, b_n \geqslant 0$, thus $\overline{\lim}\, b_n = \underline{\lim}\, b_n$ and the sequence $\{b_n\}$ is convergent.

(12) Use an argument similar to (11).

(13) $$a_{3n} = 1, \quad a_{3n-1} = a_{3n-2} = \frac{1}{n} \quad (n = 1, 2, \ldots),$$

$$b_{3n} = \frac{1}{n}, \quad b_{3n-1} = 1, \quad b_{3n-2} = \frac{1}{n} \quad (n = 1, 2, \ldots),$$

$$c_{3n} = c_{3n-1} = \frac{1}{n}, \quad c_{3n-2} = 1.$$

(14) $\forall \epsilon > 0 \, \exists N$ such that $a_n > \underline{\lim}\, a_n - \epsilon$ for all $n > N$. Thus

$$\frac{a_1 + \ldots + a_n}{n} > (\underline{\lim}\, a_n - \epsilon)\frac{n-N}{n} + \frac{a_1 + \ldots + a_N}{n} \quad \text{for} \quad n > N$$

and $$\underline{\lim}\, \frac{a_1 + \ldots + a_n}{n} \geqslant (\underline{\lim}\, a_n - \epsilon) . \lim \frac{n-N}{n}$$

$$= \underline{\lim}\, a_n - \epsilon.$$

True all $\epsilon > 0$. Hence result.

The relation $\underline{\lim}\, (1/a_n) = 1/(\overline{\lim}\, a_n)$ may be proved by an argument similar to the above or deduced from (3) with $b_n = 1/a_n$.

(15) Either $a_n = 0$ for infinitely many n in which case the subsequence formed by these n gives us a decreasing subsequence of

$\{a_n\}$ or $\exists\, N$ such that $a_n > 0$ for $n > N$. In this second case select $n_1 > N$. When $n_1, \ldots, n_k$ have been selected so that

$$a_{n_1} > a_{n_2} > \ldots > a_{n_k}$$

choose $n_{k+1} > n_k$ so that $a_{n_{k+1}} < a_{n_k}$. This is possible because $a_n > 0$ for $n > N$ and $\underline{\lim}\, a_n = 0$.

Let $\{a_n\}$ be a bounded real sequence. If $a_n \geqslant \underline{\lim}\, a_n$ for infinitely many n apply the result of the first part to this subsequence of $\{a_n - \underline{\lim}\, a_n\}$. Otherwise $\exists\, N$ such that $a_n < \underline{\lim}\, a_n$ for $n > N$. But this means that $\overline{\lim}\, a_n \leqslant \underline{\lim}\, a_n$. Hence $\overline{\lim}\, a_n = \underline{\lim}\, a_n$ and the sequence $\{a_n\}$ converges.

(16) By (15) we can find a convergent sequence whose members belong to A. The limit of this sequence has the properties required of λ.

(17) Since $b_n - a_n \geqslant 0$, $\underline{\lim}\,(b_n - a_n) \geqslant 0$, i.e. $\underline{\lim}\, b_n - \overline{\lim}\, a_n \geqslant 0$ and as b_n is convergent, $\lim b_n \geqslant \overline{\lim}\, a_n$. Thus

$$\inf_B (\lim b_n) \geqslant \overline{\lim}\, a_n.$$

Next let $\{\epsilon_i\}$ be a positive null sequence and N_i such that

$$a_n < \overline{\lim}\, a_n + \epsilon_i \quad \text{for} \quad n > N_i.$$

We can define the N_i so that $N_{i+1} > N_i$. Define b_n by

$$b_n = \overline{\lim}\, a_n + \epsilon_i \quad \text{for} \quad N_i < n \leqslant N_{i+1} \quad (i = 1, \ldots)$$

and $b_n = a_n$ for $n = 1, 2, \ldots, N_1$. Then $b_n \geqslant a_n$ and $b_n \to \overline{\lim}\, a_n$. Thus $\inf_B (\lim b_n) \leqslant \overline{\lim}\, a_n$.

(18) $$\overline{\lim}\, a_n = \max_{j=1\ldots k} (\overline{\lim}\, a_n^{(j)}) \leqslant a_m \quad (m = 1, 2, \ldots).$$

Thus $\overline{\lim}\, a_n \leqslant \underline{\lim}\, a_n$.

(19) $$a_{ql+r} < qa_l + a_r, \quad \frac{a_{ql+r}}{ql+r} < \frac{qa_l + a_r}{ql+r} \to \frac{a_l}{l} \quad \text{as} \quad q \to \infty.$$

$$\therefore \overline{\lim_{q\to\infty}}\, \frac{a_{ql+r}}{ql+r} \leqslant \frac{a_l}{l} \quad (r = 0, \ldots, l-1).$$

As in (18)

$$\overline{\lim}\, \frac{a_n}{n} = \max_r \left(\overline{\lim_{q\to\infty}}\, \frac{a_{ql+r}}{ql+r} \right) \leqslant \frac{a_l}{l} \quad \text{and} \quad \left\{\frac{a_n}{n}\right\} \text{ is convergent.}$$

(20) By (3) $\overline{\lim}\,(a_n \,.\, a_{n+p_n}) \geqslant \overline{\lim}\, a_n \,.\, \underline{\lim}\, a_{n+p_n}$

$$\geqslant \overline{\lim}\, a_n \,.\, \underline{\lim}\, a_n,$$

$$\underline{\lim}\,(a_n \,.\, a_{n+p_n}) \leqslant \overline{\lim}\, a_{n+p_n} \,.\, \underline{\lim}\, a_n$$

$$\leqslant \overline{\lim}\, a_n \,.\, \underline{\lim}\, a_n,$$

but $\{a_n \,.\, a_{n+p_n}\}$ converges. Hence result.

Chapter 7

(1) If Σa_n and $\Sigma(a_n + b_n)$ converge so does $\Sigma((a_n+b_n)-a_n) = \Sigma b_n$. Contradiction. If Σc_n converges then $\sum_1^{2n} c_r = \sum_1^n a_r + \sum_1^n b_r$, tends to a finite limit as $n \to \infty$. But Σa_n converges and thus $\sum_1^n a_r$ tends to a finite limit. Hence so also does $\sum_1^n b_r$. But this is in contradiction with the fact that Σb_n diverges.

(3) $b_n = (-1)^n$.

(5) Since Σa_n converges absolutely $|a_n| \to 0$ as $n \to \infty$. Thus for sufficiently large n,

$$|a_n| < \tfrac{1}{2} \quad \text{and} \quad |b_n| = \left|\frac{a_n}{1+a_n}\right| \leqslant 2\,|a_n|.$$

(6) Since $\Sigma\,|a_n|$ converges $a_n \to 0$ and thus for large n, $|a_n|^2 < |a_n|$. Hence $\Sigma\,|a_n|^2$ converges and thus Σa_n^2 converges. Further since $a_n \to 0$ $|a_n|/|1-a_n| < 2\,|a_n|$ for large n (i.e. n so large that $|a_n| < \tfrac{1}{2}$).

(7) (i) False. Define a_n and b_n by

$$a_{2n} = -a_{2n-1} = 1/n^{\frac{1}{2}} \quad (n = 1, 2, \ldots),$$

$$b_{2n} = 1/n^{\frac{1}{2}}, \quad b_{2n-1} = -1/n^{\frac{1}{2}} + 1/n^{\frac{2}{3}} \quad (n = 1, 2, \ldots).$$

Then $\sum_1^{2n} a_n = 0$ and $\sum_1^{2n+1} a_n = -1/(n+1)^{\frac{1}{2}}$. Thus $\sum_1^N a_n \to 0$ as $N \to \infty$, hence Σa_n converges.

$$\text{But } \sum_1^{2N} b_n = \sum_1^N 1/n^{\frac{2}{3}} > N\,.\,(1/N^{\frac{2}{3}}) = N^{\frac{1}{3}} \to \infty \quad \text{as} \quad N \to \infty.$$

Hence it follows that Σb_n cannot converge i.e. Σb_n diverges.

But $$b_n/a_n = \to 1 \text{ as } n \to \infty,$$

therefore $a_n \sim b_n$.

(ii) False. Define a_n by: $a_{2n} = (\frac{1}{2})^n$, $a_{2n-1} = (\frac{1}{3})^n$. $\Sigma(\frac{1}{2})^n$ and $\Sigma(\frac{1}{3})^n$ both converge and thus $\sum_1^N a_n$ is bounded above. By theorem 7, Σa_n converges. But $a_{2n}/a_{2n-1} = (\frac{3}{2})^n \geqslant \frac{3}{2} > 1$ for all $n = 1, 2, \ldots$. Thus, for example, $a_{n+1}/a_n > k > 1$ for infinitely many integers n where say $k = \frac{5}{4}$.

(iii) False. Define a_n by $a_n = 1/n$.

(8) (i) In place of $\Sigma(-1)^n/(n+x)$ consider the series

$$\Sigma\left(\frac{-1}{2n-1+x} + \frac{1}{2n+x}\right)$$

obtained by bracketing together pairs of consecutive summands in the given series. The sequence of partial sums of this series forms a subsequence of the sequence of partial sums of the original series. If the original series converges then so also does this second series. We should like to use the converse implication but this converse is only true because the series satisfy a certain additional condition.

First, consider

$$\sum_1^N \left(\frac{1}{2n-1+x} - \frac{1}{2n+x}\right) = \sum_1^N \frac{1}{(2n+x)(2n-1+x)}.$$

Now for n sufficiently large $1/(2n+x)(2n-1+x) > 0$ and since $1/(2n+x)(2n-1+x) = O(1/n^2)$ and $\Sigma(1/n^2)$ converges, it follows from the comparison principle that

$$\Sigma\left(\frac{1}{2n-1+x} - \frac{1}{2n+x}\right)$$

converges. Thus

$$\sum_1^{2N} \frac{(-1)^n}{n+x}$$

tends to a finite limit as $N \to \infty$. Since

$$\sum_1^{2N-1} \frac{(-1)^n}{n+x} - \sum_1^{2N} \frac{(-1)^n}{n+x} = -\frac{1}{2N+x} \to 0,$$

it follows that $$\sum_1^{2N-1} \frac{(-1)^n}{n+x}$$

tends to the same finite limit as $N \to \infty$. Thus the sequence of partial sums of $\Sigma[(-1)^n/(n+x)]$ is made up of two subsequences each tending to the same limit. Hence the whole sequence of partial sums of $\Sigma[(-1)^n/(n+x)]$ tends to this finite limit, i.e. $\Sigma[(-1)^n/(n+x)]$ converges. The assumption that x is not a negative integer is merely to ensure that each summand is defined. It has no inherent connexion with the convergence of the series.

(ii) Consider the series

$$\Sigma\left(\frac{a_{3n-2}}{3n-2+x}+\frac{a_{3n-1}}{3n-1+x}+\frac{a_{3n}}{3n+x}\right).$$

(9) $2^k a_{2^{k+1}} \leqslant \sum_{2^k+1}^{2^{k+1}} a_n \leqslant 2^k a_{2^k}$. Use theorem 7.

Write,

$$a_n = \frac{1}{n}\left(1+\tfrac{1}{2}+\ldots+\frac{1}{n}\right)^{-1}, \quad b_n = 2^n a_{2^n}.$$

Now $\{a_n\}\downarrow$ and the first part of the question can be applied. Then by the given inequality, $b_n > 1/(n+2^{-n}) > 1/2n$. Thus Σb_n diverges and hence Σa_n diverges.

(10) (ii) $\exists\, N$ such that $a_n > \frac{1}{2}(\mu+1)\,a_{n+1}$ if $n > N$. Thus

$$a_n < a_N/(\tfrac{1}{2}(\mu+1))^{n-N} \quad \text{and} \quad \overline{\lim}\, a_n^{1/n} \leqslant 1/\tfrac{1}{2}(\mu+1) < 1.$$

(11) $a_n^{\frac{1}{2}} b_n^{\frac{1}{2}} \leqslant \frac{1}{2}(a_n+b_n)$.

(12) Take λ satisfying $\overline{\lim}\,(a_{n+r}/a_n) < \lambda < 1$. $\exists\, N$ such that $a_{n+r}/a_n < \frac{1}{2}(\lambda+1)$ for $n \geqslant N$. Thus,

$$a_n < \tfrac{1}{2}(\lambda+1)\,a_{n-r} < [\tfrac{1}{2}(\lambda+1)]^2\,a_{n-2r}\ldots.$$

Let $K = \max_{N \leqslant n < N+r} a_n$. Then $a_n < [\frac{1}{2}(\lambda+1)]^p . K$, where p is the integral part of $(n-N)/r$. Thus $a_n < [\frac{1}{2}(\lambda+1)]^{n/r} . M$ where M is a constant independent of n.

(13) (i) If $\overline{\lim}\, a_n > 0\, \exists\, \delta > 0$ such that $a_n > \delta$ for infinitely many n. But then $a_n/(1+a_n) > \delta/(1+\delta)$ for infinitely many n and $\Sigma a_n/(1+a_n)$ diverges. If $\overline{\lim}\, a_n = 0$, $\exists\, N$ such that $a_n < 1$ for $n > N$. Then $a_n/(1+a_n) > \frac{1}{2}a_n$ for $n > N$ and $\Sigma a_n/(1+a_n)$ diverges.

(ii) $\Sigma a_n/(1+na_n)$ may either converge or diverge. Suppose, for example, that $a_n = 1$, then $a_n/(1+a_n) = 1/(1+n)$ and the series diverges. If, however, $a_n = 1$ when n is a perfect square and

$a_n = 1/n^2$ when n is not a perfect square, then Σa_n diverges but $\Sigma a_n/(1+na_n)$ converges. For we have

$$\frac{a_n}{1+na_n} = \frac{1}{1+n} < \frac{1}{n} \quad (n \text{ is a perfect square}),$$

$$\frac{a_n}{1+na_n} < a_n = \frac{1}{n^2} \quad (n \text{ not a perfect square}),$$

thus
$$\sum_1^N \frac{a_n}{1+na_n} < 2\sum_1^\infty \frac{1}{n^2}.$$

(iii)
$$\frac{a_n}{1+n^2a_n} < \frac{a_n}{n^2a_n} = \frac{1}{n^2}, \quad \Sigma\frac{1}{n^2} \text{ converges.}$$

(iv) If $M > a_n > \delta > 0$ for infinitely many n, then

$$\frac{a_n}{1+a_n^2} > \min\left(\frac{M}{1+M^2}, \frac{\delta}{1+\delta^2}\right)$$

for infinitely many n and the series diverges.

If $a_n \to 0$ then $\exists\, N$ such that $a_n/(1+a_n^2) > \frac{1}{2}a_n$ for $n > N$ and the series diverges.

(14) $(a_n a_{n+1})^{\frac{1}{2}} \leqslant \frac{1}{2}(a_n + a_{n+1})$. $\quad (a_n n^{-1-\delta})^{\frac{1}{2}} \leqslant \frac{1}{2}(a_n + n^{-1-\delta})$.

(15) $a_{n+1} \leqslant (a_n a_{n+1})^{\frac{1}{2}}$. Take $b_{2n} = \dfrac{1}{2n}$, $b_{2n-1} = \dfrac{1}{(2n-1)^2}$.

(16) $\forall \epsilon > 0 \,\exists\, N$ such that $\sum\limits_N^\infty a_r < \epsilon$. Then if $n > N$

$$\begin{aligned} 0 &\leqslant \frac{p_1a_1 + \ldots + p_na_n}{p_n} \\ &\leqslant \frac{p_1a_1 + \ldots + p_Na_N}{p_n} + a_{N+1} + \ldots + a_n = O\left(\frac{1}{p_n}\right) + \epsilon. \end{aligned}$$

(17) If Σa_n diverges, let $n_1 = 1$ and n_2, n_3, ... be such that $\sum\limits_{n_i}^{n_{i+1}-1} a_r > 1$. Define p_n by $p_n = i$ for $n_i \leqslant n < n_{i+1}$. Then

$$\frac{p_1a_1 + \ldots + p_{n_{i+1}-1}a_{n_{i+1}-1}}{p_{n_{i+1}-1}} > 1.$$

(18) Write $S_N = \sum\limits_1^N a_n$, then we are given that

$$S_N - N(S_N - S_{N-1}) \leqslant K,$$

for some appropriate K. Thus, $NS_{N-1} - (N-1)S_N \leqslant K$, i.e.

$$\frac{S_{N-1}}{N-1} - \frac{S_N}{N} \leqslant \frac{K}{N.(N-1)}.$$

As $N \to \infty$, $a_N \to 0$

$$\therefore \quad \frac{S_N}{N} = \frac{a_1 + \ldots + a_N}{N} \to 0, \quad \therefore \frac{S_{N-1}}{N-1} \leqslant K \sum_N^\infty \frac{1}{n(n-1)} = \frac{K}{N-1}.$$

Thus $S_{N-1} \leqslant K$ and since $a_n \geqslant 0$ this implies that Σa_n converges.

(19) Since Σa_n converges there exists a strictly increasing sequence of integers $\{N_k\}$ such that $N_1 = 1$ and $\sum_{N_k}^\infty a_n < \frac{1}{2^k}$. Define b_n by $b_n = k a_n$ for $N_k \leqslant n < N_{k+1}$.

Chapter 8

(1) Using the notation of theorem 8 (c) and (*)

$$\overline{\lim} \sum_1^N a_n b_n \leqslant \overline{\lim} \sum_1^N s_n(b_n - b_{n+1}) + \overline{\lim}\, s_N b_{N+1}.$$

If $\sum_1^N a_n < K$ for all N then

$$\sum_1^N s_n(b_n - b_{n+1}) < K \sum_1^N |b_n - b_{n+1}| < K \sum_1^\infty |b_n - b_{n+1}|$$

and $$\overline{\lim}\, s_N b_{N+1} \leqslant K . \lim |b_{N+1}| = K \left| b_1 - \sum_1^\infty (b_n - b_{n+1}) \right|$$

and these imply $$\overline{\lim} \sum_1^N a_n b_n < \infty.$$

(2) By theorem 8 (c) $\Sigma a_{nr+l} b_{nr+l}$ converges for $l = 1, 2, \ldots, r$.

(3) Theorem 8 (a) with a_n replaced by $n^{\frac{1}{2}} a_n$ and b_n by $n^{-\frac{1}{2}}$.

(4) (i) $(n^2+1)^{\frac{1}{2}} - n = 1/[(n^2+1)^{\frac{1}{2}} + n] \sim 1/2n$. Not absolutely convergent. Conditionally convergent by alternating series test.

(ii) $(2n + (-1)^{n+1})^{-\frac{1}{2}} \sim 1/(2n)^{\frac{1}{2}}$. Not absolutely convergent. Conditionally convergent by alternating series test.

(5) Use (*) and note that $b_n - b_{n+1} \geqslant 0$.

(6) Use (*). By conditions (iii) and (iv), $s_N b_{N+1} \to 0$ and by (iii) and (ii) $\Sigma s_n(b_n - b_{n+1})$ converges.

(7) Repeat the process by which (*) was obtained, we have

$$\sum_1^N a_n b_n = \sum_1^N s_n(b_n - b_{n+1}) + s_N b_{N+1}$$

$$= \sum_1^N (s_1 + \ldots + s_n)(b_n - 2b_{n+1} + b_{n+2})$$

$$+ (s_1 + \ldots + s_N)(b_{N+1} - b_{N+2}) + s_N b_{N+1}.$$

(8) (*a*) If Σb_n diverges, we can define a strictly increasing sequence of positive integers, $N_1, N_2, \ldots$ such that $\sum_{N_k+1}^{N_{k+1}} b_n > k$.

Write $\qquad a_n = 1 \quad \text{if} \quad n \leqslant N_1$

$$= 1/k \quad \text{if} \quad N_k < n \leqslant N_{k+1}.$$

Then $a_n \to 0$ as $n \to \infty$ and $\Sigma a_n b_n$ diverges.

(*b*) Follows from (*a*).

(*c*) If it is not true that $b_n = O(1)$ then there exists a strictly increasing sequence of integers $\{N_k\}$ such that $|b_{N_k}| > k^2$. Write

$$a_{N_k} = (x/k^2), \quad \text{where} \quad x = 1 \text{ if } b_{N_k} > 0, \quad x = -1 \text{ if } b_{N_k} < 0.$$

$$a_n = 0 \quad \text{any} \quad n \neq N_k \quad (k = 1, 2, \ldots).$$

Then Σa_n converges but $\Sigma a_n b_n$ diverges. Use (*b*).

Chapter 9

(4) (i) False. Take $u_{2n} = 0$, $u_{2n-1} = \frac{1}{2}$.

(ii) True,

$$\sum_1^N v_n = \sum_1^N u_{p_n} \leqslant \sum_1^{p_N} u_n \leqslant \sum_1^\infty u_n.$$

(iii) True that Σv_n converges. For by (ii) applied to $|u_n|$ and $|v_n|$, Σv_n is absolutely convergent. False that $\Sigma v_n \leqslant \Sigma u_n$, for example, take $u_1 = -1$, $u_n = 0$ $n > 1$, $p_n = n + 1$.

(5) In the first series consider $2N$ terms and bracket these terms in pairs, whilst in the second series, consider $3N$ terms bracketed in triplets.

(6) In the notation of worked example (ii)

$$\frac{C_1 + \ldots + C_N}{N} = \frac{A_1 B_N + \ldots + A_N B_1}{N} \to AB$$

by chapter 4, worked example (iv). The final conclusion follows from chapter 4, worked example (iii).

Chapter 10

(4) Write $$\frac{1}{i^2-j^2} \quad \text{as} \quad \frac{1}{2i}\left(\frac{1}{j+i}-\frac{1}{j-i}\right).$$
Then
$$\sum_{j=1}^{\infty} a_{ij} = -\frac{1}{2i^2} \quad \text{and} \quad \sum_{i=1}^{\infty}\left(\sum_{j=1}^{\infty} a_{ij}\right) = -\frac{1}{2}\sum_{i=1}^{\infty}\frac{1}{i^2} < 0.$$
Since $a_{ji} = -a_{ij}$, the sum by rows and the sum by columns could be equal only if both were zero for
$$\sum_{j=1}^{\infty}\left(\sum_{i=1}^{\infty}\frac{1}{i^2-j^2}\right) = -\sum_{i=1}^{\infty}\left(\sum_{j=1}^{\infty}\frac{1}{i^2-j^2}\right).$$
Alternatively evaluate the sum by columns writing
$$\frac{1}{i^2-j^2} \quad \text{as} \quad \frac{1}{2j}\left(-\frac{1}{i+j}+\frac{1}{i-j}\right).$$

(5) Write $$\frac{nb^n}{a^n-b^n} \quad \text{as} \quad \sum_{m=0}^{\infty}\frac{nb^n}{a^n}\frac{b^{nm}}{a^{nm}}$$
and consider the double series $\Sigma\Sigma i(b/a)^{ij}$. Use theorem 13.

(6) Express as the double series $\Sigma\Sigma a_{ij}$ where a_{ij} is the (ij)th summand in the double array
$$\begin{vmatrix} 1 & 0 & 0 & 0 & \cdots \\ q & q^2 & q^3 & q^4 & \cdots \\ 2q^2 & -2q^4 & 2q^6 & -2q^8 & \cdots \\ 3q^3 & 3q^6 & 3q^9 & 3q^{12} & \cdots \\ . & . & . & . & \cdots \end{vmatrix}$$
and use theorem 13.

(7) Express as a double array and use theorem 13.

(8) In each case $a_{ij} > 0$, and convergence and absolute convergence are equivalent.

(i) By theorem 13, the double series converges if and only if the sum by rows exists, i.e. if and only if the series $\sum_{j=1}^{\infty}\frac{1}{i^\alpha j^\beta}$ converges for $i = 1, 2, 3, \ldots$ and if the series $\sum_{i=1}^{\infty}\left(\sum_{j=1}^{\infty}\frac{1}{i^\alpha j^\beta}\right)$ converges. This is so if and only if $\alpha > 1$, $\beta > 1$.

(ii) Since $hi+kj \geqslant 2(hkij)^{\frac{1}{2}}$, it follows from (i) that $\Sigma\Sigma(hi+jk)^{-\alpha}$ converges if $\alpha > 2$. If $\alpha > 0$ and $i \geqslant j$, then
$$\frac{1}{(hi+kj)^\alpha} \geqslant \frac{1}{(h+k)^\alpha i^\alpha}.$$

Hence $$\sum_{j=1}^{i} \frac{1}{(hi+kj)^\alpha} \geqslant \frac{1}{(h+k)^\alpha} i^{1-\alpha}.$$

Thus $$\sum_{i=1}^{M} \sum_{j=1}^{i} \frac{1}{(hi+kj)^\alpha} \geqslant \frac{1}{(h+k)^\alpha} \sum_{i=1}^{M} i^{1-\alpha} \to \infty \quad \text{if} \quad \alpha \leqslant 2.$$

Thus divergence if $0 < \alpha \leqslant 2$. If $\alpha = 0$,

$$\sum_{i=1}^{N} \sum_{j=1}^{M} \frac{1}{(hi+kj)^\alpha} = M.N.$$

If $\alpha < 0$,

$$\sum_{i=1}^{N} \sum_{j=1}^{N} \frac{1}{(hi+kj)^\alpha} \geqslant \frac{1}{(hN+kN)^\alpha} \to \infty \quad \text{as} \quad N \to \infty.$$

Thus divergence if $\alpha \leqslant 0$.

(iii) $$\frac{1}{i^2+j^2} > \frac{1}{(i+j)^2},$$

$$\sum_{i=1}^{N} \sum_{j=1}^{N} \frac{1}{(i+j)^2} > \sum_{k=1}^{N-1} \sum_{i+j=k} \frac{1}{(i+j)^2} \geqslant \sum_{k=1}^{N-1} \frac{k-1}{k^2} \to \infty \quad \text{as} \quad N \to \infty.$$

Divergence.

(iv) Convergence, see (i) above,

$$\frac{1}{i^3+j^3} \leqslant \frac{1}{2i^{\frac{3}{2}}j^{\frac{3}{2}}}.$$

(10) Convergence and absolute convergence are the same if $x > 0$.

(11) Consider the sum by rows and the sum by columns of the double array

$$\left|\begin{matrix} \frac{a_1}{1.2} & \frac{a_1}{2.3} & \frac{a_1}{3.4} & \dots \\ 0 & \frac{2a_2}{2.3} & \frac{2a_2}{3.4} & \dots \\ 0 & 0 & \frac{3a_3}{3.4} & \dots \\ . & . & . & \dots \end{matrix}\right|.$$

(12) The first series is the sum by rows of the absolutely convergent double series $\Sigma\Sigma x^{ij}$. The second series is obtained by forming the sum of terms a_{ij} of the form $i = m$, $j \geqslant m$, or $i \geqslant m$, $j = m$ in this double series. This gives

$$x^{m^2} + 2x^{m(m+1)} + 2x^{m(m+2)} + \dots = x^{m^2}(1+2x^m+2x^{2m}+\dots) = x^{m^2}\frac{(1+x^m)}{1-x^m}$$

and it can be proved that these sums form a series which converges to the sum of the double series. The last series $\sum_{1}^{\infty} a_n x^n$ is obtained by adding together all terms x^{ij} for which $ij = n$. This process can be justified by the absolute convergence of the double series.

(13) Use the identity

$$\frac{1}{x^2} = \frac{1}{x(x+1)} + \frac{1!}{x(x+1)(x+2)} + \dots$$
$$+ \frac{(n-1)!}{x(x+1)\dots(x+n)} + \frac{n!}{x^2(x+1)\dots(x+n)}.$$

Since $x > 1$,

$$\frac{n!}{x^2(x+1)\dots(x+n)} < \frac{1}{n} \to 0 \quad \text{as} \quad n \to \infty$$

and the first result is proved.

Use the first part of the question and consider the double series

$$\Sigma\Sigma \frac{(j-1)!}{(x+i-1)(x+i)\dots(x+i+j-1)}.$$

Chapter 11

(1) Apply D'Alembert's test on p. 42 to $\Sigma|a_{n-1}z^{n-1}|$.

(3) (*a*) a_n/a_{n+1} is equal to

$$\frac{(2n)!}{(n!)^2}\frac{((n+1)!)^2}{(2(n+1))!} = \frac{(n+1)^2}{(2n+1)(2n+2)} \to \frac{1}{4}.$$

Radius of convergence is $\frac{1}{4}$.

(*b*) $a_n^{1/n}$ is equal to $r^n \to 0$, $r < 1$; $= 1$, if $r = 1$; $\to \infty$ if $r > 1$. Radius of convergence is 1 if $r = 1$. The series diverges everywhere except at $z = 0$ if $r > 1$ and converges everywhere if $r < 1$.

(*c*) a_{n-1}/a_n is equal to

$$\frac{n-1!}{n^n}\frac{(n+1)^{n+1}}{n!} = \left(1+\frac{1}{n}\right)^{n+1} \to e.$$

Radius of convergence is e.

(*d*) a_{n-1}/a_n is equal to

$$\frac{(3n)!}{2^3.4^3.\dots.(2n)^3}\frac{2^3.4^3.\dots.(2(n+1))^3}{(3(n+1))!}$$
$$= \frac{(2(n+1))^3}{(3n+1)(3n+2)(3n+3)} \to \frac{8}{27}.$$

(5) We have to show that $(p_n)^n \to 0$ as $n \to \infty$. If for an integer $p > 0$ $\frac{1}{2}p(p-1) < n \leqslant \frac{1}{2}p(p+1)$ then

$$\frac{1}{p+1} \leqslant p_n \leqslant \frac{p}{p+1}.$$

Thus

$$(p_n)^n < \left(\frac{p}{p+1}\right)^{\frac{1}{2}p(p-1)} = O\left(\left(1-\frac{1}{p}\right)^{\frac{1}{2}p^2}\right) \to 0 \quad \text{as} \quad p \to \infty.$$

(6) $0 < a_n < a_0$ thus $|a_n z^n| < a_0 |z^n|$ and $\Sigma |z^n|$ converges if $|z| < 1$.

$$\begin{aligned} |(1-z)\,\mathfrak{f}(z) - a_0| &= \left|(1-z)\sum_0^\infty a_n z^n - a_0\right| \\ &= |(a_1-a_0)z + (a_2-a_1)z^2 + \dots| \\ &\leqslant (a_0-a_1)|z| + (a_1-a_2)|z^2| + \dots \\ &< a_0 \quad \text{if} \quad |z| < 1. \end{aligned}$$

(7) $$\overline{\lim}\,(|a_n b_n^2|)^{1/n} \leqslant \overline{\lim}\,|a_n b_n|^{1/n}.\overline{\lim}\,|b_n|^{1/n}.$$

$$\overline{\lim}\,|a_n b_n|^{1/n} = \overline{\lim}\,|a_n^2 b_n^2|^{1/n} \leqslant \overline{\lim}\,|a_n b_n^2|^{1/n}.\overline{\lim}\,|a_n|^{1/n}.$$

Thus $\overline{\lim}\,|a_n b_n^2|^{1/n} = 1$.

Chapter 12

(1) (a) If $x \in \overline{X}_1\ \exists\ \{y_n\}$ such that $y_n \in X_1, y_n \to x$ as $n \to \infty, y_n \neq x$. Since $X_1 \subset X_2$, $y_n \in X_2$, therefore $x \in \overline{X}_2$. Thus $\overline{X}_1 \subset \overline{X}_2$.

(e) $X_1^0 \subset X_1 \subset \overline{X}_1$, $\mathscr{F}r(X_1) \subset \overline{X}_1$, therefore $X_1^0 \cup \mathscr{F}r X_1 \subset \overline{X}_1$. If $x \in \overline{X}_1$ then either $x \in X_1^0$ or for every $\delta > 0$, $\mathscr{C}(X_1) \cap \mathscr{U}(x, \delta) \neq \phi$. Thus $\exists\, y_n \in \mathscr{C}(X_1)$ with $y_n \to x$ as $n \to \infty$: if $x \in X_1$, $y_n \neq x$ thus $x \in \overline{\mathscr{C}(X_1)}$; if $x \in \mathscr{C}(X_1)$ then $x \in \overline{\mathscr{C}(X_1)}$. Thus either $x \in X_1^0$ or $x \in \overline{\mathscr{C}(X_1)}$. The latter alternative implies $x \in \mathscr{F}r X_1$ (since we are given $x \in \overline{X}_1$), therefore $\overline{X}_1 \subset X_1^0 \cup \mathscr{F}r X_1$.

(h) $\mathscr{F}r(\mathscr{F}r(X_1)) = \overline{\mathscr{F}r(X_1)} \cap \overline{\mathscr{C}(\mathscr{F}r(X_1))} \subset \overline{\mathscr{F}r(X_1)} = \mathscr{F}r(X_1)$ by theorem 19 (c).

(2) Let K be the intersection of all closed sets that contain X. Then since $K \supset X$, we have $\overline{K} \supset \overline{X}$. By theorem 17 (ii), K is closed, therefore $K = \overline{K}$. Thus $K \supset \overline{X}$. On the other hand $\overline{X}$ is closed, $\overline{X} \supset X$ and thus $\overline{X}$ is one of the sets whose intersection is K. Thus $\overline{X} \supset K$. Hence $K = \overline{X}$.

Let G be the union of all open sets contained in X. Then G is open by theorem 17 (i). Also X^0 is an open set contained in X.

Thus $G \supset X^0$. If $x \in G$, then x belongs to an open set say T, $T \subset X$. Since T is open $\exists\, \delta > 0$ such that $\mathscr{U}(x, \delta) \subset T$. Thus $\mathscr{U}(x, \delta) \subset X$ and $x \in X^0$. True all $x \in G$, therefore $G \subset X^0$ and finally $G = X^0$.

(3) If t is not an upper bound of X then $\exists\, x \in X$ with $x > t$ and thus for some $\delta > 0$ no point of $\mathscr{U}(t, \delta)$ is an upper bound of X. The set of points that are not upper bounds is open: thus the set of points that are upper bounds is closed (by complements).

(4) $y \in \mathscr{C}(X)$ an open set, therefore $\exists\, \delta > 0$ such that

$$\mathscr{U}(y, \delta) \subset \mathscr{C}(X).$$

Hence $\inf_{x \in X} |x - y| \geqslant \delta > 0$.

(5) (*a*) $\mathscr{F}r(X) \cap X^0 = \phi$, $X \subset X^0 \cup \mathscr{F}r(X)$, thus $X = X^0$ if and only if $X \cap \mathscr{F}r(X) = \phi$.

(*b*) If X is closed then $X = \bar{X}$ and

$$X \cap \mathscr{F}r(X) = \bar{X} \cap \mathscr{F}r(X) = \mathscr{F}r(X).$$

If $\mathscr{F}r(X) \subset X$ then $\bar{X} = X^0 \cup \mathscr{F}r(X) \subset X$. Thus $\bar{X} = X$ and X is closed.

(6) $x \in \bigcup_{i=1}^{\infty} X_i'$ implies $x \in X_i'$ for some i, therefore $\exists\, \{y_n\}$ such that $y_n \in X_i$, $y_n \to x$, $y_n \neq x$. But then $y_n \in \bigcup_{i=1}^{\infty} X_i$, therefore $x \in \left(\bigcup_{i=1}^{\infty} X_i\right)'$.

(7) If $\mathbf{R}$ were not connected consider sets X_1, X_2 such that $\mathbf{R} = X_1 \cup X_2$, $X_1 \neq \phi$, $X_2 \neq \phi$, $X_1 \cap \bar{X}_2 = \phi$, $\bar{X}_1 \cap X_2 = \phi$. Select points a, b one from each set X_1, X_2, say $a \in X_1$, $b \in X_2$, $a < b$. For the sets $Y_1 = X_1 \cap \{x \mid a \leqslant x \leqslant b\}$, $Y_2 = X_2 \cap \{x \mid a \leqslant x \leqslant b\}$ $\sup Y_1 \in \bar{Y}_1 \cap \bar{Y}_2$ and since $\sup Y_1 \in X_1 \cup X_2$ it follows that either $X_1 \cap \bar{X}_2 \neq \phi$ or $\bar{X}_1 \cap X_2 \neq \phi$. Contradiction.

(8) There are the sets of (v) together with $\mathbf{R}$ and sets of the form $\{x \mid x > a\}$, $\{x \mid x \geqslant a\}$, $\{x \mid x < a\}$, $\{x \mid x \leqslant a\}$.

(9) If $X_1 \cup X_2$ is not connected, then $X_1 \cup X_2 = Y_1 \cup Y_2$, where $Y_1 \neq \phi$, $Y_2 \neq \phi$, $Y_1 \cap \bar{Y}_2 = \bar{Y}_1 \cap Y_2 = \phi$. Consider X_1 and write $Z_1 = X_1 \cap Y_1$, $Z_2 = X_1 \cap Y_2$. Since X_1 is connected, it follows that either $Z_1 = \phi$ or $Z_2 = \phi$. Suppose that $Z_2 = \phi$. Then $X_1 \subset Y_1$. But $X_1 \cap X_2 \neq \phi$ thus $X_2 \cap Y_1 \neq \phi$. Now $X_2 \cap Y_2 \neq \phi$ since $\phi \neq Y_2$ and $Y_2 \subset X_1 \cup X_2$. But then we have a contradiction with the connectedness of the set X_2.

(10) $\bar{X}_1 \supset X_2$, $X_1 \subset X_2$ and $\bar{X}_2 \supset X_3$, $X_2 \subset X_3$. Hence by exercise 2 $\bar{X}_1 \supset \bar{X}_2 \supset X_3$, $X_1 \subset X_2 \subset X_3$.

(11) (a) If false $\exists x$ and $\delta > 0$ such that $\mathscr{U}(x, \delta)$ contains no rational point. Choose N so large that $1/N < \delta$ and let p be the largest integer such that $p/N \leqslant x$. Then $p/N \in \mathscr{U}(x, \delta)$ and it follows that (a) is true.

(b) Consider irrationals of form $p/N + \sqrt{2}$ and argue as in (a). If N is such that $1/N < \delta$ then the largest p for which $p/N + \sqrt{2} \leqslant x$ is necessarily such that $x - \delta < p/N + \sqrt{2}$.

(c) Argue as in (a).

Chapter 13

(1) Suppose that $\{a_n\}\uparrow$. Either a_n is the same number for $n > N$ and some appropriate N (in which case the sequence converges), or there are infinitely many distinct points amongst the a_n and then by the Bolzano–Weierstrass theorem there is a limit point. It can be shown that $\{a_n\}$ converges to this limit point.

(2) Let G_i $(i = 1, 2, 3, \ldots)$ be open sets with $\bigcup G_i \supset X$. Write $F_n = X \cap \mathscr{C}\left(\bigcup_{i=1}^{n} G_i\right)$, then F_n is compact and $F_n \supset F_{n+1}$. Also

$$\bigcap_{n=1}^{\infty} F_n \subset \mathscr{C}\left(\bigcup_{i=1}^{\infty} G_i\right) = \phi.$$

By Cantor's theorem $\exists N$ such that $F_N = \phi$, i.e. such that $\bigcup_{i=1}^{N} G_i \supset X$.

(3) Let X be a bounded infinite set. If $X' = \phi$ (i.e. if we assume Bolzano–Weierstrass's theorem to be false) then X and every subset of X is compact. Let $x_1, x_2, \ldots$ be a sequence of distinct points of X and let $G_N = \mathscr{C}\left(\bigcup_{i=N}^{\infty} x_i\right)$, G_N is open, $\bigcup_{N=1}^{\infty} G_N \supset X$ but for no finite M is it true that $\bigcup_{N=1}^{M} G_N \supset X$. This contradicts the Heine–Borel theorem. Thus the assumption is false and $X' \neq \phi$.

(4) $x \in \mathscr{C}(F)$ an open set, therefore $\exists \delta > 0$ such that

$$\mathscr{U}(x, \delta) \subset \mathscr{C}(F).$$

Hence $\inf_{y \in F} |x - y| \geqslant \delta > 0$. Write $\eta = \inf_{y \in F} |x - y|$. Then write

$$F_n = \{z \mid z \in F \text{ and } |x - z| \leqslant \eta + 1/n\}.$$

F_n is compact. $F_n \supset F_{n+1}$ and $F_n \neq \phi$. By Cantor's theorem

$$\bigcap_{n=1}^{\infty} F_n \neq \phi \quad \text{if} \quad z \in \bigcap_{n=1}^{\infty} F_n \quad \text{then} \quad z \in F \quad \text{and} \quad |x - z| = \eta.$$

(5) (i) $X_n = (0, 1/n)$ $(n = 1, 2, 3, \ldots)$.

(ii) $X_n = (1/n, 1)$ $(n = 1, 2, \ldots)$.

(iii) $X_{n+1} = [1/(n+1), 1/n]$ $(n = 1, 2, \ldots)$, $X_1 = [-1, 0]$.

(iv) $X_n = (n-2, n)$ $(n = 1, 2, \ldots)$.

(6) By the Heine–Borel theorem $\exists$ a finite number of points $z_1, z_2, \ldots, z_k$ such that the corresponding neighbourhoods $\mathscr{U}(z_i, \delta_i)$ $(i = 1, \ldots, k)$ cover X. Then $|\mathfrak{f}(x)|$ is less than the largest of the corresponding k values for K, i.e. $\mathfrak{f}$ is bounded over X.

Chapter 14

(1) $\forall \epsilon > 0$ and $x_0 \in \mathbf{S} \, \exists \, y_0 \in Y$ such that $|x_0 - y_0| - \epsilon \leqslant \mathfrak{f}(x_0)$. If $|x - x_0| < \epsilon$, then

$$|x - y_0| \leqslant |x - x_0| + |x_0 - y_0| \leqslant \mathfrak{f}(x_0) + 2\epsilon,$$

therefore $\mathfrak{f}(x) \leqslant \mathfrak{f}(x_0) + 2\epsilon$. But similarly $\mathfrak{f}(x_0) \leqslant \mathfrak{f}(x) + 2\epsilon$. Thus $|\mathfrak{f}(x) - \mathfrak{f}(x_0)| \leqslant 2\epsilon$, i.e. $\mathfrak{f}(x) \to \mathfrak{f}(x_0)$ as $x \to x_0$.

(2) $\mathfrak{f}(x)$ is defined since $|x - y| \leqslant |x - x_0| + \delta$. Thus

$$\mathfrak{f}(x) \leqslant |x - x_0| + \delta.$$

$\forall \epsilon > 0$ and $x_1 \in \mathbf{S} \, \exists \, y_1 \in Y$ such that $\mathfrak{f}(x_1) \leqslant |x_1 - y_1| + \epsilon$. If $|x - x_1| < \epsilon$, then

$$\mathfrak{f}(x) \geqslant |x - y_1| \geqslant |x_1 - y_1| - |x - x_1| \geqslant \mathfrak{f}(x_1) - 2\epsilon.$$

Similarly $$\mathfrak{f}(x_1) \geqslant \mathfrak{f}(x) - 2\epsilon.$$

Thus $$|\mathfrak{f}(x) - \mathfrak{f}(x_1)| < 2\epsilon, \quad \text{i.e.} \quad \mathfrak{f}(x) \to \mathfrak{f}(x_1) \quad \text{as} \quad x \to x_1.$$

(3) In $\mathbf{R}$ or $\mathbf{Z}$ as the case may be $\forall \epsilon > 0$ $\mathscr{U}(\mathfrak{f}(y), \epsilon)$ is open. Thus y belongs to the open set $\mathfrak{f}^{-1}(\mathscr{U}(\mathfrak{f}(y), \epsilon))$. Hence $\exists \, \delta > 0$ such that $\mathscr{U}(y, \delta) \subset \mathfrak{f}^{-1}(\mathscr{U}(\mathfrak{f}(y) \, \epsilon))$. That is to say $\forall \epsilon > 0 \, \exists \, \delta > 0$ such that $|\mathfrak{f}(x) - \mathfrak{f}(y)| < \epsilon$ if $|x - y| < \delta$.

(4) If $\mathfrak{f}(y) > \overline{\lim}_{x \to y} \mathfrak{f}(x) = \inf_{\delta > 0} (\sup_{0 < |x-y| < \delta} \mathfrak{f}(x))$, then $\exists \, \delta_0 > 0$ such that $\mathfrak{f}(y) > \sup_{0 < |x-y| < \delta} \mathfrak{f}(x)$, i.e. $\mathfrak{f}(y) = \mathfrak{g}(\delta)$ for $0 < \delta < \delta_0$. If

$$\mathfrak{f}(y) \leqslant \overline{\lim_{x \to y}} \mathfrak{f}(x) = \inf_{\delta > 0} (\sup_{0 < |x-y| < \delta} \mathfrak{f}(x)), \quad \text{then} \quad \mathfrak{f}(y) \leqslant \sup_{0 < |x-y| < \delta} (\mathfrak{f}(x))$$

every $\delta > 0$, i.e.

$\sup_{0<|x-y|<\delta} \mathfrak{f}(x) = \mathfrak{g}(\delta)$. But we have $\inf_{\delta>0} \mathfrak{g}(\delta) = \lim_{\delta\to 0,\, \delta>0} \mathfrak{g}(\delta)$, i.e. $\mathfrak{g}(\delta) \to \overline{\lim}\, \mathfrak{f}(x)$ as $\delta \to 0$, $\delta > 0$.

(5) $\mathfrak{f}(y) > \inf_{\delta>0} (\sup_{0<|x-y|<\delta} \mathfrak{f}(x))$, therefore $\exists\, \delta > 0$ such that

$$\mathfrak{f}(y) > \sup_{0<|x-y|<\delta} \mathfrak{f}(x),$$

i.e. if $x \in \mathscr{V}(y, \delta)$ then $\mathfrak{f}(x) < \mathfrak{f}(y)$. Let X_n be the subset of points y of X at which $\mathfrak{f}(y) > \mathfrak{f}(x)$ provided $x \in \mathscr{V}(y, 1/n)$. Then by the above $X = \bigcup_{n=1}^{\infty} X_n$. If $X'_k \neq \phi$ let $p \in X'_k$. Then $\exists \{p_n\}$ such that $p_n \in X_k$, $p_n \to p$ as $n \to \infty$, $p_n \neq p$. Now $\exists\, N$ such that if $n > N$, $|p_n - p| < 1/2k$ and then $|p_n - p_m| < 1/k$ if n, $m > N$. Thus if we select two distinct points p_n, p_m satisfying n, $m > N$ we have $p_n \in \mathscr{V}(p_m, 1/k)$, $p_m \in \mathscr{V}(p_n, 1/k)$ and thus $\mathfrak{f}(p_n) > \mathfrak{f}(p_m)$ and $\mathfrak{f}(p_n) > \mathfrak{f}(p_m)$. This contradiction shows that $X'_k = \phi$.

(6) Write

$$\inf_{0<|x-y|<\delta} \mathfrak{f}(x) = \mathfrak{a}(\delta), \qquad \sup_{0<|x-y|<\delta} \mathfrak{f}(x) = \mathfrak{A}(\delta),$$

$$\inf_{0<|x-y|<\delta} \mathfrak{g}(x) = \mathfrak{b}(\delta), \qquad \sup_{0<|x-y|<\delta} \mathfrak{g}(x) = \mathfrak{B}(\delta),$$

$$\inf_{0<|x-y|<\delta} \mathfrak{f}(x) \,.\, \mathfrak{g}(x) = \mathfrak{c}(\delta), \qquad \sup_{0<|x-y|<\delta} \mathfrak{f}(x) \,.\, \mathfrak{g}(x) = \mathfrak{C}(\delta).$$

Then, $\quad \mathfrak{a}(\delta) \,.\, \mathfrak{b}(\delta) \leqslant \mathfrak{c}(\delta) \leqslant \mathfrak{a}(\delta) \,.\, \mathfrak{B}(\delta) \leqslant \mathfrak{C}(\delta) \leqslant \mathfrak{A}(\delta) \,.\, \mathfrak{B}(\delta)$

and the result follows since

$$\overline{\lim_{x\to y}}\, \mathfrak{f}(x) = \lim_{\delta\to 0} \mathfrak{A}(\delta), \quad \text{etc.}$$

(7) $$\underline{\lim_{x\to y}}\, \mathfrak{f}(x) = \sup_{\delta>0} (\inf_{0<|x-y|<\delta} \mathfrak{f}(x)) = \lim_{\delta\to 0} (\inf_{0<|x-y|<\delta} \mathfrak{f}(x)).$$

$\exists\, x_n$ such that $0 < |x - x_n| < 1/n$ and

$$\inf_{0<|x-x_n|<1/n} \mathfrak{f}(x) \leqslant \mathfrak{f}(x_n) < \inf_{0<|x-x_n|<1/n} \mathfrak{f}(x) + 1/n.$$

Then $x_n \to x$ and $\mathfrak{f}(x_n) \to \underline{\lim_{x\to y}}\, \mathfrak{f}(x)$.

(9) If E is not closed $\exists\, y$ and a sequence $\{y_n\}$ such that $y \notin E$, $y_n \in E$, $y_n \to y$ as $n \to \infty$, $y_n \neq y$. Define $\mathfrak{f}(x)$ by $\mathfrak{f}(x) = |x - y|$.

Then $\mathfrak{f}(E)$ does not contain the point 0, but $\mathfrak{f}(E)$ does contain the points $|y-y_n|$ and $|y-y_n| \to 0$ as $n \to \infty$. Thus $\mathfrak{f}(E)$ is not closed. This contradiction with given hypotheses show that E is closed.

Chapter 15

(1) (*a*) Let $x_n = 1/(n+\frac{1}{2})$, then

$$x_n \to 0, \quad x_n > 0, \quad \mathfrak{f}(x_n) = \tfrac{1}{2} \not\to \mathfrak{f}(0) = 0.$$

(*b*) $|\mathfrak{f}(x)| \leqslant \frac{1}{2}$ thus

$$|\mathfrak{h}(x).\mathfrak{f}(x)| \leqslant \tfrac{1}{2}\mathfrak{h}(x) \to 0 \quad \text{if} \quad \mathfrak{h}(0) = 0.$$

$\mathfrak{h}(x)$ is continuous at 0. If

$$\mathfrak{h}(0) \neq 0 \quad \text{then} \quad \mathfrak{h}(x_n).\mathfrak{f}(x_n) \to \tfrac{1}{2}\mathfrak{h}(0) \neq 0.$$

x_n defined as in (*a*).

(2) (*a*) $\mathfrak{f}$, $x.\mathfrak{f}(x)$, $\mathfrak{f}(x^2)$ are all continuous at x if $x \neq 0$. Consider then $x = 0$. If $x > 0$, $\mathfrak{f}(x) = -\frac{1}{2}x$. If $x < 0$, $\mathfrak{f}(x) = -1-\frac{1}{2}x$. Thus $\mathfrak{f}(x)$ is discontinuous at $x = 0$: $x\mathfrak{f}(x)$ is continuous at $x = 0$, and $\mathfrak{f}(x^2) = -\frac{1}{2}x$ is continuous at $x = 0$.

(*b*) $x > 1$, $\mathfrak{g}(x) = x$; $\mathfrak{g}(1) = 1$; $-1 < x < 1$, $\mathfrak{g}(x) = x^2$, $\mathfrak{g}(-1) = 0$; $x < -1$, $\mathfrak{g}(x) = x^2$. Thus continuous at $x = 1$; discontinuous at $x = -1$.

(3) $\overline{\lim\limits_{x\to a}}\, \mathfrak{f}(x) = \lim\limits_{\delta\to 0} \Big(\sup\limits_{a-\delta<x<a+\delta} \mathfrak{f}(x)\Big)$. Thus for each integer n, $\exists\, x_n$ such that

$$a-1/n < x_n < a+1/n \quad \text{and} \quad \Big|\overline{\lim_{x\to a}}\, \mathfrak{f}(x) - \mathfrak{f}(x_n)\Big| < 1/n,$$

i.e. $x_n \to a$ and $\mathfrak{f}(x_n) \to \overline{\lim\limits_{x\to a}}\, \mathfrak{f}(x)$.

(4) By condition (*a*), $\sup\limits_n \mathfrak{f}_n(x)$ is defined for all x. By (*b*) $\forall\, \epsilon > 0\, \exists\, \delta > 0$ such that

$$|\mathfrak{f}_n(x) - \mathfrak{f}_n(y)| < \epsilon \quad (n = 1, 2, \ldots)$$

if $|x-y| < \delta$. Thus

$$\mathfrak{f}_n(y) < \mathfrak{f}_n(x) + \epsilon \leqslant \sup_n \mathfrak{f}_n(x) + \epsilon,$$

therefore $$\sup_n \mathfrak{f}_n(y) \leqslant \sup_n \mathfrak{f}_n(x) + \epsilon.$$

Similarly $$\sup_n \mathfrak{f}_n(x) \leqslant \sup_n \mathfrak{f}_n(y) + \epsilon,$$

i.e. $$\Big|\sup_n \mathfrak{f}_n(x) - \sup_n \mathfrak{f}_n(y)\Big| \leqslant \epsilon.$$

(5) $\mathfrak{f}(x) = \mathfrak{f}(x+0) = \mathfrak{f}(x)+\mathfrak{f}(0)$, therefore $\mathfrak{f}(0) = 0$. Induction shows that $\mathfrak{f}(nx) = n\mathfrak{f}(x)$. n a positive integer.

$$0 = \mathfrak{f}(0) = \mathfrak{f}(x-x) = \mathfrak{f}(x)+\mathfrak{f}(-x),$$

therefore $\mathfrak{f}(-x) = -\mathfrak{f}(x)$. $x = p/q$, p, q integers,

$$\mathfrak{f}(x) = \mathfrak{f}(p.1/q) = p\mathfrak{f}(1/q).$$

$\mathfrak{f}(1) = \mathfrak{f}(q.1/q) = q\mathfrak{f}(1/q)$, therefore $\mathfrak{f}(p/q) = p/q\mathfrak{f}(1)$. $\mathfrak{f}(x) = x\mathfrak{f}(1)$ by continuity.

(6) Method of theorem 26 (i) applies to both parts of the question. The method of theorem 26 (ii) cannot be applied to the second part.

(7) $\{\mathfrak{f}_n(x)\}\downarrow \geqslant 0$, therefore $\mathfrak{f}_n(x) \to \mathfrak{f}(x)$ as $n \to \infty$.

$$\mathfrak{f}(y) = \lim_{n\to\infty} \mathfrak{f}_n(y) \geqslant \lim_{n\to\infty} \overline{\lim_{x\to y}} \mathfrak{f}_n(x)$$

since $\mathfrak{f}_n(x)$ is upper semi-continuous. But $\mathfrak{f}_n(x) \geqslant \mathfrak{f}(x)$ thus

$$\lim_{n\to\infty} \overline{\lim_{x\to y}} \mathfrak{f}_n(x) \geqslant \lim_{n\to\infty} \overline{\lim_{x\to y}} \mathfrak{f}(x) = \overline{\lim_{x\to y}} \mathfrak{f}(x).$$

Hence $\mathfrak{f}(y) \geqslant \overline{\lim_{x\to y}} \mathfrak{f}(x)$.

(8) (a) $|\mathfrak{f}(x)-\mathfrak{g}(x)| = |\mathfrak{f}(x)|.|1-\mathfrak{g}(x)/\mathfrak{f}(x)|$ if $\mathfrak{f}(x) \neq 0$. Since $\mathfrak{f}(x) = O(1)$ and $\mathfrak{f} \sim \mathfrak{g}$ it follows that $\forall \epsilon > 0 \, \exists X$ such that if $\mathfrak{f}(x) \neq 0$, then $|\mathfrak{f}(x)-\mathfrak{g}(x)| < \epsilon$ $(x > X)$ and $\mathfrak{f}-\mathfrak{g} = o(1)$.

(9) $\mathfrak{g}(x)/[\mathfrak{f}(x)+\mathfrak{g}(x)] \to 0$ as $x \to \infty$, i.e. $\forall \epsilon > 0 \, \exists X$ such that

$$\frac{\mathfrak{g}(x)}{\mathfrak{f}(x)+\mathfrak{g}(x)} < \epsilon \quad (x > X),$$

i.e. $\mathfrak{g}(x) < \epsilon(\mathfrak{f}(x)+\mathfrak{g}(x))$, i.e. $\mathfrak{g}(x) < \epsilon/(1-\epsilon)\,\mathfrak{f}(x)$. Hence $\mathfrak{g}(x)/\mathfrak{f}(x) \to 0$, i.e. $\mathfrak{g} = o(\mathfrak{f})$.

(10) Otherwise for every $x \in [a, b] \, \exists \, \delta > 0$ such that if

$$x-\delta < y < x+\delta, \quad a \leqslant y \leqslant b \quad \text{then} \quad \mathfrak{f}(x) = \mathfrak{f}(y).$$

By the Heine–Borel theorem a finite number of such intervals cover $[a, b]$. Suppose that they are intervals $(x_i-\delta_i, x_i+\delta_i)$, where $i = 1, \ldots, k$, $x_1 < x_2 < \ldots < x_k$ and $x_i+\delta_i > x_{i+1}-\delta_{i+1}$. It follows that $\mathfrak{f}(x)$ is a constant.

Chapter 16

(1) $\overline{\lim_{n\to\infty}} \mathfrak{f}_n(x) = \lim_{n\to\infty} (\sup_{p\geqslant 1} \mathfrak{f}_{n+p}(x))$. If $x_1 > x_2$ then $\mathfrak{f}_n(x_1) \geqslant \mathfrak{f}_n(x_2)$. Thus

$$\sup_{p\geqslant 1} \mathfrak{f}_{n+p}(x_1) \geqslant \sup_{p\geqslant 1} \mathfrak{f}_{n+p}(x_2)$$

and hence

$$\overline{\lim_{n\to\infty}} \mathfrak{f}_n(x_1) \geqslant \overline{\lim_{n\to\infty}} \mathfrak{f}_n(x_2).$$

(2) Use worked example (iii).

(3) Write $a_n^+ = \max(a_n, 0)$, $a_n^- = \min(a_n, 0)$. Then

$$\sum_0^\infty a_n z^n = \sum_0^\infty a_n^+ z^n + \sum_0^\infty a_n^- z^n \quad |z| < R$$

(since the last two series have radius of convergence at least equal to R). But for $0 < z < R$, $\sum_0^\infty a_n^+ z^n \uparrow$, $\sum_0^\infty a_n^- z^n \downarrow$. Thus the given power series is of bounded variation over $[0, R]$ where $R_1 < R$.

Similarly, $\sum_0^\infty a_n(-z)^n = \sum_0^\infty ((-1)^n a_n) z^n \in \mathsf{B}[0, R_1]$.

Thus $\sum_0^\infty a_n z^n \in \mathsf{B}[-R_1, 0]$. Combine this with the above, choose R_1 so that $-R_1 < a < b < R_1$.

(5) Worked example (i) and theorem 29.

(6) If $0 < x < 1$ then by theorems 27 and 29, $\lim_{y\to x-} \mathfrak{f}(y)$ and $\lim_{x\to y+} \mathfrak{f}(y)$ exist. By Darboux continuity they must equal $\mathfrak{f}(x)$. Similarly $\lim_{y\to 0+} \mathfrak{f}(y) = \mathfrak{f}(0)$ and $\lim_{y\to 1-} \mathfrak{f}(y) = \mathfrak{f}(1)$.

(7) Let $\mathfrak{V}(x)$ be the total variation of $\mathfrak{f}$ over $[0, x]$. Then $\mathfrak{V}(x)\uparrow$. Write $\lim_{x\to y+} \mathfrak{V}(x) = \mathfrak{V}^+(y)$. Then $\mathfrak{V}^+(y) \geqslant \mathfrak{V}(y)$. If $\mathfrak{V}^+(y) > \mathfrak{V}(y) + \eta$ with $\eta > 0$, $\forall \epsilon > 0\ \exists$ points $p_1 \dots p_k$ such that

$$\sum_{i=1}^{k-1} |\mathfrak{f}(p_{i+1}) - \mathfrak{f}(p_i)| > \eta, \quad y = p_1 < p_2 < \dots < p_k < y + \epsilon.$$

Since $\mathfrak{f}(x)$ is continuous at y, we can find such a set of points $p_1^{(1)} \dots p_k^{(1)}$ with $y < p_1^{(1)} < p_2^{(1)} < \dots < p_k^{(1)} < y + \epsilon$. Write $q_1 = p_1^{(1)}$, $r_1 = p_k^{(1)}$, then we have shown that $\forall \epsilon > 0\ \exists q_1, r_1$ such that $y < q_1 < r_1 < y + \epsilon$ and $\mathfrak{V}(r_1) - \mathfrak{V}(q_1) > \eta$. Similarly, $\exists q_2, r_2$ such that $y < q_2 < r_2 < q_1$ and $\mathfrak{V}(r_2) - \mathfrak{V}(q_2) > \eta$ and so on. After n stages we have (since $\mathfrak{V}(x)\uparrow$)

$$\mathfrak{V}(1) \geqslant \sum_{i=1}^n (\mathfrak{V}(r_i) - \mathfrak{V}(q_i)) > \eta . n.$$

But this is to be true for all integers n and since $\mathfrak{V}(1)$ is finite implies $\eta = 0$. Hence $\mathfrak{V}^{+}(y) = \mathfrak{V}(y)$. Similarly $\mathfrak{V}(y)$ is continuous on the left.

(8) Consider the sequence of functions $\{\mathfrak{f}_n(x)\}$, where if x is a rational number that can be written as p/q, where p, q are integers and $q < n$, then $\mathfrak{f}_n(x) = 1$; otherwise $\mathfrak{f}_n(x) = 0$.

(9) If $V^{\mathfrak{f}} > 1$ $\exists$ points $a = x_0 < x_1 < \ldots < x_k = b$ such that

$$\sum_{i=0}^{k-1} |\mathfrak{f}(x_{i+1}) - \mathfrak{f}(x_i)| > 1.$$

But then for n sufficiently large

$$\sum_{i=0}^{k-1} |\mathfrak{f}_n(x_{i+1}) - \mathfrak{f}_n(x_i)| > 1$$

and this implies $V^{\mathfrak{f}_n} > 1$. This is false thus $V^{\mathfrak{f}} \leqslant 1$.

(10) Let $\mathfrak{g}(x)$ be the function for which $\mathfrak{g}(x) = 0$ if $x < 0$ and $\mathfrak{g}(x) = 1$ if $x \geqslant 0$. Let $\{r_n\}$ be the rational numbers between 0 and 1 enumerated in some sequence. The function

$$\mathfrak{f}(x) = \sum_{1}^{\infty} \frac{1}{2^n} \mathfrak{g}(x - r_n)$$

has the required properties.

(11) Since $\mathfrak{f} \uparrow$ the one-sided limits of $\mathfrak{f}(x)$ exist. Write

$$\mathfrak{f}_{+}(x) = \lim_{y \to x+} \mathfrak{f}(y), \quad \mathfrak{f}_{-}(x) = \lim_{y \to x-} \mathfrak{f}(y).$$

Then in the given relation keep x_1 fixed and let $x_2 \to x_1$, $x_2 > x_1$. The given relation together with $\mathfrak{f} \uparrow$, $x_2 > x_1$ implies

$$\mathfrak{f}(x_1) < \mathfrak{f}(x_3) < \mathfrak{f}(x_2),$$

i.e. $x_1 < x_3 < x_2$ and thus $x_3 \to x_1$, $x_3 > x_1$. Hence

$$(1-\alpha)\mathfrak{f}(x_1) + \alpha\mathfrak{f}_{+}(x_1) \leqslant \mathfrak{f}_{+}(x_1) \leqslant \alpha\mathfrak{f}(x_1) + (1-\alpha)\mathfrak{f}_{+}(x_1),$$

i.e. $\quad (1-\alpha)\mathfrak{f}(x_1) \leqslant (1-\alpha)\mathfrak{f}_{+}(x_1) \quad$ and $\quad \alpha\mathfrak{f}_{+}(x_1) \leqslant \alpha\mathfrak{f}(x_1)$.

Thus $\mathfrak{f}(x_1) = \mathfrak{f}_{+}(x_1)$. Similarly keeping x_2 fixed and letting $x_1 \to x_2$, $x_1 < x_2$ we deduce that $\mathfrak{f}(x_2) = \mathfrak{f}_{-}(x_2)$.

(13) Assume statement false and use the Heine–Borel theorem to cover $[c, d]$ with intervals in each of which $\mathfrak{f}(x)$ is of bounded variation.

Chapter 17

(1) Mean-value theorem applied to $\dfrac{\mathfrak{f}(x)-\mathfrak{f}(a)}{x-a}$.

(2) Mean-value theorem applied to $\dfrac{\mathfrak{f}(x)-\mathfrak{f}(c)}{x-c}$.

(3) Trivial if $\mathfrak{f}$ is a constant. Otherwise $\exists\, c$ with $\mathfrak{f}(c) \neq \mathfrak{f}(a)$. Suppose $\mathfrak{f}(c) > \mathfrak{f}(a)$. Take y satisfying $\mathfrak{f}(a) < y < \mathfrak{f}(c)$. By Darboux continuity $\exists\, x_1$, $a < x_1 < c$ with $\mathfrak{f}(x_1) = y$. Similarly $\exists\, x_2$, $c < x_2 < b$ with $\mathfrak{f}(x_2) = y$. Apply Rolle to $[x_1, x_2]$. Continuity over this interval follows from differentiability.

(4) $\dfrac{a_0}{n+1}x^{n+1}+\dfrac{a_1}{n}x^n+\ldots+a_n x$ is zero at $x = 0$, $x = 1$. Rolle's theorem.

(5) Suppose $\mathfrak{f}(x_1) = \mathfrak{f}(x_2) = 0$, $0 < x_1 < x_2 < 1$. Since $\mathfrak{f}(x) \geqslant 0$, $\mathfrak{f}'(x_1) = \mathfrak{f}'(x_2) = 0$. Rolle applied to $[x_1, x_2]$ shows that $\exists\, x_3$ with $x_1 < x_3 < x_2$ and $\mathfrak{f}'(x_3) = 0$. Apply Rolle successively to $\mathfrak{f}'$, $\mathfrak{f}''$.

(6) Write $\mathfrak{g}(x) = \mathfrak{f}(x)-\mathfrak{f}(x-h)$. Then

$$\begin{aligned}\mathfrak{f}(a+2h)-3\mathfrak{f}(a)+2\mathfrak{f}(a-h)&\\ &= [\mathfrak{g}(a+2h)-\mathfrak{g}(a+h)]+2[\mathfrak{g}(a+h)-\mathfrak{g}(a)]\\ &= hD\mathfrak{g}(\xi_1)+2hD\mathfrak{g}(\xi_2) = h^2(D^2\mathfrak{f}(\eta_1)+2D^2\mathfrak{f}(\eta_2)).\end{aligned}$$

But $D^2\mathfrak{f}$ is Darboux continuous, therefore $\exists\, c$ between η_1 and η_2 such that $D^2\mathfrak{f}(c) = \frac{1}{3}(D^2\mathfrak{f}(\eta_1)+2D^2\mathfrak{f}(\eta_2))$.

(7) $\mathfrak{c}$ is single valued as a function of x: since given x

$$(\mathfrak{f}(x+h)-\mathfrak{f}(x))/h$$

is uniquely defined and $D\mathfrak{f}\Uparrow$ implies that $D\mathfrak{f}(\mathfrak{c}_1) = D\mathfrak{f}(\mathfrak{c}_2)$ only if $\mathfrak{c}_1 = \mathfrak{c}_2$. Write $\mathfrak{g}(x) = \mathfrak{f}(x+h)-\mathfrak{f}(x)$. Then

$$\mathfrak{g}(x+\delta)-\mathfrak{g}(x) = \delta D\mathfrak{f}(\xi_1)-\delta D\mathfrak{f}(\xi_2),$$

where $x+h < \xi_1 < x+h+\delta$ and $x < \xi_2 < x+\delta$. Thus if $\delta < h$, $\xi_1 > \xi_2$ and $\mathfrak{g}(x+\delta) > \mathfrak{g}(x)$. Thus $\mathfrak{g}(x)\Uparrow$. Thus $D\mathfrak{f}(\mathfrak{c})\Uparrow$ as a function of x but $D\mathfrak{f}\Uparrow$, therefore $\mathfrak{c}\Uparrow$. If we write $\mathfrak{c} = \mathfrak{c}(x)$ and $\lim\limits_{x\to x_0+}\mathfrak{c}(x) > \mathfrak{c}(x_0)$ then $\lim\limits_{x\to x_0+} D\mathfrak{f}(\mathfrak{c}(x)) > D\mathfrak{f}(\mathfrak{c}(x_0))$. Contradiction.

(8) We must show that if $x_1 > x_2$, then

$$\frac{\mathfrak{f}(x_1)-\mathfrak{f}(a)}{x_1-a} > \frac{\mathfrak{f}(x_2)-\mathfrak{f}(a)}{x_2-a},$$

i.e. writing $\mathfrak{g}(x) = \mathfrak{f}(x) - \mathfrak{f}(a)$ (so that $D^2\mathfrak{g} = D^2\mathfrak{f} > 0$),

$$(x_2 - a)\,\mathfrak{g}(x_1) > (x_1 - a)\,\mathfrak{g}(x_2),$$

$$(x_2 - a)\,(\mathfrak{g}(x_1) - \mathfrak{g}(x_2)) > (x_1 - x_2)\,\mathfrak{g}(x_2),$$

i.e.
$$\frac{\mathfrak{g}(x_1) - \mathfrak{g}(x_2)}{x_1 - x_2} > \frac{\mathfrak{g}(x_2)}{x_2 - a}.$$

By the mean-value theorem ($\mathfrak{g}(a) = 0$),

$$\frac{\mathfrak{g}(x_1) - \mathfrak{g}(x_2)}{x_1 - x_2} = D\mathfrak{g}(\xi_1), \quad \frac{\mathfrak{g}(x_2)}{x_2 - a} = D\mathfrak{g}(\xi_2) \quad \text{and} \quad x_1 > \xi_1 > x_2 > a.$$

Since $D^2\mathfrak{g} > 0$, $D\mathfrak{g} \Uparrow$ and the result follows.

(9) (i) $\forall \epsilon > 0 \,\exists\, x_0$ such that for $x > x_0$, $|D\mathfrak{f}(x) - a| < \epsilon$. Then

$$\frac{\mathfrak{f}(x) - \mathfrak{f}(x_0)}{x - x_0} = D\mathfrak{f}(\xi) = a + \theta\epsilon, \quad |\theta| < 1.$$

Thus $\mathfrak{f}(x) = (a + \theta\epsilon)(x - x_0) + \mathfrak{f}(x_0)$, and

$$\frac{\mathfrak{f}(x)}{ax} = 1 + \frac{\theta\epsilon}{a} - \frac{(a+\theta\epsilon)}{a}\frac{x_0}{x} + \frac{\mathfrak{f}(x_0)}{ax} \quad \text{and} \quad \overline{\lim_{x\to\infty}} \left|\frac{\mathfrak{f}(x)}{ax} - 1\right| \leqslant \epsilon/|a|.$$

True all $\epsilon > 0$. Hence $\mathfrak{f}(x)/ax \to 1$ as $x \to \infty$.

(ii) $\forall \epsilon > 0 \,\exists\, x_0$ such that for $x > x_0$, $|D\mathfrak{f}(x)| < \epsilon$. Then

$$\frac{\mathfrak{f}(x) - \mathfrak{f}(x_0)}{x - x_0} = D\mathfrak{f}\xi = \theta\epsilon, \quad \text{where} \quad |\theta| < 1.$$

(iii) Argue as above.
(iv) By (i).

(10) $\forall \epsilon > 0 \,\exists\, X$ such that if $x \geqslant X$ then

$$a - \epsilon < \mathfrak{f}(x) + D\mathfrak{f}(x) < a + \epsilon.$$

Given $X_1 > X \,\exists\, c$, $X \leqslant c \leqslant X_1$ such that $\mathfrak{f}(c) = \sup\limits_{X \leqslant c \leqslant X_1} \mathfrak{f}(x)$. If ($a$) $X < c < X_1$, then $D\mathfrak{f}(c) = 0$, therefore $\mathfrak{f}(c) < a + \epsilon$, i.e.

$$\mathfrak{f}(x) < a + \epsilon, \quad X \leqslant c \leqslant X_1.$$

If (b) $c = X$, $\mathfrak{f}(x) \leqslant \mathfrak{f}(X)$; if ($c$) $c = X_1$, $\sup \mathfrak{f}(x) = \mathfrak{f}(X_1)$. If the alternative ($a$) holds for a sequence of points X_1 tending to infinity, then $\mathfrak{f}(x) < a + \epsilon$ all $x \geqslant X$. Otherwise $\mathfrak{f}(x) \leqslant \mathfrak{f}(X)$ all $x \geqslant X$, or $\exists\, X_2$ and $\mathfrak{f}(x) \uparrow$ for $x > X_2$. Similarly, either $\mathfrak{f}(x) > a - \epsilon$ all $x \geqslant X_3$

or $\mathfrak{f}(x) \geqslant \mathfrak{f}(X)$, all $x \geqslant X_3$ or $\exists\, X_4$ and $\mathfrak{f}(x)\downarrow$ for $x > X_4$. Thus either (*) $a-\epsilon \leqslant \underline{\lim}\, \mathfrak{f}(x) \geqslant \overline{\lim}\, \mathfrak{f}(x) \leqslant a+\epsilon$ or $\mathfrak{f}(x)$ is monotonic for sufficiently large x. If $\mathfrak{f}(x)\uparrow$, $D\mathfrak{f} \geqslant 0$ and thus $\mathfrak{f}(x) \leqslant a$ all $x > X_5$, say. Hence $\mathfrak{f}(x) \to l$, $l \leqslant a$ and by $\mathfrak{f}+D\mathfrak{f} \to a$, it follows that $D\mathfrak{f} \to l-a$. By exercise 9 (iv) $l = a$. Similarly if $\mathfrak{f}(x)\downarrow$.

If $\mathfrak{f}(x)$ is not monotonic then (*) holds for all $\epsilon > 0$, therefore $\lim \mathfrak{f}(x)$ exists and is equal to a.

(11) $$\frac{\mathfrak{f}(x_1)-\mathfrak{f}(x_0)}{\mathfrak{g}(x_1)-\mathfrak{g}(x_0)} = \frac{D\mathfrak{f}(c)}{D\mathfrak{g}(c)} \quad (x_0 < c < x_1).$$

$\forall \epsilon > 0\, \exists\, X$ such that

$$\frac{D\mathfrak{f}(c)}{D\mathfrak{g}(c)} > \underline{\lim}\, \frac{D\mathfrak{f}}{D\mathfrak{g}} - \epsilon \quad \text{all} \quad c \geqslant X,$$

$$\therefore \quad \frac{\dfrac{\mathfrak{f}(x_1)}{\mathfrak{g}(x_1)} - \dfrac{\mathfrak{f}(x_0)}{\mathfrak{g}(x_1)}}{1-\dfrac{\mathfrak{g}(x_0)}{\mathfrak{g}(x_1)}} > \underline{\lim}\, \frac{D\mathfrak{f}}{D\mathfrak{g}} - \epsilon \quad (x_1 \geqslant X).$$

Let $x_1 \to \infty$ then $\underline{\lim}\, \mathfrak{f}/\mathfrak{g} \geqslant \underline{\lim}\, D\mathfrak{f}/D\mathfrak{g} - \epsilon$. True for all $\epsilon < 0$ and therefore $\underline{\lim}\, \mathfrak{f}/\mathfrak{g} \geqslant \underline{\lim}\, D\mathfrak{f}/D\mathfrak{g}$.

(12) Apply Cauchy's formula to prove

$$\lim_{h\to 0} \frac{\mathfrak{f}(h)-\mathfrak{f}(0)-\dfrac{h}{1!}D\mathfrak{f}(0)-\ldots-\dfrac{h^n}{n!}D^n\mathfrak{f}(0)}{h^{n+1}} = \lim_{h\to 0} \frac{D^n\mathfrak{f}(h)-D^n\mathfrak{f}(0)}{h}.$$

(13) Let E be the set of values of $(\mathfrak{f}(x)-\mathfrak{f}(y))/(x-y)$, $a < x < y < b$ and F the set of values of $D\mathfrak{f}(x)$, $a < x < b$. Since

$$\frac{\mathfrak{f}(x)-\mathfrak{f}(y)}{x-y} = D\mathfrak{f}(c) \quad (x < c < y),$$

we have $E \subset F$ and thus $\bar{E} \subset \bar{F}$. On the other hand given c, $x < c < y$, $D\mathfrak{f}(c) = \lim\limits_{h\to 0}[\mathfrak{f}(x+h)-\mathfrak{f}(x)]/h$ and thus $D\mathfrak{f}(c) \in \bar{E}$, i.e. $F \subset \bar{E}$, and $\bar{F} \subset \bar{E}$ and finally $\bar{F} = \bar{E}$.

(14) (i) Mean-value theorem.
(ii) Given $\epsilon > 0$, $x \in [a, b]$, $\exists\, \delta > 0$ such that

$$\left|\frac{\mathfrak{f}(x+h)-\mathfrak{f}(x)}{h}\right| < \epsilon \quad (0 < h \leqslant \delta).$$

Since $\mathfrak{f}(x)$ is continuous $\exists\, \delta_1 > 0$ such that if $x_1,\ x_2 \in (x-\delta_1, x+\delta)$ then $|\mathfrak{f}(x_1)-\mathfrak{f}(x_2)| < 2\epsilon(\delta_1+\delta_2)$. By the Heine–Borel theorem a finite number of these intervals cover $[a, b]$, and if we omit from this covering any interval completely contained in all the other intervals, we obtain a covering such that each point of $[a, b]$ belongs to at most two intervals of the covering. But then for any two points y_1, y_2 of $[a, b]$, $|\mathfrak{f}(y_1)-\mathfrak{f}(y_2)| < 4\epsilon(b-a)$.

(15) See exercise (8).

Chapter 18

(1) If $D^n\mathfrak{f}(x)$ is continuous then so is $D^n\mathfrak{F}(x)$ which therefore takes all values between h and H, i.e. $\exists\, \xi$ such that

$$\mathfrak{F}(x) = \frac{(x-a)^n}{n!} D^n\mathfrak{F}(\xi) = \frac{(x-a)^n}{n!} D^n\mathfrak{f}(\xi).$$

(2) $$\mathfrak{f}(a+h)-\mathfrak{f}(a) = hD\mathfrak{f}(a)+\frac{h^2}{2!}D^2\mathfrak{f}(\xi) = hD\mathfrak{f}(a+\theta h).$$

But $D\mathfrak{f}(a+\theta h) = D\mathfrak{f}(a)+\theta h D^2\mathfrak{f}(\eta)$. As $h \to 0$, $\xi \to a$, $\eta \to a$ and thus from $\theta D^2\mathfrak{f}(\eta) = \frac{1}{2}D^2\mathfrak{f}(\xi)$, we conclude that $\theta \to \frac{1}{2}$.

(3) We have to show that $\mathfrak{f}(x) = \mathfrak{f}(0)+xD\mathfrak{f}(x)-\frac{1}{2}x^2D^2\mathfrak{f}(\xi)$, i.e. $\mathfrak{f}(0) = \mathfrak{f}(x)+(-x)D\mathfrak{f}(x)+\frac{1}{2}(-x)^2D^2\mathfrak{f}(\xi)$ and this is the second mean-value theorem.

(*a*) $D\mathfrak{f}(x) > \mathfrak{f}(x)/x$. Thus

$$D\left(\frac{\mathfrak{f}(x)}{x}\right) = \frac{D\mathfrak{f}(x)}{x} - \frac{\mathfrak{f}(x)}{x^2} > 0, \quad \text{i.e.} \quad \frac{\mathfrak{f}(x)}{x}\Uparrow.$$

(*b*) $\mathfrak{f}(0) \geqslant 0$ and $D^2\mathfrak{f} < 0$ implies $D\mathfrak{f}(x) < \mathfrak{f}(x)/x$ and thus as above $\mathfrak{f}(x)/x \Downarrow$.

(4) Write $\mathfrak{g}(x) = \mathfrak{f}(x)-\mathfrak{f}(x+h)$ and

$$\sum_{r=0}^{n}(-1)^r\binom{n}{r}\mathfrak{f}(x+rh) = \sum_{r=0}^{n-1}(-1)^r\binom{n-1}{r}\mathfrak{g}(x+rh).$$

Induction.

Chapter 19

(1) Take two points y_1, y_2 of $[\mathfrak{f}(a), \mathfrak{f}(b)]$. If $\mathfrak{f}(x_1) = y_1, \mathfrak{f}(x_2) = y_2$ and $\mathfrak{g}$ is the inverse function of $\mathfrak{f}$, then as $\mathfrak{f}$ is convex,

$$\mathfrak{f}(\lambda x_1 + \mu x_2) \leqslant \lambda \mathfrak{f}(x_1) + \mu \mathfrak{f}(x_2) \quad (\lambda \geqslant 0,\ \mu \geqslant 0,\ \lambda + \mu = 1).$$

But $\mathfrak{g} \Uparrow$, thus $\mathfrak{g}(\mathfrak{f}(\lambda x_1 + \mu x_2)) \leqslant \mathfrak{g}(\lambda \mathfrak{f}(x_1) + \mu \mathfrak{f}(x_2))$, i.e.

$$\lambda x_1 + \mu x_2 \leqslant \mathfrak{g}(\lambda y_1 + \mu y_2) \quad \text{and} \quad \lambda \mathfrak{g}(y_1) + \mu \mathfrak{g}(y_2) \leqslant \mathfrak{g}(\lambda y_1 + \mu y_2).$$

(2) Take x_1, x_2 with $a < x_1 < x_2 < b$ and λ, μ such that $\lambda + \mu = 1$, $\lambda \geqslant 0, \mu \geqslant 0$. Then $\mathfrak{f}(\lambda x_1 + \mu x_2) \leqslant \lambda \mathfrak{f}(x_1) + \mu \mathfrak{f}(x_2)$. As $\mathfrak{g} \Uparrow$, and convex

$$\mathfrak{g}(\mathfrak{f}(\lambda x_1 + \mu x_2)) \leqslant \mathfrak{g}(\lambda \mathfrak{f}(x_1) + \mu \mathfrak{f}(x_2)) \leqslant \lambda \mathfrak{g}\mathfrak{f}(x_1) + \mu \mathfrak{g}\mathfrak{f}(x_2).$$

(3) Take x, y such that $a < x < c < y < b$. Then

$$\mathfrak{f}(x) \leqslant \frac{c-x}{c-a}\mathfrak{f}(a) + \frac{x-a}{c-a}\mathfrak{f}(c) = \mathfrak{f}(a).$$

Similarly $\mathfrak{f}(y) \leqslant \mathfrak{f}(a)$. Also

$$\mathfrak{f}(c) \leqslant \frac{y-c}{y-x}\mathfrak{f}(x) + \frac{c-x}{y-x}\mathfrak{f}(y).$$

And if either $\mathfrak{f}(x) < \mathfrak{f}(a)$ or $\mathfrak{f}(y) < \mathfrak{f}(a)$, we conclude that $\mathfrak{f}(c) < \mathfrak{f}(a)$. This is not so and hence $\mathfrak{f}(x) = \mathfrak{f}(y) = \mathfrak{f}(a)$.

(4) If $\mathfrak{f}$ is not a constant since $\mathfrak{f} \uparrow \exists\, x_1, x_2$ such that $x_1 < x_2$ and $\mathfrak{f}(x_1) < \mathfrak{f}(x_2)$. Write $x_n = x_1 + (n-1)(x_2 - x_1)$. By the convexity condition,

$$\mathfrak{f}(x_n) - \mathfrak{f}(x_{n-1}) \geqslant \mathfrak{f}(x_{n-1}) - \mathfrak{f}(x_{n-2}) \quad (n = 3, \ldots),$$

$$\therefore\ \mathfrak{f}(x_n) - \mathfrak{f}(x_1) \geqslant n(\mathfrak{f}(x_2) - \mathfrak{f}(x_1)),$$

$$\therefore\ \frac{\mathfrak{f}(x_n)}{x_n} > \frac{n(\mathfrak{f}(x_2) - \mathfrak{f}(x_1)) + \mathfrak{f}(x_1)}{(n-1)(x_2 - x_1) + x_1} \sim \frac{\mathfrak{f}(x_2) - \mathfrak{f}(x_1)}{x_2 - x_1} > 0,$$

$$\therefore\ \mathfrak{f}(x) \neq o(x).$$

(5) Jensen's inequality applied to x^α.

(6) An example is given by $\mathfrak{f}_{2n}(x) = |x|$, $\mathfrak{f}_{2n-1}(x) = |x-1|$. $\underline{\lim}\, \mathfrak{f}_n(x)$ is not convex over $(-1, 2)$.

Chapter 20

(1) Consider differential coefficients.

(2) Consider differential coefficients. For the second part of the question use

$$D\left(\cos x - 1 + \frac{x^2}{2!} - \dots (-1)^{m+1}\frac{x^{2m}}{(2m)!}\right)$$
$$= -\sin x + x - \dots (-1)^{m+1}\frac{x^{2m-1}}{(2m-1)!},$$
$$D\left(\sin x - x + \frac{x^3}{3!} - \dots (-1)^{m+1}\frac{x^{2m+1}}{(2m+1)!}\right)$$
$$= \cos x - 1 + \frac{x^2}{2!} - \dots (-1)^{m+1}\frac{x^{2m}}{(2m)!}.$$

Induction.

(3) Compare differential coefficients. Induction.

(4) $D(\sin \pi x - \pi x(1-x))$ is a strictly concave function of x zero at $x = 0$, $x = 1$ and thus positive for $0 < x < 1$. Hence

$$\sin \pi x > \pi x(1-x) \quad \text{for} \quad 0 < x < 1.$$

(5) Addition formula for $\sin((\lambda - 1)x + x)$.

(6) (i) Take logs of the given function and apply theorem 35. Answer $e^{-\frac{1}{2}a^2/b^2}$.

(ii) $|\sin 1/x| < 1$ and $x/(e^x - 1) \to 1$. Answer 0.

(iii) Answer $-12/5$.

(iv) $-8/(15.4!)$.

(8) (*a*) Since $|n^{i-x}| = |(-1)^n n^{i-x}| = n^{-x}$ both series are absolutely convergent if $x > 1$ and are not absolutely convergent if $x \leqslant 1$.

(*b*) $n^{i-x} = n^{-x}\sin(\log n) + in^{-x}\cos(\log n)$ and Σn^{i-x} converges only if $\Sigma n^{-x}\sin(\log n)$ converges. Suppose $x \leqslant 1$. If

$$(2k+1)\tfrac{1}{2}\pi - \tfrac{1}{4}\pi \leqslant \log n \leqslant (2k+1)\tfrac{1}{2}\pi + \tfrac{1}{4}\pi,$$

k an integer, then $\sin\log n \geqslant 1/\sqrt{2}$. Let this range of n be from N_k to M_k so that $N_k - 1 < e^{(2k+1)\frac{1}{2}\pi - \frac{1}{4}\pi}$ and $M_k + 1 > e^{(2k+1)\frac{1}{2}\pi + \frac{1}{4}\pi}$. Then

$$\sum_{N_k}^{M_k} n^{-x}\sin(\log n) > \frac{1}{\sqrt{2}}\sum_{N_k}^{M_k} n^{-x}$$
$$\geqslant \frac{1}{\sqrt{2}}(M_k - N_k)M_k^{-x} \quad \text{if} \quad x \geqslant 0,$$
$$\geqslant \frac{1}{\sqrt{2}}(e^{(2k+1)\frac{1}{2}\pi})^{1-x}\left(e^{\frac{1}{4}\pi} - e^{-\frac{1}{4}\pi} - \frac{2}{e^{(2k+1)\frac{1}{2}\pi}}\right).$$

Since $x \leqslant 1$ this last expression does not tend to zero as $k \to \infty$. Thus Σn^{i-x} diverges. The same result holds if $x \leqslant 0$ since the individual terms do not then tend to zero.

Consider next $\Sigma(-1)^n n^{i-x}$. We have to consider two series, a sine series and a cosine series. We consider only the sine series. If $x \leqslant 0$, then as above the individual terms do not tend to zero and $\Sigma(-1)^n n^{i-x}$ diverges. Suppose then that $0 < x \leqslant 1$, and consider two consecutive terms of $\Sigma(-1)^n \sin(\log n)\, n^{-x}$, say

$$(2n)^{-x}\sin(\log 2n) - (2n-1)^{-x}\sin(\log(2n-1))$$
$$= (+1)\,[-x(\xi)^{-x-1}\sin(\log\xi) + \xi^{-x-1}\cos(\log\xi)],$$

where $2n-1 < \xi < 2n$ by the mean-value theorem. Thus

$$|(2n)^{-x}\sin(\log 2n) - (2n-1)^{-x}\sin\log(2n-1)| < \frac{1+x}{(2n-1)^{1+x}}.$$

Since $1+x > 1$ it follows by comparing with $\Sigma\, 1/(2n-1)^{1+x}$ that

$$\Sigma((2n)^{-x}\sin(\log 2n) - (2n+1)^{-x}\sin(\log(2n+1)))$$

converges. Thus if the partial sums of $\Sigma(-1)^n n^{-x}\sin\log n$ are $S_1, S_2, \ldots, S_n, \ldots$ we have shown that $S_{2n} \to$ a finite limit S as $n \to \infty$. But

$$|S_{2n+1} - S_{2n}| = |(2n+1)^{-x}\sin\log(2n+1)|$$
$$\leqslant (2n+1)^{-x} \to 0 \quad \text{as} \quad n \to \infty.$$

Hence $S_{2n+1} \to S$ as $n \to \infty$, i.e. $\Sigma n^{-x}\sin(\log n)$ converges to S. A similar argument applies to $\Sigma n^{-x}\cos(\log n)$. Thus $\Sigma(-1)^n n^{i-x}$ converges if $0 < x \leqslant 1$.

(9) It is easy to see that the radii of convergence of $\Sigma e^{\sqrt{n}} z^n$ and of $\Sigma e^{-\sqrt{n}} z^n$ are both equal to 1. If $|z| = 1$, $|e^{\sqrt{n}} z^n| > 1$, therefore $\Sigma e^{\sqrt{n}} z^n$ diverges. If $|z| = 1$,

$$|e^{-\sqrt{n}} z^n| = \frac{1}{e^{\sqrt{n}}} < \frac{1}{1+\sqrt{n}+\dfrac{(\sqrt{n})^2}{2!}+\ldots} < \frac{6}{n^{\frac{3}{2}}},$$

and $\Sigma e^{-\sqrt{n}} z^n$ converges absolutely.

(10) x^p, $\cos x$, $1/x^q$ are all continuous for $0 \leqslant x \leqslant 1$, except possibly at $x = 0$. Thus $\mathfrak{f}(x)$ is continuous at each x, except possibly $x = 0$. $x^p \to 0$ as $x \to 0+$ if and only if $p > 0$ and it is not true that $\cos 1/x^q \to 0$ as $\to 0+$ for any q. Thus $\mathfrak{f}(x)$ is continuous at $x = 0$ in $[0, 1]$ if and only if $p > 0$. Since $D\mathfrak{f}(x)$ is bounded in $\delta \leqslant x \leqslant 1$ for any $\delta > 0$ we have $\mathfrak{f}$ is of bounded variation in $[\delta, 1]$. If $q \leqslant 0$,

$\lim_{x\to 0+} D\mathfrak{f}(x)$ exists and $\mathfrak{f}(x)$ is monotonic in $(0, \delta)$ if and only if it is bounded in this interval, i.e. if and only if $p \geqslant 0$. If next $q > 0$, $\mathfrak{f}(x)$ for

$$\left(\frac{1}{2(k+1)\pi}\right)^{1/q} \leqslant x \leqslant \left(\frac{1}{2k\pi}\right)^{1/q}$$

oscillates between

$$\left(\frac{1}{2k\pi}\right)^{p/q} \quad \text{and} \quad -\left(\frac{1}{(2k+1)\pi}\right)^{p/q}$$

and it can be shown that $\mathfrak{f}(x)$ is of bounded variation over $[0, 1]$ if and only if $\Sigma((2k)^{-p/q} - (2k+1)^{-p/q})$ converges. Since

$$(2k+1)^{-p/q} - (2k)^{-p/q} = -\frac{p}{q}\xi^{-p/q-1}, \quad 2k < \xi < 2k+1$$

(by the mean-value theorem) this is equivalent to $p/q > 0$.

For Darboux continuity we need simply $\overline{\lim}_{x\to 0+} \mathfrak{f}(x) \geqslant 0 \geqslant \underline{\lim}_{x\to 0+} \mathfrak{f}(x)$, i.e. $p > 0$ or $q > 0$.

Thus the results are

(*a*) Continuity in $[0, 1]$ if and only if $p > 0$.

(*b*) of bounded variation in $[0, 1]$ if and only if $p \geqslant 0$, $q \leqslant 0$ or $p > 0$, $q > 0$.

(*c*) Darboux continuity if $p > 0$ or $q > 0$.

Chapter 21

(1) As a function of λ the right-hand side is least when

$$\lambda = |z_2|/|z_1|$$

and its value then is $(|z_1| + |z_2|)^2$. Alternatively write $|z_1|^2 = z_1\overline{z_1}$ and reduce the inequality to $(\lambda\overline{z_1} - \overline{z_2})(\lambda z_1 - z_2) \geqslant 0$.

(2) Consider the various possibilities with respect to sign of a, b, c.

(3) Induction.

(4) Arithmetic–Geometric mean applied to $a_1 + \ldots + a_n$ and to $(a_1^{-1} + \ldots + a_n^{-1})$.

(5) Write $\lambda_i = a_i/(a_i + b_i)^{\frac{1}{2}}$, $\mu_i = b_i/(a_i + b_i)^{\frac{1}{2}}$. The inequality to be established reduces to

$$(\Sigma\lambda_i(\lambda_i + \mu_i))(\Sigma\mu_i(\lambda_i + \mu_i)) \geqslant \Sigma(\lambda_i + \mu_i)^2 . \Sigma\lambda_i\mu_i,$$

i.e. $$(\Sigma\lambda_i^2).(\Sigma\mu_i^2) \geqslant (\Sigma\lambda_i\mu_i)^2$$

and this is Cauchy's inequality.

(6) Cauchy's inequality.

(7) Write $p_i/(p_1+\ldots+p_n) = q_i$. Then the inequality to be proved becomes $a_1^{q_1}.a_2^{q_2}\ldots a_n^{q_n} \leqslant a_1q_1+\ldots+a_nq_n$, where $q_1+\ldots+q_n = 1$ and $0 \leqslant q_i$. Take logs. This is equivalent to

$$q_1\log a_1+\ldots+q_n\log a_n \leqslant \log(a_1q_1+\ldots+a_nq_n)$$

and this follows from Jensen's inequality applied to the convex function $-\log x$.

Take $a_n = p_n = n$.

(8) Write

$$A_i = a_i/(a_i+b_i+c_i), \quad B_i = b_i/(a_i+b_i+c_i), \quad C_i = c_i/(a_i+b_i+c_i).$$

Then the inequality to be proved becomes

$$A_1^{q_1}A_2^{q_2}A_3^{q_3}+B_1^{q_1}B_2^{q_2}B_3^{q_3}+C_1^{q_1}C_2^{q_2}C_3^{q_3} \leqslant 1.$$

By (7), $A_1^{q_1}A_2^{q_2}A_3^{q_3} \leqslant q_1A_1+q_2A_2+q_3A_3$, etc. Using $q_1+q_2+q_3 = 1$, $A_i+B_i+C_i = 1$ then gives the required result.

(9) Minkowski's inequality applied twice—say to $a_i+(b_i+c_i)$ and to b_i+c_i.

(10) (9) extended to m instead of three sums.

Chapter 22

(1) If $\mathfrak{f}\uparrow$ then $\mathfrak{f}(\mathfrak{g})\uparrow$, therefore R-integrable. Similarly if $\mathfrak{f}\downarrow$, $\mathfrak{f}(\mathfrak{g})\downarrow$, therefore R-integrable. If $\mathfrak{f}$ is continuous $\forall\,\epsilon > 0\;\exists\,\delta > 0$ such that $|\mathfrak{f}(x_1)-\mathfrak{f}(x_2)| < \epsilon$ provided that $|x_1-x_2| < \delta$. Since $\mathfrak{g}(x)\Uparrow$ there are at most a finite number of points say the points $p_1, \ldots, p_N$ such that (*) $\overline{\lim\limits_{x\to p_i}}\, \mathfrak{g}(x) \geqslant \underline{\lim\limits_{x\to p_i}}\, \mathfrak{g}(x)+\delta$ (chapter 16, worked example (i)). Let $p_0 = a$, $p_{N+1} = b$. Let X be the union of the open intervals $(p_i-\epsilon/N, p_i+\epsilon/N)$ $(i = 0, 1, \ldots, N+1)$. We can find $\delta_1 > 0$ such that if x_1, x_2 belong to $\mathscr{C}(X)\cap[a,b]$ and $|x_1-x_2| < \delta_1$, then $|\mathfrak{g}(x_1)-\mathfrak{g}(x_2)| < \delta$. (Otherwise by the Bolzano–Weierstrass theorem $\mathscr{C}(X)$ will contain a point satisfying (*).) Consider a dissection $\mathscr{D}$ of $[a,b]$ which contains the points $p_i \pm \epsilon/N$ and has mesh less than δ_1. Then combining the above results,

$$S_{\mathscr{D}}^{\mathfrak{f}(\mathfrak{g})} - s_{\mathscr{D}}^{\mathfrak{f}(\mathfrak{g})} < 2M\epsilon+\epsilon(b-a), \quad \text{where} \quad M = \sup_{a\leqslant x\leqslant b} |\mathfrak{f}(\mathfrak{g}(x))|.$$

(2) (*a*) Continuous $\exp(\sin x) \in \mathsf{R}(0, 1)$.

(*b*) Continuous except at $x = 0$; bounded; $\exp(\sin 1/x) \in \mathsf{R}(0, 1)$.

(*c*) Unbounded. $\operatorname{Exp} 1/x \notin \mathsf{R}(0, 1)$.

(*d*) Continuous except at $x = 0$; bounded; $\sin(\exp 1/x) \in \mathsf{R}(0, 1)$.

(3) $\Sigma \mathfrak{f}(x - r_n)/n^2$ converges for all x of $[0, 1]$ since $|\mathfrak{f}(x - r_n)| \leqslant 1$. Write $\mathfrak{Q}_N(x) = \sum_1^{N-1} \mathfrak{f}(x - r_n)/n^2$, $\mathfrak{P}_N(x) = \sum_N^{\infty} \mathfrak{f}(x - r_n)/n^2$.

$\forall \epsilon > 0 \, \exists N$ such that $|\mathfrak{P}_N(x)| \leqslant 1/N < \epsilon$. With this N, since $\mathfrak{Q}_N(x)$ has at most a finite number of discontinuities $\exists \delta > 0$ such that for any dissection $\mathscr{D}$ of mesh less than δ, $S_{\mathscr{D}}^{\mathfrak{Q}_N} - s_{\mathscr{D}}^{\mathfrak{Q}_N} < \epsilon$. Since also $|S_{\mathscr{D}}^{\mathfrak{P}_N} - s_{\mathscr{D}}^{\mathfrak{P}_N}| \leqslant 2 \sup_x |\mathfrak{P}_N(x)| \leqslant 2\epsilon$. It follows that for

$$\mathfrak{f}(x) = \mathfrak{Q}_N(x) + \mathfrak{P}_N(x)$$

we have $S_{\mathscr{D}}^{\mathfrak{f}} - s_{\mathscr{D}}^{\mathfrak{f}} < 3\epsilon$. Hence $\mathfrak{f} \in \mathsf{R}(0, 1)$.

(4) Suppose $v_N(\alpha, \beta)/N \to (\beta - \alpha)/(b - a)$. If $\mathfrak{f}$ is a step-function say $\mathfrak{f}(x) = \Sigma k_i \mathfrak{g}_i(x)$, where $\mathfrak{g}_i(x) = 0$, $x \leqslant a_i$; $\mathfrak{g}_i(x) = 1$, $a_i < x < b_i$; $\mathfrak{g}_i(x) = 0$, $x \geqslant b_i$, then $\sum_{n=1}^{N} \mathfrak{g}_i(x_n)$ is the number of x_n, $1 \leqslant n \leqslant N$ for which $a_i < x_n < b_i$, i.e. $\sum_{n=1}^{N} \mathfrak{g}_i(x_n) = v_N(a_i, b_i)$. Thus

$$\frac{\sum_1^N \mathfrak{g}_i(x_n)}{N} \to \frac{b_i - a_i}{b - a} = \frac{\int_a^b \mathfrak{g}_i}{b - a}.$$

By addition
$$\frac{\sum_1^N \mathfrak{f}}{N} \to \frac{\int_a^b \mathfrak{f}}{b - a}.$$

True for general $\mathfrak{f} \in \mathsf{R}(0, 1)$ by an approximation argument.

(5) For x_1 satisfying $0 < x_1 < 2x_1 < b$ we have $\int_{x_1}^{2x_1} \mathfrak{f} > x_1 \mathfrak{f}(2x_1)$. As $\lim_{x \to 0+} \int_x^b \mathfrak{f}$ exists we have from the general principle of convergence $\forall \epsilon > 0 \, \exists X$ such that

$$\left| \int_{X_1}^{X_2} \mathfrak{f} \right| < \epsilon \quad \text{if} \quad 0 < X_1 < X_2 < X.$$

Thus if $2x_1 < X$ we have $2x_1 \mathfrak{f}(2x_1) < 2\epsilon$. Since $\mathfrak{f} \geqslant 0$ for x sufficiently small this implies that $x\mathfrak{f}(x) \to 0$.

$$\int_{b/n}^{b} \mathfrak{f} \leqslant \frac{b}{n} \sum_1^n \mathfrak{f}\left(\frac{rb}{n}\right) \leqslant \int_{b/n}^{b} \mathfrak{f} + \frac{b}{n} \mathfrak{f}\left(\frac{b}{n}\right),$$

(since $\mathfrak{f}\downarrow$ we have

$$\frac{b}{n}\mathfrak{f}\left(\frac{rb}{n}\right) \leqslant \int_{b(r-1)/n}^{br/n} \mathfrak{f} \leqslant \frac{b}{n}\mathfrak{f}\left(\frac{(r-1)b}{n}\right).$$

(6) $\forall \epsilon > 0 \,\exists\, \delta > 0$ such that $\mathfrak{f} < M-\epsilon$ if $x < \delta$ or $x > \delta$ and as $\epsilon \to 0$ we can suppose that $\delta \to 0$. (Since $\mathfrak{f}(x) < M$ except at $x = 0$.) Then

$$\int_{-1}^{+1} \exp n\mathfrak{f}(x)\,dx \leqslant 2\exp n(M-\epsilon) + 2\delta \exp(nM)$$
$$= o(\exp(nM)).$$

(7) Use $\mathfrak{f}(x)-\mathfrak{f}(x_0) = (x-x_0)\,\mathfrak{f}'(x_0)+\frac{1}{2}[(x-x_0)^2]\,\mathfrak{f}''(\eta)$. η varies with x and $\mathfrak{f}''(\eta)$ is an R-integrable function of x. For the identity above shows that $(x-x_0)^2$. $\mathfrak{f}''(\eta)$ is R-integrable and $\mathfrak{f}''(\eta)$ is bounded. Hence $\mathfrak{f}''(\eta)$ is R-integrable. Apply the integral mean-value theorem.

(8) Write $M = \sup|D\mathfrak{f}(x)|$. By the mean-value theorem

$$\mathfrak{f}(x) = (x-a)\,D\mathfrak{f}(\xi) \leqslant M(x-a), \quad a \leqslant x \leqslant \tfrac{1}{2}(a+b);$$
$$\mathfrak{f}(x) = (x-b)\,D\mathfrak{f}(\eta) \leqslant M(b-x), \quad \tfrac{1}{2}(a+b) \leqslant x \leqslant b;$$

for suitable ξ, η satisfying $a < \xi < x$, $x < \eta < b$. Equality cannot occur in both the above inequalities for all x concerned or $\mathfrak{f}$ would not be differentiable at $x = \frac{1}{2}(a+b)$. Hence

$$\int_a^b \mathfrak{f} < \tfrac{1}{4}M(b-a)^2.$$

Chapter 23

(1) Use $\displaystyle\int_x^{x+h} \mathfrak{f}\mathfrak{g} = \mathfrak{g}(x)\left(\int_x^{x+h} (\mathfrak{f}+o(1))\right)$ as $h \to 0$.

(2) Use (1).

(3) $$\int_n^{n+1} [t]\,D\mathfrak{f}(t)\,dt = n(\mathfrak{f}(n+1)-\mathfrak{f}(n)).$$

(4) $D\mathfrak{g} = \mathfrak{f} \geqslant \mathfrak{g}^{\frac{1}{2}}$. Thus for $x \geqslant a$, since

$$\mathfrak{g} = \int \mathfrak{f} > 0, \quad \frac{1}{\mathfrak{g}^{\frac{1}{2}}}D\mathfrak{g} \geqslant 1, \quad \text{i.e.} \quad (\mathfrak{g}(x))^{\frac{1}{2}} \geqslant \tfrac{1}{2}(x-a).$$

(5) Approximate to $\mathfrak{f}$ by step-functions.

(6) Integrate by parts. Induction.

(7) (See worked example (i), chapter 22.) Integrate by parts in $\displaystyle\int_0^a \mathfrak{g}\,D\mathfrak{f}$.

(8) Integrate $\int_0^1 (\mathfrak{f}(x)-\mathfrak{f}(0))$ by parts in the form

$$\Big[(x-1)(\mathfrak{f}(x)-\mathfrak{f}(0))\Big]_0^1 - \int_0^1 (x-1)\,D\mathfrak{f}(x).$$

Chapter 24

(1) $\mathfrak{f}(x)$ is a sum of functions $\mathfrak{k}(x)$ of the form $\mathfrak{k}(x) = 0$, $x < c$; $\mathfrak{k}(x) = l$, $c \leqslant x \leqslant d$; $\mathfrak{k}(x) = 0$, $x > d$; where c, d are discontinuities of $\mathfrak{f}$. It is sufficient to show that $\mathfrak{k} \in \mathsf{RS}_{\mathfrak{g}}(a, b)$. For any dissection $\mathscr{D}$ of $[a, b]$ say $x_0 < x_1 < \ldots < x_k$, suppose that p and q are chosen so that $x_p \leqslant c < x_{p+1}$ and $x_q \leqslant d < x_{q+1}$. Then if $\tau(\mathscr{D}) < d-c$

$$\Sigma\mathfrak{k}(\xi_i)\,(\mathfrak{g}(x_{i+1})-\mathfrak{g}(x_i)), \quad \text{where} \quad x_i < \xi_i < x_{i+1}$$

is equal to

$$\mathfrak{k}(\xi_p)\,(\mathfrak{g}(x_{p+1})-\mathfrak{g}(x_p)) + l(\mathfrak{g}(x_q)-\mathfrak{g}(x_{p+1})) + \mathfrak{k}(\xi_q)\,(\mathfrak{g}(x_{q+1})-\mathfrak{g}(x_q)).$$

Since $\mathfrak{g}$ is continuous at c and at d, the first and last expressions tend to zero and the middle expression to $l(\mathfrak{g}(d)-\mathfrak{g}(c))$ as the mesh of $\mathscr{D}$ tends to zero.

(2) $\mathfrak{f} \in \mathsf{RS}_{\mathfrak{g}}(0, 1)$ because $\mathfrak{g}$ is of bounded variation.

(3) A convex function is both continuous and of bounded variation.

(4), (5) Compare with the analogous result for Riemann integration.

Chapter 25

(1) (*a*) $\mathfrak{f}$ is continuous except possibly at $x = 0$. Thus $\mathfrak{f} \in \mathsf{R}(0, 1)$ is equivalent to $\mathfrak{f}$ is bounded. Since

$$\mathfrak{f}(x) = kx^{k-1}\sin(\pi/x) - \pi x^{k-2}\cos(\pi/x)$$

this is equivalent to $k \geqslant 2$.

(*b*) $\mathfrak{f} \in \mathsf{R}(\epsilon, 1)$ for every $\epsilon > 0$ and

$$\int_\epsilon^1 \mathfrak{f} = \Big[x^k \sin(\pi/x)\Big]_\epsilon^1 = -\epsilon^k \sin \pi/\epsilon$$

tends to a finite limit as $\epsilon \to 0+$ if and only if $k > 0$.

(2) $(\log x)^n/(1+x) \in \mathsf{R}(\epsilon, 1)$ if $\epsilon > 0$, and $\log x < 0$ if $0 < x < 1$. It can be shown that the required result follows from

$$\sup_{0<\epsilon<1} \left| \int_\epsilon^1 \frac{(\log x)^n}{1+x}\,dx \right| < \infty.$$

This last inequality follows from the fact that if $0 < \lambda < 1$ then

$$\left|\int_{\lambda^{r+1}}^{\lambda^r} \frac{(\log x)^n}{1+x}\,dx\right| \leqslant \lambda^r(1-\lambda)\,|\log(\lambda)^{r+1}|^n$$

$$\leqslant Kr^n\lambda^r \quad \text{and} \quad \Sigma r^n\lambda^r \text{ converges.}$$

(3) Evaluate the integral directly writing

$$\frac{1-x^n}{1-x} = 1+x+\ldots+x^{n-1}$$

and evaluate the integral again after transforming from x to y, where $x = 1-y$.

Integrate $x^a(1-(1-x)^n)/x$.

(4) (a) $$\sin(\pi/x) > 1/\sqrt{2}$$

if $$(2k+\tfrac{1}{4})\pi < \pi/x < (2k+\tfrac{3}{4})\pi \quad (k = 1, 2, \ldots),$$

i.e. if $(2k+\frac{3}{4})^{-1} < x < (2k+\frac{1}{4})^{-1}$. Thus

$$\int_{(2N+\frac{3}{4})^{-1}}^{1} \left|\frac{1}{x}\sin\frac{\pi}{x}\right| dx > \frac{1}{\sqrt{2}}\sum_{1}^{N}(2k+\tfrac{1}{4}).\left(\frac{1}{2k+\frac{1}{4}}-\frac{1}{2k+\frac{3}{4}}\right)$$

$$\to \infty \quad \text{as} \quad N \to \infty.$$

(b) Integrate.

(5) $$\left|\int_{k-\frac{1}{2}}^{k+\frac{1}{2}} \frac{1}{x}dx - \frac{1}{k}\right| = \left|\int_{k-\frac{1}{2}}^{k+\frac{1}{2}} \frac{k-x}{kx}dx\right|$$

$$= \int_{k}^{k+\frac{1}{2}} \frac{k-x}{kx}dx + \int_{k}^{k-\frac{1}{2}} \frac{x-k}{xk}dx$$

$$= \int_0^{\frac{1}{2}} \frac{-y\,dy}{k(y+k)} + \int_0^{\frac{1}{2}} \frac{-y\,dy}{k(y-k)}$$

$$= \int_0^{\frac{1}{2}} \frac{-y}{k}\cdot\frac{2y}{y^2-k^2}dy$$

$$= O\left(\frac{1}{k^3}\right),$$

$$\sum_{n+1}^{2n}\frac{1}{n} = \int_{n+\frac{1}{2}}^{2n+\frac{1}{2}}\frac{dx}{x} + \sum_{n+1}^{2n} O\left(\frac{1}{n^3}\right)$$

$$= \log\left(\frac{2n+\frac{1}{2}}{n+\frac{1}{2}}\right) + O\left(\frac{1}{n^2}\right).$$

(6) (a) Converges if $\alpha < 1$. Diverges if $\alpha \geqslant 1$.
(b) Converges if $\alpha > 1$. Diverges if $\alpha \leqslant 1$.
(c) Diverges. For large x, $|\log x| < x$. Use (b).

(d) $$\frac{1}{x^{\alpha}(1+\log 1/x)} \in \mathsf{R}(\epsilon, 1)$$

and putting $x = 1/y$

$$\int_{\epsilon}^{1} \frac{dx}{x^{\alpha}(1+\log(1/x))} = \int_{1}^{1/\epsilon} \frac{dy}{y^{2-\alpha}(1+\log y)}.$$

The integral $$\int_{1}^{\infty} \frac{dy}{y^{2-\alpha}(1+\log y)}$$

converges if $2-\alpha > 1$ and diverges if $2-\alpha < 1$ (by comparison with the integral $\int_{1}^{\infty} \frac{dy}{y^{\beta}}$). If $2-\alpha = 1$ we have

$$\frac{1}{y(1+\log y)} > \frac{1}{2y\log y} \quad \text{if} \quad y > e$$

and $$\int_{e}^{Y} \frac{dy}{y(1+\log y)} > \tfrac{1}{2}\log\log Y \to \infty \quad \text{as} \quad Y \to \infty.$$

Thus the given integral diverges if $2-\alpha = 1$.

(7) (a) The integral behaves like that of $x^{k-l}/(\log x)^2$. Converges if $l \geqslant k+1$. Diverges if $l < k+1$.

(b) If $(2k+\frac{1}{2})\pi \leqslant x \leqslant (2k+\frac{3}{4})\pi$, then $\sin x(1-\cos x) > 1/\sqrt{2}$ and the integral diverges when $k \leqslant 1$. On the other hand

$$|\sin x(1-\cos x)| < 2$$

and the integral converges when $k > 1$.

(8) By the second mean-value theorem

$$\int_{Y}^{X} \frac{\mathfrak{f}(x)\sin nx}{x}\,dx = \frac{\mathfrak{f}(Y)}{Y}\int_{Y}^{\xi} \sin nx\,dx,$$

where ξ may depend on n. Now

$$\int_{2r\pi/n}^{(2r+2)\pi/n} \sin nx\,dx = 0,$$

thus $$\left|\int_{Y}^{\xi} \sin nx\,dx\right| \leqslant 2\pi/n \to 0 \quad \text{as} \quad n \to \infty.$$

(9) Show that

$$u_n - u_{n-1} = 0, \quad v_n = \int_0^{n\pi} \frac{\sin y}{y}\, dy,$$

$$\begin{aligned} u_n - v_n &= \int_0^{\frac{1}{2}\pi} \sin 2nx(\cot x - 1/x)\, dx \\ &= \left[\frac{\cos 2nx}{2n}(\cot x - 1/x)\right]_0^{\frac{1}{2}\pi} + \int_0^{\frac{1}{2}\pi} \frac{\cos 2nx}{2n}\left(\frac{1}{x^2} - \operatorname{cosec}^2 x\right) dx, \end{aligned}$$

$$\cot x - \frac{1}{x} = \frac{x\cos x - \sin x}{x \sin x} \to 0 \quad \text{as} \quad x \to 0,$$

$$\frac{1}{x^2} - \operatorname{cosec}^2 x = \frac{\sin^2 x - x^2}{x^2 \sin^2 x} \to -\tfrac{1}{3} \quad \text{as} \quad x \to 0.$$

(10)

$$\int_\epsilon^X \frac{\cos ax}{x}\, dx = \int_{a\epsilon}^{aX} \frac{\cos y}{y}\, dy, \quad \int_\epsilon^X \frac{\cos bx}{x}\, dx = \int_{b\epsilon}^{bX} \frac{\cos y}{y}\, dy,$$

$$\int_\epsilon^X \frac{\cos ax - \cos bx}{x}\, dx = \int_{a\epsilon}^{b\epsilon} \frac{\cos y}{y}\, dy - \int_{aX}^{bX} \frac{\cos y}{y}\, dy.$$

As $\epsilon \to 0$ $\quad \displaystyle\int_{a\epsilon}^{b\epsilon}\left(\frac{\cos y}{y} - \frac{1}{y}\right) dy = O\left(\int_{a\epsilon}^{b\epsilon} y\, dy\right) = o(1)$

and thus $\quad \displaystyle\int_{a\epsilon}^{b\epsilon} \frac{\cos y}{y}\, dy \to \log \frac{b}{a}.$

(11) Consider integrals over ranges $(k\pi, (k+1)\pi)$. Second part take $X = N\pi$ and use

$$\sum_1^N (n+N)^{-2} = \sum_1^N (n+N)^{-1}(n+N+1)^{-1} + O(N^{-2}) = 1/2N + O(N^{-2}).$$

(12) $\displaystyle\mathfrak{F}(x) = \int_0^X \mathfrak{f}$ then $\displaystyle\int_X^Y \frac{\mathfrak{f}}{\mathfrak{F}} \geqslant \frac{\mathfrak{F}(Y) - \mathfrak{F}(X)}{\mathfrak{F}(Y)} \to 1$ as $Y \to \infty$.

By the general principle of convergence, $\displaystyle\int^\infty \frac{\mathfrak{f}}{\mathfrak{F}}$ does not converge.

Take $\epsilon_n = 1 \Big/ \sum_{r=1}^n a_r$.

(13) Write the integrand in the form $e^{\sin x} \cos x \,.\, 2x^{-n} \sin x$ and integrate by parts twice.

Chapter 26

(1) $$\int_{r_1}^{r_N} \frac{\mathfrak{n}(u)}{u^{\rho+1}}\, du = \sum_1^{N-1} \int_{r_n}^{r_{n+1}} \frac{n}{u^{\rho+1}}\, du$$

$$= \sum_1^{N-1} \frac{n}{\rho}(-r_{n+1}^{-\rho} + r_n^{-\rho}) = \rho^{-1}\left(\sum_1^N r_n^{-\rho} - (N-1) r_N^{-\rho}\right).$$

(2) Take $h = (\log(1/t))^{\frac{1}{2}}$ and use $\log 1/t \sim 1-t$ as $t \to 1-$.

(3) $\mathfrak{f}(x)-\mathfrak{f}(n+\frac{1}{2}) = (x-n-\frac{1}{2})\,D\mathfrak{f}(n+\frac{1}{2})+\dfrac{(x-n-\frac{1}{2})^2}{2!}\,D^2\mathfrak{f}(\xi),$
where ξ lies between $n+\frac{1}{2}$ and x. Since $\alpha > 1$,

$$\Sigma\int_n^{n+1}(\mathfrak{f}(x)-\mathfrak{f}(n+\tfrac{1}{2}))\,dx$$

converges say to C. Then

$$\sum_1^N \mathfrak{f}(m+\tfrac{1}{2}) = \int_1^{N+1}\mathfrak{f}+\sum_{N+1}^{\infty}\int_n^{n+1}(\mathfrak{f}(x)-\mathfrak{f}(n+\tfrac{1}{2}))\,dx-C$$

and
$$\sum_{N+1}^{\infty}\int_n^{n+1}(\mathfrak{f}(x)-\mathfrak{f}(n+\tfrac{1}{2}))\,dx = O(n^{1-\alpha}).$$

(4) Theorem 55 applied to $(a+bx)^{-s}$,

$$\begin{aligned}\mathfrak{f}(n)-\mathfrak{f}(n+1) &= \frac{(a+(n+1)\,b)^{1-s}-(a+nb)^{1-s}}{b(1-s)}-(a+(n+1))\,b^{-s}\\ &= (a+\xi b)^{-s}-(a+(n+1)\,b)^{-s} \qquad (n<\xi<n+1)\\ &= (n+1-\xi)\,(-s).b.(a+\eta b)^{-1-s} \qquad (\xi<\eta<n+1)\\ &= O(n^{-1-s})\end{aligned}$$

by two applications of the mean-value theorem.

$$\mathfrak{f}(n) = \sum_n^{\infty}(\mathfrak{f}(r)-\mathfrak{f}(r+1)).$$

(5) $$1+\tfrac{1}{2}+\ldots+\frac{1}{n}-\log n\downarrow \quad\text{and}\quad \to\gamma.$$

Thus
$$0<1+\tfrac{1}{2}+\ldots+\frac{1}{n}-\log n<1.$$

Consider
$$\frac{1}{N}\sum_1^N\left(1+\tfrac{1}{2}+\ldots+\frac{1}{n}-\log n\right).$$

(6) Theorem 55 applied to $s/x-1/x^s$.

(7) Gauss's test. Diverges.

(8) Induction on s assuming

$$\frac{n}{p^s}\leqslant n^{(s)}<\frac{n}{p^s}+1+\ldots+\frac{1}{p^{s-1}}.$$

If $\lim a_n/a_{n(1)} < \lambda < 1/p$ choose N so that for $n > N$, $a_n < \lambda a_{n(1)}$. If $Np^s < n \leqslant Np^{s+1}$, then

$$n^{(s)} \geqslant n/p^s > N \quad \text{and} \quad n^{(s)} \leqslant Np + \frac{p}{p-1} = N_1.$$

Write $M = \max\limits_{N<n<N_1} a_n$. Then

$$\sum_{Np^s<n\leqslant Np^{s+1}} a_n \leqslant \lambda^s . M . Np^s(p-1).$$

Since $\lambda p < 1$ the series $\sum\limits_{s=1}^{\infty} (\lambda p)^s$ converges. Hence result.

(9) $$\left|\log\left(\frac{a_{n+1}}{b_{n+1}}\right) - \log\left(\frac{a_n}{b_n}\right)\right| = |\log(1+c_n)| < 2\,|c_n|$$

for n sufficiently large.

$$\sum_1^N \left(\log\frac{a_{n+1}}{b_{n+1}} - \log\frac{a_n}{b_n}\right) = \log\frac{a_{N+1}}{b_{N+1}}\frac{b_1}{a_1}.$$

(10) Integrate by parts

$$I_{n+\frac{1}{2}}(x) - I_{n-\frac{1}{2}}(x) = \frac{1}{2}\int_x^\pi (\sin(n+\tfrac{1}{2})t - \sin(n-\tfrac{1}{2})t)\operatorname{cosec}\tfrac{1}{2}t\,dt$$

$$= \int_x^\pi \cos nt\,dt$$

$$= -\frac{\sin nx}{n},$$

$$\sum_1^\infty \frac{\sin nx}{n} = +I_{\frac{1}{2}}(x).$$

(11) Let a_n be the nth summand of the rearranged series. Then

$$\sum_1^N a_n = \sum_1^{\mathfrak{f}(N)} \frac{1}{2r-1} - \sum_1^{\mathfrak{g}(N)} \frac{1}{2r},$$

$$\sum_1^N a_n \sim \log(2\mathfrak{f}(N)-1) - \tfrac{1}{2}\log(\mathfrak{f}(N)-1) - \tfrac{1}{2}\log\mathfrak{g}(N),$$

$$\left(\sum_1^N \frac{1}{n} = \log N + \gamma + o(1)\right),$$

$$\sum_1^N a_n = \log(2\mathfrak{f}(N)-1) - \tfrac{1}{2}\log(\mathfrak{f}(N)-1) - \tfrac{1}{2}\log\mathfrak{g}(N) + o(1)$$

$$= \tfrac{1}{2}\log\left(4\frac{\mathfrak{f}(N)}{\mathfrak{g}(N)}\right) + o(1).$$

Second part

$\lambda = \frac{1}{2}$, $\mu = 2$ limits of oscillation are $\frac{1}{2}\log\frac{1}{2}$ and $\frac{1}{2}\log 8$, i.e. $\log 2\sqrt{2}$ and $-\log\sqrt{2}$.

(12) Integrate by parts twice the integrals

$$\int_0^{\frac{1}{2}} (1-2t)^2 D^2\mathfrak{f}(n+t)\,dt \quad \text{and} \quad \int_{-\frac{1}{2}}^0 (1+2t)^2 D^2\mathfrak{f}(n+t)\,dt.$$

Write $\mathfrak{f}(x) = e^{i\theta x^{\frac{1}{2}}} x^{-\frac{1}{2}}$. Then $|D^2\mathfrak{f}(x)| = O(1/x^{\frac{3}{2}})$. Thus the integral $\int^X \mathfrak{f}$ and the series $\sum^N \mathfrak{f}(n)$ converge and diverge together. Convergence if $\beta > \frac{1}{2}$, divergence if $\beta \leqslant \frac{1}{2}$.

(13) $\forall \delta > 0$, $\exists\, l$, X such that $|\mathfrak{f}(x) - l| < \delta$ if $x > X$. Then if $M = |\sup \mathfrak{f}(t)|$

$$\left|\epsilon \int_0^\infty e^{-\epsilon t}\mathfrak{f}(t)\,dt - l\right| \leqslant \epsilon \int_0^\infty e^{-\epsilon t}|\mathfrak{f}(t) - l|\,dt$$

$$\leqslant (M + |l|)(1 - e^{-\epsilon X}) + \delta(e^{-\epsilon X}) \to \delta$$

$$\text{as} \quad \epsilon \to 0+.$$

Chapter 27

(1) (*a*) The sum of $\Sigma[x/(1+x)^n]$ is discontinuous since it is 0 at $x = 0$ and 1 when $x > 0$. Hence the convergence is not uniform.

(*b*) By Abel's lemma, p. 47, with

$$a_n = (-1)^n \quad \text{and} \quad b_n = 1/n^x \log(1+n)$$

we have

$$\left|\sum_N^M \frac{(-1)^n}{n^x \log(1+n)}\right| \leqslant \frac{1}{N^x \log(1+N)}$$

$$\leqslant \frac{1}{\log(1+N)} \quad \text{since } x \geqslant 0.$$

By theorem 57*a* we have uniform convergence for $x \geqslant 0$.

Alternatively consider the series obtained by bracketing together pairs of consecutive terms. These terms are of opposite signs and we can apply the Mean Value Theorem. Hence we can prove the uniform convergence of the bracketed series and then deduce the same result for the original series since its summands tend uniformly to zero.

(c) Take $x = 1/N$ then

$$\sum_{N}^{2N} \frac{x}{(nx+1)((n+2)x+1)} > \frac{1}{15}$$

and the convergence is non-uniform.

(2) For $0 \leqslant x \leqslant 1-\delta$, $|x^n/(x^{2n}+1)| \leqslant (1-\delta)^n$ and $\exists N$ independent of x such that $(1-\delta)^n < \epsilon$ if $n > N$. Hence $x^n/(x^{2n}+1) \to 0$ uniformly over $[0, 1-\delta]$. For

$$1+\delta \leqslant x \leqslant 2, \quad \left|\frac{x^n}{x^{2n}+1}\right| \leqslant \left|\frac{1}{x^n}\right| \leqslant \frac{1}{(1+\delta)^n}$$

and again we have uniform convergence to zero. But at $x = 1$, $x^n/(x^{2n}+1) = \frac{1}{2}$ thus the limit function is discontinuous at $x = 1$. Hence convergence cannot be uniform over $[0, 2]$. For any x of the interval $[0, 2]$ we have $x^n/(x^{2n}+1) \leqslant \frac{1}{2}$;

$$\forall \epsilon > 0 \quad \left|\frac{(1-x)\,x^n}{x^{2n}+1}\right| < \epsilon \quad \text{if} \quad |1-x| < 2\epsilon.$$

If $\quad 0 \leqslant x \leqslant 1-2\epsilon, \quad \left|\frac{(1-x)\,x^n}{x^{2n}+1}\right| \leqslant x^n \leqslant (1-2\epsilon)^n.$

If $\quad 1+2\epsilon \leqslant x \leqslant 2, \quad \left|\frac{(1-x)\,x^n}{x^{2n}+1}\right| \leqslant \frac{1}{x^n} \leqslant \frac{1}{(1+2\epsilon)^n}.$

Thus $\exists N$ independent of x such that

$$\left|\frac{(1-x)\,x^n}{x^{2n}+1}\right| < \epsilon \quad \text{if} \quad n \geqslant N, \quad \text{i.e.} \quad \frac{(1-x)\,x^n}{x^{2n}+1} \to 0$$

uniformly over $[0, 2]$.

(3) If $x = 0$ the series is $\Sigma 0$ and converges. If $x > 0$

$$\frac{n^\alpha x^\beta}{1+n^2x^2} \sim n^{\alpha-2}.x^{\beta-2}$$

and the series converges if and only if $\alpha < 1$. If $\beta \geqslant 2$, $x \leqslant 1$,

$$\frac{n^\alpha x^\beta}{1+n^2x^2} \leqslant n^{\alpha-2}.$$

If $\beta < 2$, and n is sufficiently large, by considering $x < 1/n$ and $x > 1/n$ separately

$$\frac{n^\alpha x^\beta}{1+n^2x^2} \leqslant Kn^{\alpha-\beta},$$

where K depends only on α and β. Thus for n sufficiently large

$$\frac{n^\alpha x^\beta}{1+n^2x^2} \leqslant n^{\alpha-2} + Kn^{\alpha-\beta} = a_n.$$

Since $\alpha-2 < -1$ and $\alpha-\beta < -1$, Σa_n converges and thus by the Weierstrass M test $\Sigma[n^\alpha x^\beta/(1+n^2x^2)]$ is uniformly convergent. For convergence and therefore for uniform convergence we must have $\alpha < 1$. If now $\alpha+1 \geqslant \beta$ it follows that $\beta < 2$ and for

$$x = \frac{1}{N}$$

and N sufficiently large we have

$$\sum_{N}^{2N} \frac{n^\alpha x^\beta}{1+n^2x^2} > \sum_{N}^{2N} N^{\alpha-\beta}K_1$$

$$> N^{\alpha-\beta+1}K_1,$$

where K_1 depends only on α and β. Since $\alpha+1-\beta \geqslant 0$, $N^{\alpha-\beta+1}$ does not tend to zero as $N \to \infty$. Hence the convergence cannot be uniform. For the range $x \geqslant 1$,

$$\frac{n^\alpha x^\beta}{1+n^2x^2} \sim n^{\alpha-2}x^{\beta-2} \quad \text{as} \quad x \to \infty$$

n fixed. For uniform convergence and n sufficiently large $n^{\alpha-2}x^{\beta-2}$ must be small for all x, $x \geqslant 1$. Hence $\beta \leqslant 2$. If $\beta \leqslant 2$, $0 \leqslant \alpha < 1$

$$\left|\frac{n^\alpha x^\beta}{1+n^2x^2}\right| \leqslant n^{\alpha-2}x^{\beta-2} \leqslant n^{\alpha-2}$$

and we have uniform convergence by the Weierstrass M test.

(4) The sum of the series is 1 if $x \neq 0$, 0 if $x = 0$.

(5) Let $\mathfrak{g}(x)$ be the limit of the sequence. Then $\mathfrak{g}(x) = 0, 0 \leqslant x < 1$ and $\mathfrak{g}(1) = \mathfrak{f}(1)$. If the convergence is uniform $\mathfrak{g}(1) = 0$, i.e. $\mathfrak{f}(1) = 0$. On the other hand if $\mathfrak{f}(1) = 0$ $\forall \epsilon > 0 \, \exists \, \delta$ such that $|\mathfrak{f}(x)| < \epsilon$ if $1-\delta < x \leqslant 1$ and thus $|x^n\mathfrak{f}(x)| < \epsilon$ if $1-\delta < x \leqslant 1$. Next since $\mathfrak{f}(x)$ is continuous over $[0, 1-\delta]$ it is also bounded. Thus $\exists \, M$ such that $|\mathfrak{f}(x)| < M$, $0 \leqslant x < 1-\delta$. Hence $|x^n\mathfrak{f}(x)| \leqslant M(1-\delta)^n$, $0 \leqslant x \leqslant 1-\delta$. $\forall \epsilon > 0 \, \exists \, N$ such that $M(1-\delta)^n < \epsilon$ for $n > N$ and thus finally $|x^n\mathfrak{f}(x)| < \epsilon$ if $0 \leqslant x \leqslant 1$ and $n > N$. Since the definition of N does not depend on x it follows that $x^n\mathfrak{f}(n) \to 0$ uniformly over $0 \leqslant x \leqslant 1$.

(6) $\mathfrak{s}_n(0) = 2\cos 1/n \to 2$ as $n \to \infty$. $|\mathfrak{s}_n(x)| \leqslant \dfrac{2\sin^2 1/n}{x\cos 1/n} \to 0$ as $n \to \infty$, $x > 0$. $\mathfrak{s}_n(x)$ converges for $x \geqslant 0$ but not uniformly. $\mathfrak{s}_n(x)$ converges uniformly for $x \geqslant \delta > 0$.

(7) If $M_n \to 0$, $\forall \epsilon > 0 \exists N$ such that $|M_n| < \epsilon$ for $n > N$. Hence $|\mathfrak{s}_n(x) - \mathfrak{s}(x)| < \epsilon$ for $n > N$, $x \in E$. Since N is independent of x it follows that $\mathfrak{s}_n(x) \to \mathfrak{s}(x)$ uniformly over E.

If M_n does not converge to $0 \exists \epsilon > 0$ and a sequence of integers $\{n_i\}$ such that $M_{n_i} > \epsilon$. Hence $\exists x \in E$ say x_i such that

$$(1) \qquad |\mathfrak{s}_{n_i}(x_i) - \mathfrak{s}(x_i)| > \epsilon \quad (i = 1, 2, \ldots).$$

But if $\mathfrak{s}_n(x) \to \mathfrak{s}(x)$ uniformly over E $\forall \epsilon > 0 \exists N$ such that

$$(2) \quad |\mathfrak{s}_n(x) - \mathfrak{s}(x)| < \epsilon \quad \text{for all } n > N \quad \text{and} \quad \text{all } x \in E.$$

Since (1) and (2) are contradictory we see that $\mathfrak{s}_n(x) \to \mathfrak{s}(x)$ not uniformly over E.

$$\sin\frac{m.\pi}{2n} \geqslant \frac{m}{n} \quad \text{for} \quad 1 \leqslant m \leqslant n.$$

Thus $$\mathfrak{s}_n\left(\frac{\pi}{2n}\right) \geqslant \frac{1}{n^3}\sum_1^n m^2 = \frac{1}{6}\left(2+\frac{1}{n}\right)\left(1+\frac{1}{n}\right) > \frac{1}{3}.$$

$k > 1$, $|\sin^3 mx/m^k| \leqslant 1/m^k$. Uniform convergence over $[0, \frac{1}{2}\pi]$.

$k \leqslant 0$, $\sin^3 mx/m^k$ does not tend to zero as $m \to \infty$ unless $x = 0$. Not convergent for any x over $[0, \frac{1}{2}\pi]$ except $x = 0$.

$0 < k \leqslant 1$, use $\sin^3 mx = \frac{3}{4}\sin mx - \frac{1}{4}\sin 3mx$ to show that $\sum_1^N \sin^3 mx$ is bounded as N varies (x fixed $0 < x \leqslant \frac{1}{2}\pi$). Then theorem 8 establishes convergence (convergence at $x = 0$ is trivial).

The inequality $\mathfrak{s}_n(\pi/2n) > \frac{1}{3}$ is true if $0 < k < 1$ as well as for $k = 1$. If the convergence were uniform over $[0, \frac{1}{2}\pi]$ and if $\mathfrak{s}(x)$ were the sum function then $\mathfrak{s}(0) = 0$ and $\mathfrak{s}(x)$ is continuous. Thus $\exists \delta > 0$ such that $|\mathfrak{s}(x)| < \frac{1}{6}$ for $0 \leqslant x < \delta$.

By the uniformity of convergence again $\exists N$ such that

$$|\mathfrak{s}_n(x) - \mathfrak{s}(x)| < \tfrac{1}{6} \quad \text{all } n > N \quad \text{and} \quad x \in [0, \tfrac{1}{2}\pi].$$

Choose n so large that $\pi/2n < \delta$ and $n > N$, then we have both $|\mathfrak{s}(\pi/2n)| < \frac{1}{6}$ and $|\mathfrak{s}_n(\pi/2n) - \mathfrak{s}(\pi/2n)| < \frac{1}{6}$. Hence $|\mathfrak{s}_n(\pi/2n)| < \frac{1}{3}$. But this contradicts the inequality proved above and the convergence is therefore non-uniform.

(8) As $N \to \infty$, $\sum_1^N \mathfrak{f}_n(x) \to x^2 \sin\frac{1}{x} \quad (x \neq 0)$.

If $x \neq 0$, $$D\left(\sum_1^\infty \mathfrak{f}_n(x)\right) = 2x\sin\frac{1}{x} - \cos\frac{1}{x}.$$

If $x = 0$, $$D\left(\sum_1^\infty \mathfrak{f}_n(x)\right) = \lim_{x\to 0}\left(\frac{x^2 \sin 1/x}{x}\right) = 0.$$

Also

$$\sum_1^N D\mathfrak{f}_n(x) = D\left(\sum_1^N \mathfrak{f}_n(x)\right)$$

$$= \left(2+\frac{1}{2N-1}\right) x^{1+(1/2N-1)} \sin\frac{1}{x} - x^{1/(2N-1)} \cos\frac{1}{x} \quad (x \neq 0)$$

$$= 0 \quad (x = 0).$$

As $N \to \infty$,

$$\sum_1^N D\mathfrak{f}_n(x) \to D\left(\sum_1^\infty \mathfrak{f}_n(x)\right), \quad \text{i.e.} \quad \sum_1^\infty D\mathfrak{f}_n(x) = D\left(\sum_1^\infty \mathfrak{f}_n(x)\right).$$

The convergence of $\Sigma D\mathfrak{f}_n(x)$ cannot be uniform since the limit function $2x \sin 1/x - \cos 1/x$ is not continuous at $x = 0$, whereas the functions $\sum_1^N D\mathfrak{f}_n(x)$ are continuous for all x.

(9) $$\frac{t^{\alpha-1}}{1+t} = t^{\alpha-1} \sum_1^\infty (-t)^{n-1} \quad (0 < t < 1).$$

If $0 < x < 1$ and $0 \leqslant t \leqslant x$, then $\sum_1^\infty (-1)^{n-1} t^{\alpha+n-2}$ converges uniformly and thus

$$\int_0^x \frac{t^{\alpha-1}}{1+t} dt = \sum_1^\infty (-1)^{n-1} \int_0^x t^{\alpha+n-2}\, dt = \sum_1^\infty (-1)^{n-1} \frac{x^{\alpha+n-1}}{\alpha+n-1}.$$

The results hold when $x = 1$ because the series

$$\Sigma(-1)^{n-1} \frac{x^{\alpha+n-1}}{\alpha+n-1}$$

is uniformly convergent for $0 \leqslant x \leqslant 1$ and thus has a continuous sum. That is

$$\lim_{x\to 1-} \Sigma(-1)^{n-1} \frac{x^{\alpha+n-1}}{\alpha+n-1} = \Sigma(-1)^{n-1} \frac{1}{\alpha+n-1}.$$

Hence both $\lim_{x\to 1-} \int_0^x \frac{t^{\alpha-1}}{1+t} dt$ exists and its value is $\Sigma(-1)^{n-1} \frac{1}{\alpha+n-1}$. The fact that the series $\Sigma(-1)^{n-1}[x^{\alpha+n-1}/(\alpha+n-1)]$ is uniformly convergent over $[0, 1]$ can be seen by following a procedure used in question 1 (*b*).

(10) For $0 \leqslant x \leqslant 1$, $0 \leqslant \sin(\pi x^n) \leqslant \pi x^n$, and for $0 \leqslant x \leqslant 1-\delta$,

$$\left|\frac{\sin(\pi x^n)}{n}\right| \leqslant (1-\delta)^n \frac{\pi}{n}$$

Weierstrass's M test. By uniform convergence

$$\sum_1^\infty \frac{1}{n}\int_0^{1-\delta} \sin(\pi x^n)\,dx = \int_0^{1-\delta} \mathfrak{f}(x)\,dx.$$

Also

$$\int_{1-\delta}^1 \sin(\pi x^n)\,dx \leqslant \int_{1-\delta}^1 \pi x^n\,dx = \frac{\pi}{n+1}(1-(1-\delta)^{n+1}) \leqslant \frac{\pi}{n+1}.$$

Thus
$$\sum_1^\infty \frac{1}{n}\int_0^{1-\delta} \sin(\pi x^n)\,dx$$

converges uniformly in δ, where $0 \leqslant \delta \leqslant 1$. Hence

$$\lim_{\delta\to 0+}\int_0^{1-\delta} \mathfrak{f}(x)\,dx \quad \text{exists and equals} \quad \sum_1^\infty \frac{1}{n}\int_0^1 \sin(\pi x^n)\,dx.$$

(11) Since both $\quad \sum_1^\infty \mathfrak{f}_n(x) \quad$ and $\quad \sum_{-\infty}^0 \mathfrak{f}_n(x)$

converge uniformly it follows that

$$D\left(\sum_{-\infty}^\infty \mathfrak{f}_n(x)\right) = \sum_{-\infty}^\infty \mathfrak{f}_n(x).$$

Hence
$$D\left(e^{-x}\sum_{-\infty}^\infty \mathfrak{f}_n(x)\right) = 0.$$

Thus $e^{-x}\sum_{-\infty}^\infty \mathfrak{f}_n(x)$ is a constant and its value is $\sum_{-\infty}^\infty \mathfrak{f}_n(0)$.

(12)
$$\frac{d}{dx}\left(\frac{x}{1+n^2x^p}\right) = \frac{1+n^2x^p-pn^2x^p}{(1+n^2x^p)^2}$$
$$= \frac{1+n^2(x^p-px^p)}{(1+n^2x^p)^2} = \frac{1+n^2x^p(1-p)}{(1+n^2x^p)^2}.$$

If $p \leqslant 1$ this is positive. If $p > 1$ this is positive until

$$n^2x^p(1-p) = -1, \quad x^p = \frac{1}{(p-1)\,n^2} \leqslant 1,$$

and then negative. If $p \leqslant 1$,

$$0 \leqslant \mathfrak{s}_n(x) \leqslant \mathfrak{s}_n(1) = \frac{n}{1+n^2}.$$

If $p > 1$,

$$0 < \mathfrak{s}_n(x) < \mathfrak{s}_n\left(\left(\frac{1}{(p-1)\,n^2}\right)^{1/p}\right) = \frac{n^{1-2/p}(p-1)^{1-1/p}}{p}.$$

Thus uniform convergence if $p < 2$. Not uniform convergence if $p \geqslant 2$. In the case $p = 2$,

$$\int_0^1 \frac{nx}{1+n^2x^2}\,dx = \frac{1}{2n}\Big[\log(1+n^2x^2)\Big]_0^1 = \frac{\log(1+n^2)}{2n} \to 0 \quad \text{as} \quad n \to \infty.$$

In the case $p = 4$,

$$2\int_0^1 \frac{nx}{1+n^2x^4}\,dx = \Big[\tan^{-1} nx^2\Big]_0^1 = \tan^{-1} n \to \tfrac{1}{2}\pi \quad \text{as} \quad n \to \infty.$$

(13) Write $\mathfrak{g}(x)$ for $\sum_{n=1}^{\infty} 2^{-n+1}\mathfrak{f}(2^{n-1}x)$. Each function $\mathfrak{f}(2^{n-1}x)$ is continuous at every x except possibly x of the form $p/2^{n-1}$ where p is an integer. By uniform convergence $\mathfrak{g}(x)$ is continuous except at such points. (Compare with (i) p. 183.) If x is itself an odd integer p then

$$\lim_{x\to p+} \mathfrak{f}(x) = -1, \quad \lim_{x\to p-} \mathfrak{f}(x) = 1,$$

$$\lim_{x\to p+} \mathfrak{f}(2^{n-1}x) = 1, \quad \lim_{x\to p-} \mathfrak{f}(2^{n-1}x) = -1 \quad (n = 2, 3, \ldots).$$

Thus $$\lim_{x\to p+} \sum_{1}^{N} 2^{-n+1}\mathfrak{f}(2^{n-1}x) = -\tfrac{1}{2} + \sum_{2}^{N} 2^{-n} = -2^{-N},$$

and $$\overline{\lim_{\delta\to 0+}}\, |\mathfrak{g}(p) - \mathfrak{g}(p+\delta)| \leqslant 2^{-N+1}.$$

This is true for every integer N. Hence $\mathfrak{g}(p+\delta) \to \mathfrak{g}(p)$ as $\delta \to 0+$. Similarly $\mathfrak{g}(p+\delta) \to \mathfrak{g}(p)$ as $\delta \to 0-$ and $\mathfrak{g}(x)$ is continuous at p.

Finally for any x $$\mathfrak{g}(\tfrac{1}{2}x) = \mathfrak{f}(\tfrac{1}{2}x) + \tfrac{1}{2}\mathfrak{g}(x).$$

Now if x is not an even integer $\mathfrak{f}(\tfrac{1}{2}x)$ is continuous and thus in this case $\mathfrak{g}(x)$ is continuous if and only if $\mathfrak{g}(\tfrac{1}{2}x)$ is continuous. Thus $\mathfrak{g}(x)$ is continuous for all x of the form $p/2^{n-1}$ where p is an odd integer and $n \geqslant 1$.

Answers to further examples

(14) Write $\mathfrak{f}(x) = \sum_{r=1}^{\infty} \mathfrak{f}_n(x)$. Let M be a large integer. Use the convergence of the series at each point i/M to make

$$\left|\sum_{r=1}^{n} \mathfrak{f}_r(i/M) - \mathfrak{f}(i/M)\right|$$

small and then use the Mean Value Theorem. Both convergent and uniformly convergent over $[0, 1]$.

(18) Uniformly convergent over $[\tfrac{1}{2}\pi, \pi]$. Not uniformly convergent over $[0, \tfrac{1}{2}\pi]$.

(19) (i) $\int_0^X \sum_{n=1}^{\infty} \mathfrak{f}_n(x)\,dx = -(b-c)\log(1-X^a)$

$$-(c-a)\log(1-X^b)-(a-b)\log(1-X^c)$$

$$= \sum_{n=1}^{\infty} \int_0^X \mathfrak{f}_n(x)\,dx \quad (0 < X < 1).$$

(ii) $\int_0^1 \sum_{n=1}^{\infty} \mathfrak{f}_n(x)\,dx = -(b-c)\log a-(c-a)\log b-(a-b)\log c.$

(Writing $(1-X^a)$ as $(1-X)(1+X+\ldots+X^{a-1})$)

$$\sum_{n=1}^{\infty} \int_0^1 \mathfrak{f}_n(x)\,dx = 0.$$

Chapter 28

(1) (*a*) Not continuous or differentiable.

(*b*) Continuous; not differentiable since the partial derivatives do not exist.

(*c*) Continuous and differentiable. Near (0, 0).

$$\mathfrak{f}(0,y) = |iy-1| = \sqrt{(1+y^2)}. \quad \text{Thus } \mathfrak{f}_2(0,0) = 0.$$

$$\mathfrak{f}(x,0) = |x-1| = 1-x. \quad \text{Thus } \mathfrak{f}_1(0,0) = -1.$$

Now $|\mathfrak{f}(x,y)-\mathfrak{f}(0,0)+x| = \big|\,|x+iy-1|-1+x\big|$

$$= \sqrt{[(1-x)^2+y^2]}-(1-x)$$

$$= (1-x)\,O\left(\frac{y^2}{(1-x)^2}\right) = O(y^2) = o[\sqrt{(x^2+y^2)}] \quad \text{as} \quad \sqrt{(x^2+y^2)} \to 0.$$

(2) (iii) $\mathfrak{f}_1(0,0) = \mathfrak{f}_2(0,0) = 0$. Thus if differentiable we must have $|\mathfrak{f}(x,y)| = o[\sqrt{(x^2+y^2)}]$. But $\mathfrak{f}(y^2,y) = y \neq o[\sqrt{(y^2+y^4)}]$ as $y \to 0$. Hence not differentiable at (0, 0).

(3) (*a*) (i) Continuous for each $p > 0$.

(ii) Continuous for each $p > 0$.

(iii) Continuous for each $p > 0$ (by uniform convergence).

(*b*) (i) $\mathfrak{f}(x,y) = (|x|^p+|y|^p)^{1/p}$, $(x,y) \neq (0,0)$, $\mathfrak{f}(0,0) = 0$.

$\partial\mathfrak{f}/\partial x$ does not exist at (0, 0), therefore $\mathfrak{f}(x,y)$ is not differentiable at (0, 0).

(ii) $\mathfrak{f}(x,y) = |xy|^p$, $(x,y) \neq (0,0)$, $\mathfrak{f}(0,0) = 0$.

$\partial\mathfrak{f}/\partial x = \partial\mathfrak{f}/\partial y = 0$ at (0, 0) and thus $\mathfrak{f}$ is differentiable at (0, 0) if and only if $|xy|^p = o(x^2+y^2)^{\frac{1}{2}}$ as $(x^2+y^2)^{\frac{1}{2}} \to 0$, i.e. if and only if $p > \frac{1}{2}$.

(iii) $\mathfrak{f}(x,y) = \sum_{1}^{\infty} a_n |xy|^{p/n}$, $(x,y) \neq (0,0)$. $\mathfrak{f}(0,0) = 0$.

Again as in (ii) the necessary and sufficient condition for differentiability is $\mathfrak{f}(x,y) = o(x^2+y^2)^{\frac{1}{2}}$ as $(x^2+y^2)^{\frac{1}{2}} \to 0$. This is false for each $p > 0$. For given p choose N so that $p/N \leqslant \frac{1}{2}$, then

$$\mathfrak{f}(x,y) \geqslant a_N |xy|^{p/N} \quad \text{and} \quad \mathfrak{f}(x,x) \geqslant a_N |x| \quad \text{if} \quad |x| < 1.$$

(4) $|\mathfrak{f}(x,y)-\mathfrak{f}(x_0,y_0)| \leqslant |\mathfrak{f}(x,y_0)-\mathfrak{f}(x_0,y_0)|+|\mathfrak{f}(x,y)-\mathfrak{f}(x,y_0)|$.
For $|x-x_0|+|y-y_0|$ small the first term of the right-hand side above is small by the continuity of $\mathfrak{f}$ with respect to x and the second by the Mean Value Theorem and the fact that $\mathfrak{f}_y(x,y)$ is bounded.

(5) $\mathfrak{f}(0,y) = -y \ (y \neq 0)$, $\left(\dfrac{\partial \mathfrak{f}}{\partial y}\right)_{0,0} = -1.$

$\mathfrak{f}(x,y) = 0$ if $x = y$ or $x = -y$. The theorem 64 does not apply. Condition (i) is not satisfied. If $\partial\mathfrak{f}/\partial y \neq 0$ were continuous at $(0,0)$ it could be applied.

(6) Write $\mathfrak{f}(x,y)$ for the polynomial. Then $(\partial\mathfrak{f}/\partial y)_{0,0} = 1$ and theorem 64 can be applied to show that $\mathfrak{f}(x,y) = 0$ defines y as a unique function of x with the properties stated.

(7) $\mathfrak{F}(x,x)\uparrow$ and therefore has at most enumerably many discontinuities (see p. 99). $\therefore \exists\, x_0$ such that $\mathfrak{F}(x,x)$ is continuous at x_0, where $0 < x_0 < 1$. Then $\forall\, \epsilon > 0\, \exists\, \delta > 0$ such that

$$|\mathfrak{F}(x,x)-\mathfrak{F}(x_0,x_0)| < \epsilon \quad \text{if} \quad |x-x_0| < \delta.$$

Consider $\mathfrak{F}(x,y)$, where $|x-x_0| < \delta$, $|y-y_0| < \delta$. Then if $x \leqslant y$, $\mathfrak{F}(x,x) \leqslant \mathfrak{F}(x,y) \leqslant \mathfrak{F}(y,y)$, with reversed inequalities if $y \leqslant x$. Hence $|\mathfrak{F}(x,y)-\mathfrak{F}(x_0,x_0)| < \epsilon$ if $|x-x_0| < \delta$ and $|y-y_0| < \delta$, i.e. $\mathfrak{F}(x,y)$ is continuous at (x_0,x_0).

(8) For $(x,y) \neq (0,0)$,

$$\frac{\partial \mathfrak{f}}{\partial x} = \frac{y(x^4-y^4)+4x^2y^3}{(x^2+y^2)^2}, \quad \frac{\partial \mathfrak{f}}{\partial y} = \frac{x(x^4-y^4)-4x^3y^2}{(x^2+y^2)^2}.$$

At $(0,0)$,

$$\frac{\partial \mathfrak{f}}{\partial x} = \frac{\partial \mathfrak{f}}{\partial y} = 0, \quad \therefore \quad \frac{\partial}{\partial x}\left(\frac{\partial \mathfrak{f}}{\partial y}\right)_{(0,0)} = 1 = -\frac{\partial}{\partial y}\left(\frac{\partial \mathfrak{f}}{\partial x}\right)_{(0,0)}.$$

(9) For $(x,y) \neq (0,0)$,

$$\frac{\partial \mathfrak{f}}{\partial x} = 2x\tan^{-1}\frac{y}{x}-y, \quad \frac{\partial \mathfrak{f}}{\partial y} = x-2y\tan^{-1}\frac{x}{y}.$$

At $(0,0)$,

$$\frac{\partial \mathfrak{f}}{\partial x} = \frac{\partial \mathfrak{f}}{\partial y} = 0, \quad \therefore \quad \frac{\partial}{\partial x}\left(\frac{\partial \mathfrak{f}}{\partial y}\right)_{(0,0)} = 1 = -\frac{\partial}{\partial y}\left(\frac{\partial \mathfrak{f}}{\partial x}\right)_{(0,0)}.$$

(10) From the differentiability of $\mathfrak{f}$ and an argument similar to that at the end of theorem 64 we have

$$\frac{dy}{dx} = -\frac{(\partial\mathfrak{f}/\partial x)}{(\partial\mathfrak{f}/\partial y)}. \quad \text{Then} \quad \frac{\partial\mathfrak{f}}{\partial y}\frac{dy}{dx} + \frac{\partial\mathfrak{f}}{\partial x} = 0.$$

Repeating the process with the function $(\partial\mathfrak{f}/\partial y)(dy/dx) + \partial\mathfrak{f}/\partial x$ in place of $\mathfrak{f}$ (with dy/dx expressed as a function of x) we obtain

$$\frac{\partial}{\partial y}\left(\frac{\partial\mathfrak{f}}{\partial y}\frac{dy}{dx} + \frac{\partial\mathfrak{f}}{\partial x}\right)\frac{dy}{dx} + \frac{\partial}{\partial x}\left(\frac{\partial\mathfrak{f}}{\partial y}\frac{dy}{dx} + \frac{\partial\mathfrak{f}}{\partial x}\right) = 0,$$

i.e.
$$\frac{\partial^2\mathfrak{f}}{\partial y^2}\frac{dy}{dx} + 2\frac{\partial^2\mathfrak{f}}{\partial y\,\partial x}\frac{dy}{dx} + \frac{\partial\mathfrak{f}}{\partial y}\frac{d^2y}{dx} + \frac{\partial^2\mathfrak{f}}{\partial x^2} = 0.$$

Substitution for dy/dx gives d^2y/dx^2.

(11) Use an extension of the argument of theorem 61.

(12) The only possible points of non-differentiability lie on $x = 0, x = 1$ or $y = 0$ or $y = 1$. For continuity we need $p > 0, q > 0$.

For differentiability the partial derivatives must exist and (since $\mathfrak{f}(x, y) = 0$ if (x, y) does not lie in $0 < x < 1$, $0 < y < 1$) they must be zero on $x = 0, 1$ or $y = 0, 1$. This implies that $p > 1$, $q > 1$. On the other hand, if $p > 1$, $q > 1$, then the function is differentiable.

(14) See p. 196.

If $a \neq 0$,

$$e^x(ax^2 + bxy^2 + cy^4) = e^x\left(\frac{(ax + \tfrac{1}{2}by^2)^2}{a} + \left(c - \frac{b^2}{4a}\right)y^4\right).$$

This function is positive if $y = 0$, negative if $x = -(b/2a)\,y^2$ when $b^2 > 4ac$. Thus no maximum or minimum at $(0, 0)$. A similar argument holds if $c \neq 0$. If $a = c = 0$, then $e^x b\,xy^2$ is positive or negative according as $x > 0$, $x < 0$ or vice versa (in this case $b^2 > 4ac$ implies $b \neq 0$).

The remainder of the question is an immediate application of the criterion given.

(15) Show that the integral $\int_0^\infty \frac{\sin xy}{x}\,dx$ is uniformly convergent in $y > 0$. (The lack of definition at $x = 0$ is not important, since we can consider $\mathfrak{f}(x, y)$ instead of $\sin xy/x$, where $\mathfrak{f}(x, y) = \sin xy/x$ if $x > 0$ and $\mathfrak{f}(x, y) = y$ if $x = 0$. $\mathfrak{f}(x, y)$ is then continuous in $0 \leqslant x < \infty$, $a \leqslant y \leqslant b$) by writing

$$\int_{X_1}^{X_2} \frac{\sin xy}{x}\,dx = \int_{X_1 y}^{X_2 y} \frac{\sin u}{u}\,du$$

and observing that $\forall \epsilon > 0 \, \exists X_0$ such that

$$\left|\int_{X_1}^{X_2} \frac{\sin u}{u} du\right| < \epsilon \quad \text{if} \quad X_2 \geqslant X_1 > X_0.$$

Hence $$\left|\int_{X_1}^{X_2} \frac{\sin xy}{x} dx\right| < \epsilon \quad \text{if} \quad X_2 \geqslant X_1 > X_0/a$$

and since X_0/a does not depend on y this means that $\int_0^\infty \frac{\sin xy}{x} dx$ converges uniformly. Then by (vi), p. 200

$$\int_0^\infty \left(\int_a^b \frac{\sin xy}{x} dy\right) dx = \int_a^b \left(\int_0^\infty \frac{\sin xy}{x} dx\right) dy = \int_a^b \left(\int_0^\infty \frac{\sin u}{u} du\right) dy.$$

(16) Establish the uniform convergence of $\int_0^\infty e^{-x^2}(-2x)\sin 2xy\, dx$ by the Weierstrass M test ($|e^{-x^2}(-2x)\sin 2xy| \leqslant 2x\, e^{-x^2}$ and the integral $\int_0^\infty e^{-x^2}(-2x)\, dx$ converges). Then if $\mathfrak{J} = \int_0^\infty e^{-x^2} \cos 2xy\, dx$ obtain the relation $d\mathfrak{J}/dy = -2y . \mathfrak{J}$ and integrating $\mathfrak{J} = K . e^{-y^2}$.

(17) Induction.

APPENDIX

The number system

The whole numbers

§ 1. *The axioms*

The mathematician's idea of a whole number is based upon the notion of the physical process of counting. The two basic principles of counting are that each element is 'followed' by another unless it is a 'last' element and that each element is 'preceded' by another unless it is a 'first' element. The counting procedure 'orders' the set so that each element occupies a definite position in the order.

Both of these principles are embodied in the axioms of the whole number system except that we omit the 'last' element and assert that every element is 'followed' by another. This enables us to use one system of whole numbers for all practical counting purposes.

These two principles do not characterize a unique system and we add a third which asserts that the set of whole numbers is the smallest system satisfying the two basic principles. This third axiom enables us to make assertions about 'all whole numbers' which we could not otherwise do.

The axioms are due to Peano.

The set X and function $\mathfrak{F}$ satisfy $\mathbf{d}(\mathfrak{F}) = X$, $\mathbf{r}(\mathfrak{F}) \subset X$. We also write x^* for $\mathfrak{F}(x)$ and Y^* for $\mathfrak{F}(Y)$ where $Y \subset X$.

$\Rightarrow$ denotes logical implication.

Definition. *A non-empty subset Y of X is inductive if $Y \supset Y^*$.*

X is an inductive set.

X and $\mathfrak{F}$ define the *whole number system* if and only if the following axioms hold.

I. $\quad x^* = y^* \Rightarrow x = y$.

II. $\exists$ exactly one element of X not in $\mathbf{r}(\mathfrak{F})$. It is written 0.

III. Y is inductive and $0 \in Y \Rightarrow Y$ is X.

III is called the *Principle of Mathematical Induction*. Write 1 for 0^*, 2 for 1^*, 3 for 2^* etc. also zero for 0, one for 1, two for 2 etc.

§2. *Properties of inductive sets*

I (1). If *Y is inductive then so is Y^**

$$Y^* \subset Y \Rightarrow (Y^*)^* \subset Y^*.$$

I (2). If *Y_i is inductive and $\bigcap Y_i \neq \phi$ then $\bigcap Y_i$ is inductive*

$$\bigcap Y_i \subset Y_i \quad \text{and} \quad Y_i^* \subset Y_i \Rightarrow (\bigcap Y_i)^* \subset Y_i^* \subset Y_i \quad \text{for every } i,$$

therefore
$$(\bigcap Y_i)^* \subset (\bigcap Y_i).$$

By I (2) the intersection of all those inductive subsets of X which contain a fixed element a is an inductive set. It is denoted by $M(a)$. In particular III implies that $M(0) = X$.

I (3). If $b \in M(a)$ then $M(b) \subset M(a)$.

I (4). $a \notin M(a^*)$.

Write $A = \{a \mid a \notin M(a^*)\}$.

X^* is inductive and contains 0^*. Since $M(0^*) \subset X^*$ it follows that $0 \notin M(0^*)$. Thus $0 \in A$.

$$\begin{aligned} b \in A &\Rightarrow b \notin M(b^*) \\ &\Rightarrow b^* \notin (M(b^*))^*. \end{aligned}$$

Now
$$\begin{aligned} M(b^*) \supset b^* &\Rightarrow (M(b^*))^* \supset (b^*)^* \\ &\Rightarrow (M(b^*))^* \supset M((b^*)^*) \quad \text{from I (1)}. \end{aligned}$$

Hence $b^* \notin M((b^*)^*)$, i.e. $b^* \in A$.

By Axiom III $A = X$.

I (5). $M(a) = (a) \cup M(a^*)$.

$$a \in M(a) \Rightarrow a^* \in M(a) \Rightarrow M(a^*) \subset M(a)$$
$$\Rightarrow (a) \cup M(a^*) \subset M(a).$$

But $(a) \cup M(a^*)$ is inductive and contains a. Thus

$$(a) \cup M(a^*) \supset M(a).$$

I (6). $M(a^*) = (M(a))^*$.

$$(M(a))^* \supset a^* \Rightarrow (M(a))^* \supset M(a^*). \qquad \text{I (1)}$$

But $$(M(a))^* = (a^*) \cup (M(a^*))^* \subset M(a^*). \qquad \text{I (5)}$$

I (7). $M(a) \supset M(b) \Rightarrow$ *either* $M(a^*) \supset M(b)$ *or* $M(a^*) \subset M(b)$.

$$M(a) = (a) \cup M(a^*) \Rightarrow \text{either } a \in M(b) \text{ or } M(a^*) \supset M(b),$$
$$a \in M(b) \Rightarrow M(b) \supset M(a) \Rightarrow M(b) = M(a) \Rightarrow M(b) \supset M(a^*).$$

I (8). *Either* $M(a) \supset M(b)$ *or* $M(a) \subset M(b)$.

With fixed b write

$$A = \{a | \text{either } M(a) \supset M(b) \text{ or } M(a) \subset M(b)\}.$$

$$M(0) = X \Rightarrow M(b) \subset M(0) \Rightarrow 0 \in A,$$
$$M(a) \subset M(b) \Rightarrow M(a^*) \subset M(a) \subset M(b),$$
$$M(a) \supset M(b) \Rightarrow \text{either } M(a^*) \supset M(b) \text{ or } M(a^*) \subset M(b) \quad \text{by I (7).}$$

Thus $a \in A \Rightarrow a^* \in A$.

$A = X$ and I (8) is proved.

I (9). $M(a) = M(b)$ *if and only if* $a = b$.

$$a = b \Rightarrow M(a) = M(b) \quad \text{by definition.}$$
$$M(a) = M(b) \Rightarrow (a) \cup M(a^*) = (b) \cup M(b^*) \qquad \text{I (5)}$$
$$\Rightarrow a = b \text{ or } a \in M(b^*),$$
$$a \in M(b^*) \Rightarrow M(a) \subset M(b^*)$$
$$\Rightarrow (a) \cup M(a^*) \subset M(b^*)$$
$$\Rightarrow b \notin (a) \cup M(a^*).$$

This is false. Hence $a \notin M(b^*)$ and $a = b$.

Definition. *If $M(a) \subset M(b)$ we write $a \geqslant b$ or $b \leqslant a$.*
If in addition $a \neq b$ then we write $a > b$ or $b < a$.

The properties of inductive sets imply the following

(i)	$0 \leqslant x$ all $x \in X$.	
(ii)	Either $a \leqslant b$ or $b \leqslant a$.	I (8)
(iii)	$a \leqslant b$ and $b \leqslant a \Rightarrow a = b$.	I (9)
(iv)	$a \leqslant b$ and $b < c \Rightarrow a < c$.	
(v)	$a < b \Rightarrow a^* < b^*$.	I (5)
(vi)	$a < b \Rightarrow a^* \leqslant b$.	I (5)
(vii)	$a < b^* \Rightarrow a \leqslant b$.	
(viii)	$a < b \leqslant a^* \Rightarrow b = a^*$.	

Definition. *We call the set $Y = \{y \mid a \leqslant y, y \leqslant b\}$ the interval $[a, b]$. This symbol is used only when $a \leqslant b$.*

I (10). *If $T \neq \phi$ and $\mathscr{C}(T) \neq \phi$ then not both T and $\mathscr{C}(T)$ are inductive subsets of X.*

If $0 \in T$ and T is inductive then $T = X$ by Axiom III and $\mathscr{C}(T) = \phi$. Hence whichever of $T, \mathscr{C}(T)$ contains 0 is not inductive.

I (11). *T inductive and $a \notin T \Rightarrow M(a) \supset T$.*

For otherwise $\exists\, b \in T \cap \mathscr{C}(M(a))$. Since T is inductive $T \supset M(b)$. By I (8) since $M(a) \not\ni b$, $M(b) \supset M(a) \ni a$. Therefore $a \in T$. This is false. Hence $M(a) \supset T$.

I (12). *T inductive $\Rightarrow \exists\, a$ such that $T = M(a)$.*

Either $T = X = M(0)$ or by I (10) $\exists\, a \in T \cap (\mathscr{C}(T))^*$. For by I (10) $\mathscr{C}(T)$ is not inductive, i.e. $\exists\, a \in (\mathscr{C}(T))^*$ such that $a \notin \mathscr{C}(T)$. Then

$$a \in T \Rightarrow M(a) \subset T.$$

$a \in (\mathscr{C}(T))^* \Rightarrow a = c^*$, where $c \notin T \Rightarrow M(c) \supset T$. Thus by I (5) $(c) \cup M(a) \supset T$. But $c \notin T$, therefore $M(a) \supset T$. Hence finally $M(a) = T$.

Theorem 1. *Every non-void subset Y of X contains a least member, i.e. $\exists\, p \in Y$ such that $y \in Y \Rightarrow y \geqslant p$.*

Let T be the intersection of all inductive sets containing Y. By I (2), I (12) $T = M(p)$, say. Since $T \cap \mathscr{C}(p)$ is inductive, $p \in Y$ and since $T \supset Y$ we have $y \in Y \Rightarrow y \geqslant p$.

§ 3. 1–1 *correspondences*

If for any two sets Y, Z there exists a 1–1 correspondence between the elements of Y and those of Z we write

$$Y \leftrightarrow Z.$$

The function $\mathfrak{F}(x) = x^*$ shows that $[a, b] \leftrightarrow [a^*, b^*]$.

C (1). *If $\mathfrak{f}$ satisfies* $\mathbf{d}(\mathfrak{f}) = [a,b]$ *and* $\mathbf{r}(\mathfrak{f}) \supset [a,b]$ *then* $\mathbf{r}(\mathfrak{f}) = [a,b]$.

Let B be the set of whole numbers b for which this statement is true with fixed a. Then $a \in B$. Assume $b \in B$. If $\mathfrak{f}$ satisfies

$$\mathbf{d}(\mathfrak{f}) = [a, b^*], \quad \mathbf{r}(\mathfrak{f}) \supset [a, b^*]$$

suppose $\mathfrak{f}(b^*) = t$ and that s is such that $\mathfrak{f}(s) = b^*$. Define $\mathfrak{g}$ by $\mathfrak{g}(x) = \mathfrak{f}(x)$ if $x \neq b^*$, s and $\mathfrak{g}(s) = t$. Then $\mathbf{d}(\mathfrak{g}) = [a,b]$, $\mathbf{r}(\mathfrak{g}) \supset [a,b]$ and since $b \in B$ $\mathbf{r}(\mathfrak{g}) = [a,b]$. Hence $\mathbf{r}(\mathfrak{f}) = [a,b^*]$. Thus $b^* \in B$. B is inductive and $B = M(a)$.

C (2). *If* $[a,b] \leftrightarrow [a,c]$ *then* $b = c$.

If $\mathfrak{f}$ is a 1–1 correspondence between $[a,b]$ and $[a,c]$ then C (2) follows from C (1) applied to $\mathfrak{f}$ or $\mathfrak{f}^{-1}$.

§ 4. *Bounded, finite and infinite sets*

Definition. *A non-empty set of whole numbers Y is said to be bounded if $\exists\, k \in X$ such that $y \in Y \Rightarrow y \leqslant k$.*

The union of two bounded sets is bounded.

F (1). *A non-empty bounded set Y has a largest member.*

Write $K = \{k | y \leqslant k \text{ for all } y \in Y\}$. K is inductive and thus $\exists\, a$ such that $K = M(a)$. If $a = 0$, $Y = (0)$ and 0 is the largest member of Y. If $a \neq 0$ then $a = b^*$ say. Now $y \leqslant a$ implies $y \leqslant b$ or $y = a$ Since $b \notin M(a) = K$ $\exists\, y \in Y$ such that $y = a$ and a is the largest member of Y.

F (2). *If Y is a non-empty bounded set then $\exists\, p$ such that $Y \leftrightarrow [0,p]$.*

Let $T = \{t | \exists$ a 1–1 correspondence between Y and a subset of $[0,t]\}$. Since Y is bounded T is non-empty. Let p be the least member of T. Then $\exists\, K \subset [0,p]$ such that $Y \leftrightarrow K$, denote this correspondence by $\mathfrak{f}$ so that $\mathfrak{f}(Y) = K$. Now $K = [0,p]$ for if $\exists\, a \in [0,p]$ and $a \notin K$ define a new 1–1 correspondence $\mathfrak{g}$ by

$$\mathfrak{g}(\mathfrak{f}^{-1}(p)) = a, \quad \mathfrak{g}(x) = \mathfrak{f}(x), \quad x \neq \mathfrak{f}^{-1}(p).$$

Then $\mathfrak{g}$ sets up a 1–1 correspondence between Y and a subset of $[0,q]$, where $p = q^*$. This contradicts the fact that p is the least member of T. Thus $K = [0,p]$ and $Y \leftrightarrow [0,p]$.

By C (2) the p defined in F (2) is unique.

Definition. *If $Y \leftrightarrow [0,k]$ we say that Y has k^* members or is of cardinal k^*. Equivalently* if *$Y \leftrightarrow [1,k]$ we say Y has k members.*

Definition. *If $\exists\, k$ such that $Y \leftrightarrow [0,k]$ we say that Y is a finite set. Otherwise Y is an infinite set.*

Thus F (2) can be stated as 'Every bounded set is finite'. The converse is also true.

F(3). *If Y is a finite set of whole numbers then Y is bounded.*

Let K be the set of whole numbers k such that if Y has k^* members then Y is bounded. $0 \in K$ since a set consisting of 1 number is bounded. If $t \in K$ and $\mathfrak{f}$ is a 1–1 correspondence between Y and $[0, t^*]$ so that $\mathfrak{f}([0, t^*]) = Y$, then $\mathfrak{f}[0, t]$ is a bounded set and so is the single number $\mathfrak{f}(t^*)$. Hence Y, the union of two bounded sets, is itself bounded.

Corollary 1. *The union of two finite sets is a finite set.*

Corollary 2. *A subset of a finite set is finite.*

F(4). *If Y is finite and $\mathfrak{f}$ is a function for which $\mathbf{d}(\mathfrak{f}) = Y$ and $\mathbf{r}(\mathfrak{f}) \subset X$ then $\mathfrak{f}(Y)$ is a finite set*

From the set $\mathfrak{f}^{-1}(p)$, where $p \in \mathfrak{f}(Y)$ select the least number. This is done for each p of $\mathfrak{f}(Y)$ and the selected numbers form a subset Y_1 of Y. Then $Y_1 \leftrightarrow \mathfrak{f}(Y)$ and as Y_1 is finite $Y_1 \leftrightarrow [0, k]$ for some k, therefore $\mathfrak{f}(Y) \leftrightarrow [0, k]$ and $\mathfrak{f}(Y)$ is a finite set.

F(5). *There is no 1–1 correspondence between a finite set Y and an infinite set T.*

For a function $\mathfrak{f}$ with $\mathbf{d}(\mathfrak{f}) = Y$, by F(4), $\mathfrak{f}(Y)$ is finite and therefore cannot coincide with T.

The set X is infinite for if $X \leftrightarrow [0, k]$ $\exists$ $\mathfrak{f}$ with $\mathbf{d}(\mathfrak{f}) = [0, k]$, $\mathbf{r}(\mathfrak{f}) \supset [0, k]$. By C(1) $\mathbf{r}(\mathfrak{f}) = [0, k]$. But this is not so for $k^* \in X$ and $k^* \notin [0, k]$.

The set of all whole numbers X has the remarkable property that there exists a 1–1 correspondence between X and a proper subset of X; the function $\mathfrak{F}(x) = x^*$ sets up such a correspondence. But it is a fundamental that if m, n are whole numbers and $m \neq n$ then $[0, m]$ cannot be put into 1–1 correspondence with $[0, n]$, (as proved in C(2)). This property is the rational basis of counting finite sets, i.e. if we say that a set has five members we mean that it can be put into 1–1 correspondence with $[0, 4]$. If it were also possible to put the set into 1–1 correspondence with $[0, 3]$ the practical purpose of counting would be nullified since it would fail to reveal when an object had been removed from the set.

Any set of entities is said to be finite if and only if it can be put into 1–1 correspondence with a finite set of whole numbers. Otherwise it is infinite. If such a set can be put into 1–1 correspondence with $[0, n]$ we say that it has n^* members or is of cardinal n^*.

§ 5. *Addition*

Given a, consider the set B of numbers b for which $\exists$ x such that $[a, x] \leftrightarrow [0, b]$. Since $[a, a] \leftrightarrow [0, 0]$, $0 \in B$. If $b \in B$ and we add to the correspondence $[a,x] \leftrightarrow [0,b]$, $(x^*) \leftrightarrow (b^*)$ we see that $b^* \in B$. B is indictive and by Axiom III contains X. Thus given $b \in X$ $\exists\, x \in X$ such that

$$[a,x] \leftrightarrow [0,b]$$

x is uniquely defined by this relation (C (2)). We denote it by $a+b$, and call it the *sum* of a and b. We refer to this process as addition.

Properties of addition

A (1). $a+0 = a$. For $[a,a] \leftrightarrow [0,0]$.

A (2). $a^* = a+1$. For $[a,a^*] \leftrightarrow [0,1]$.

A (3). $a+b \geqslant a$. For $[a,a+b]$ exists.

A (4). $a^*+b = (a+b)^*$. For $[a^*,a^*+b] \leftrightarrow [0,b] \leftrightarrow [a,a+b] \leftrightarrow [a^*,(a+b)^*]$.

A (5). $a+b = b+a$. The *Communative Law*. For

$$\begin{aligned}[0,(a+b)^*] &= [0,a] \cup [a^*,(a+b)^*]\\ &\leftrightarrow [b^*,(b+a)^*] \cup [0,b]\\ &= [0,(b+a)^*]. \qquad \text{A (4)}\end{aligned}$$

Therefore $(a+b)^* = (b+a)^*$, i.e. $a+b = b+a$.

A (6). $(a+b)+c = a+(b+c)$. The *Associative Law*. For

$$\begin{aligned}[0,(a^*+b^*)+c] &= [0,a] \cup [a^*,(a^*+b)] \cup [a^*+b^*,(a^*+b^*)+c]\\ &\leftrightarrow [b^*+c^*,(b^*+c^*)+a] \cup [0,b] \cup [b^*,b^*+c]\\ &= [0,(b^*+c^*)+a].\end{aligned}$$

By A (4) applied twice and A (5) it follows that

$$(a+b)+c = a+(b+c).$$

We write $a+b+c$ for $(a+b)+c$.

A (7). $a+b = a+c \Rightarrow b = c$. The *Cancellation Law*

$$[0,b] \leftrightarrow [a,a+b] = [a,a+c] \leftrightarrow [0,c].$$

A (8). $\left.\begin{aligned}&a+b = 0 \Rightarrow a = b = 0,\\ &a+b = 1 \Rightarrow a = 1, b = 0 \text{ or } a = 0, b = 1.\end{aligned}\right\}$ A (3)

A (9). $a \geqslant b \Rightarrow a+c \geqslant b+c$. For $a \geqslant b \Rightarrow a = b+k$, where k is defined by $[0,k] \leftrightarrow [b,a]$ and thus

$$a+c = (b+k)+c = (b+c)+k \geqslant b+c.$$

Definition. *The difference between two whole numbers a, b is defined to be k, where if $a \geqslant b$ $[0, k] \leftrightarrow [b, a]$ and if $a \leqslant b$ $[0, k] \leftrightarrow [a, b]$.*

§ 6. *Multiplication*

Given $k, l \in X$ denote by $T(k, l)$ the set of all ordered pairs of whole numbers (a, b) where $1 \leqslant a \leqslant k$, $1 \leqslant b \leqslant l$. Since $T(k, 1) \leftrightarrow [1, k]$, $T(k, 1)$ is finite. Assume that $T(k, l)$ is finite. Then

$$T(k, l^*) = T(k, l) \cup P,$$

where P is the set of ordered pairs (a, l^*), $1 \leqslant a \leqslant k$. $P \leftrightarrow [1, k]$ thus P is finite, $T(k, l)$ is finite by assumption. Therefore $T(k, l^*)$ is finite. It follows from the Principle of Mathematical Induction that $T(k, l)$ is finite for all l, k.

Thus $\exists\, x$ such that $[0, x] \leftrightarrow T(k, l)$. We denote x^* by kl or $k.l$ so that $[1, kl] \leftrightarrow T(k, l)$. kl is uniquely defined by k and l.

Properties of multiplication

M (1). $k.1 = k$, $1.1 = 1$. For $T(k, 1) \leftrightarrow [1, k]$ and $T(1, 1) \leftrightarrow [1, 1]$.

M (2). $k.l = l.k$. *The Commutative Law.* For if we make the pair (a, b) correspond to (b, a) we see that $T(k, l) \leftrightarrow T(l, k)$.

M (3). $(k.l)\, m = k(l.m)$. *The Associative Law.*

$$\text{For} \quad [1, (kl)\, m] \leftrightarrow T(kl, m). \quad \text{Now} \quad [1, kl] \leftrightarrow T(k, l).$$

If $(a, b) \in T(kl, m)$ then a corresponds under the last 1–1 correspondence to a pair (a', a''), where $1 \leqslant a' \leqslant k$, $1 \leqslant a'' \leqslant l$. Thus $T(kl, m)$ is in 1–1 correspondence with the set of all ordered trials (a', a'', b), where $1 \leqslant a' \leqslant k, 1 \leqslant a'' \leqslant l, 1 \leqslant b \leqslant m$. This is also true for $T((lm)\, k)$ and the result follows. We write klm for $(kl)\, m$.

M (4). $k(l+m) = kl + km$. *The Distributive Law.* For

$$T(k(l+m)) = T(k, l) \cup T_1,$$

where T_1 is the set of ordered pairs (a, b) with $1 \leqslant a \leqslant k$ and $1 + l \leqslant b \leqslant l + m$. Thus $T_1 \leftrightarrow T(k, m)$. Hence $k(l+m) = kl + km$.

We define multiplication by 0 in such a way that M (4) is universally true. If this is to be so we must have $kl = k(l+0) = k.l + k.0$, i.e. $k.0 = 0$. Thus we *define* $k.0$ and $0.k$ to be 0.

M (5). $kl = km$ and $k \neq 0 \Rightarrow l = m$. *The Cancellation Law.* For if $l \geqslant m$, $l = m + t$ then $kl = km + kt$ and $kl > km$ unless $k.t = 0$. Since $k \neq 0$ this implies $t = 0$, therefore l is not greater than m. Similarly $m \leqslant l$, therefore $l = m$. We have

$$2.a = a + a, \quad 3.a = a + a + a, \quad \text{etc.}$$

M (6). $(a) x \geqslant y \Rightarrow x.z \geqslant y.z$; $(b) x.z \geqslant y.z$ and $z \neq 0 \Rightarrow x \geqslant y$.

(a) $x \geqslant y \Rightarrow x = y+k \Rightarrow x.z = y.z.+k.z \Rightarrow x.z \geqslant y.z.$

(b) $x < y \Rightarrow x.z < y.z$ as in (a) since $z \neq 0$, therefore

$$x.z \geqslant y.z \Rightarrow x \geqslant y.$$

Definition. *If a, b are whole numbers such that $\exists x$ for which $a = x.b$ then we say that b divides a or that b is a divisor or factor of a. If the only factors of a are a and 1 then we say that a is a prime; otherwise a is composite.*

M (7). *If $a = b+c$ where $b \neq 0$, $c \neq 0$ and p divides two of a, b, c then p also divides the third.*

Suppose that p divides b and c so that $b = pk$, $c = pl$. Then by the Distributive Law

$$a = p.k+p.l = p(k+l),$$

and p divides a.

Suppose that p divides a and b so that $b = p.k$, $a = p.m$. Now $a = b+c \Rightarrow a \geqslant b \Rightarrow m \geqslant k \Rightarrow m = k+x$ for some x. Thus

$$p.m = p.k+p.x, \quad \text{i.e.} \quad a = b+p.x.$$

Therefore by the Cancellation Law for addition $c = p.x$ and p divides c.

M (8). *If p divides a and $a \neq 0$ then $p \leqslant a$.*

For $\exists x$ such that $p.x = a$. Since $a \neq 0 \Rightarrow x \neq 0 \Rightarrow x \geqslant 1$ we have $p.x \geqslant p.1$, i.e. $a \geqslant p$.

Definition. *The largest number dividing both a, b is called the 'greatest common divisor' of a and b provided $a \neq 0$, $b \neq 0$.*

The existence of the greatest common divisor follows from M (8) and F (1).

M (9). *If $a, b \in X$ and $b \neq 0$ then $\exists t, s$ such that $a = tb+s$ and $s < b$.*

Let $T = \{x | xb \leqslant a\}$. By M (8) T is bounded and by F (1) has a largest member say t. Then since $t+1 > t$ we must have

$$tb \leqslant a < (t+1)b.$$

The first of these inequalities gives $a = tb+s$ and the second (from the Distributive Law) gives $s < b$.

M (10). *If $Y \subset X$ is such that the sum and the difference of any two members of Y also belongs to Y then Y consists of 0 and all the multiples of its least non-zero element (if it has any non-zero element).*

If Y contains any non-zero elements there is one of them say a that is least. By M (9) if $x \in Y$ and $x \neq 0$ then $x = at+s$, where $s < a$. By the definition of Y, $s \in Y$. Since $s < a$ we have $s = 0$ and $x = at$ is a multiple of a. On the other hand every multiple of a belongs to Y.

M (11). *If d is the greatest common divisor of a, b ($a \neq 0, b \neq 0$) then $\exists\, x, y$ such that either*

$$ax = by + d \quad \text{or} \quad bx = ay + d.$$

The set of all sums and differences of multiples of a and b has a least non-zero member say k and for appropriate x, y either $ax = by + k$ or $bx = ay + k$. Since d divides b and a it also divides k, therefore $d \leqslant k$ (M (8)). By M (10) a, b are multiples of k, therefore k divides both a, b, therefore $k \leqslant d$ since d is the *greatest* common divisor of a and b, therefore $k = d$.

M (12). *If c divides $a.b$ and the greatest common divisor of a and c is 1 then c divides b (assuming $a \neq 0$, $b \neq 0$).*

By M (11) $\exists\, x, y$ such that either

$$ax = cy + 1 \quad \text{or} \quad cx = ay + 1.$$

Thus either

$$abx = cby + b \quad \text{or} \quad cbx = aby + b$$

and by M (7) since c divides ab it follows that c divides b.

Powers

The set of functions $L(y, x)$ for which $\mathbf{d}(\mathfrak{f}) = [1, y]$ and $\mathbf{r}(\mathfrak{f}) \subset [1, x]$ is finite. For this is true if $y = 1$ since $L(1, x) \leftrightarrow [1, x]$. Let Y be the set of y for which $L(y, x)$ is finite for every x. Suppose $y \in Y$ and $\mathfrak{f}$ is such that $\mathbf{d}(\mathfrak{f}) = [1, y^*]$, $\mathbf{r}(\mathfrak{f}) \subset [1, x]$. Define $\mathfrak{g}$ by $\mathfrak{g}_{\mathfrak{f}}(u) = \mathfrak{f}(u)$ if $1 \leqslant u \leqslant x$. There is a 1–1 correspondence between the functions $\mathfrak{f}$ of $L(y^*, x)$ and the ordered pairs $(\mathfrak{g}_{\mathfrak{f}}, \mathfrak{f}(y^*))$. Now $\mathfrak{g}_{\mathfrak{f}} \in L(y, x)$, a finite set by assumption and the possible values of $\mathfrak{f}(y^*)$ also form a finite set. Thus the set $L(y^*, x)$ is finite and its cardinal is equal to that of $L(y, x)$ multiplied by x.

Hence $\exists\, k$ such that $[1, k] \leftrightarrow L(y, x)$. We define k to be x^y. Thus x^y is defined if $x \geqslant 1, y \geqslant 1$. We define x^0 to be 1 if $x \geqslant 1$. The symbol 0^0 is not defined and has no meaning. x^y is referred to as *x to the power y*.

It is inherent in the definition that $x^{y+1} = x^y . x$.

Properties of powers

P (1). $1^y = 1$. For $L(y, 1)$ contains only one function.

P (2). $x^y . t^y = (x.t)^y$ if $xt \neq 0$. For let Y be the set of whole numbers y for which this result is true for all non-zero x, t of X. Then $0 \in Y$ and if $y \in Y$ then

$$x^{y+1} . t^{y+1} = x^y . x . t^y . t = x^y . t^y . x . t = (xt)^y . xt = (xt)^{y+1}.$$

Thus $y \in Y \Rightarrow y + 1 \in Y$. Y is inductive, therefore $Y = X$.

P(3). $x^y . x^t = x^{y+t}$ if $x \neq 0$. Let T be the set of t for which this is true for all x, y. The $0 \in T$ and if $t \in T$ we have

$$x^y . x^{t+1} = x^y . x^t . x = x^{y+t} . x = x^{y+t+1},$$

i.e. $t+1 \in T$. Thus T is inductive and hence $T = X$.

P(4). $(x^y)^t = x^{y.t}$ if $x \neq 0$. Let T be the set of t for which this is true for all x, y. Then $0 \in T$ and if $t \in T$ we have

$$(x^y)^{t+1} = (x^y)^t . x^y = x^{y.t} . x^y = x^{y.t+y} = x^{y(t+1)},$$

i.e. $t+1 \in T$. Thus T is inductive and hence $T = X$.

§ 7. *Remarks on the whole number system*

Although we could pursue the development of the properties of whole numbers to a much greater extent we do not do so but prefer to define enlarged number systems. There is only one fundamental operation in the whole number system, that of adding a unit to a given whole number. Repetitions of this process lead to addition as defined between any two whole numbers; repetitions of additions lead to multiplication and repetitions of multiplications lead to powers of numbers. It is logically possible to base the whole of analysis upon this property. We do not need ever to define any other sort of number. But our analysis would be cumbersome and difficult to manipulate. In fact it is the difficulties of manipulation that led to the various extensions of the number system.

For example given a, b, we can always find x to satisfy the equations

$$x = a+b, \quad x = a.b, \quad x = a^b \quad (\text{if } a, b \text{ not both } 0)$$

but we can not always find x to satisfy the equations

$$a = b+x, \quad a = b.x, \quad a = x^b, \quad a = b^x$$

and the conditions under which these last equations can be solved are quite different from one another. For example, we can solve $a = b+x$ if $a \geqslant b$, i.e. given b there are only a finite number of values of a for which the equation $a = b+x$ has no solution, whereas given b there are infinitely many values of a for which $a = b.x$ has no solution for example if $a = lb+1$ for any l.

It is a matter of experience that one often needs to use 'solutions' of these equations. Of course we can never really 'solve' these equations. What we do is a typical mathematical evasion. If we consider what is the nature of our requirement for 'solutions' we find that it matters not at all whether or not the 'solutions' are whole numbers. What matters is that laws similar to those for whole numbers should hold between the solutions. For this reason we consider briefly more abstract algebraic systems defined axiomatically.

Algebraic systems†

I. Hemi-groups

A hemi-group is a set S and an operation . defined between every two elements x, y of S such that $x.y \in S$ and

(i) $x.y = y.x$ (The Commutative Law).

(ii) $(x.y).z = x.(y.z)$ (The Associative Law).

(iii) $x.y = x.z \Rightarrow y = z$ (The Cancellation Law).

(iv) $\exists\, e \in S$ such that $e.e = e$. We call e the *neutral element*. It has the property that $e.x = x = x.e$. For by (iv) $(e.e).x = e.x$; by (ii) $e.(e.x) = e.x$ and by (iii) $e.x = x$. Similarly $x.e = x$.

The whole numbers from a hemi-group with addition as . and 0 as e; they also form a hemi-group with multiplication as . and 1 as e provided that 0 is excluded.

II. Groups

A set of elements G and an operation . defined between every two elements of G so that $a.b \in G$ is a group if and only if

(i) $(a.b).c = a.(b.c)$, (Associative Law).

(ii) $\exists e \in G$ such that $e.a = a.e = a$.

(iii) $\forall\, a \in G \,\exists\, a^{-1} \in G$ such that $a.a^{-1} = a^{-1}.a = e$.

The element e in (ii) is unique for if f is such that $a.f = f.a = a$ for all a of G then we have in particular $e.f = f.e = e$. But from (ii) with a replaced by f $e.f = f$, therefore $f = e$. e is called the *neutral element* of G. Similarly in (iii) if $a.b = b.a = e$ then

$$a^{-1} = a^{-1}.e = a^{-1}.(a.b) = b.$$

Thus a^{-1} is uniquely determined by a. It is called the *inverse* of a.

If in addition we have

(iv) $a.b = b.a$, (Commutative Law)

for all a, b of G then G is called a *commutative* group.

We write $a.b.c$ instead of $a.(b.c)$ or $(a.b).c$.

Properties of groups

(α) $a.b = a.c \Rightarrow a^{-1}.a.b = a^{-1}.a.c \Rightarrow b = c$ (The Cancellation Law).

† An excellent and detailed account of the relations between the number systems and algebraic systems is to be found in *The Number-System* by H. A. Thurston.

(β) $(a.b)^{-1} = b^{-1}.a^{-1}$. For $b^{-1}.a^{-1}.a.b = b^{-1}.e.b = b^{-1}.b = e$.

From (α) every commutative group is a hemi-group.

Every finite hemi-group is a group. Let S be a hemi-group. By the Cancellation Law for each fixed y the correspondence $x \leftrightarrow x.y$ is a 1–1 correspondence of S onto itself. Therefore given $y, e \exists$ a member of S, say y^{-1}, such that $y^{-1}.y = e$. This combined with the fact that S is a hemi-group implies that S is a group.

If S is an infinite hemi-group it is no longer the case that it must be a group but it can always be embedded in a group. This may be seen as follows. We consider the elements of a new set T to be classes of ordered pairs (a, b) of elements of S. Two ordered pairs (a, b) and (c, d) belong to the same class if and only if either $c = a.k, d = b.k$ or $a = c.k$, $b = d.k$ for the same k of S. We use $[(a, b)]$ to denote this class of ordered pairs. Define . over T by

$$[(a,b)].[(c,d)] = [(a.c, b.d)].$$

To show that this is valid we must show that . does not depend upon the particular members of the classes involved. This can easily be verified. II (i) is obviously a consequence of I (ii). The neutral element is $[(e, e)]$ and the inverse of $[(a, b)]$ is $[(b, a)]$.

III. Fields

A set of entities T and two operations . and + constitute a field if and only if,

(i) *T is a commutative group with respect to* +,

(ii) *the elements of T with the neutral element of the group of* (i) *omitted, form a commutative group with respect to* .,

(iii) *for all* a, b, c *of T* $a.(b+c) = a.b+a.c$..

The neutral element of the group defined in (i) is denoted by 0 and that of the group defined in (ii) by 1. The inverse of a in (i) is written $-a$ and that of a in (ii) is written a^{-1}.

Properties of Fields

(α) $0 = -0$. For $0+0 = 0$.

(β) $a.0 = 0.a = 0$. For $a.0 = a.(0+0) = a.0+a.0$.

(γ) $a.b = a.c$ and $a \neq 0 \Rightarrow b = c$. For

$$b = 1.b = a^{-1}.a.b = a^{-1}.a.c = c.$$

(δ) $a.b = 0 \Rightarrow a = 0$ or $b = 0$. For

$$b \neq 0 \Rightarrow a = a.1 = a.b.b^{-1} = 0.b^{-1} = 0.$$

(ϵ) If $a \neq 0$ then $a.x = b$ if and only if $x = a^{-1}.b$. For if $x = a^{-1}.b$ then $a.x = b$ and conversely if $a.x = b$ and $a \neq 0$ then

$$x = a^{-1}.a.x = a^{-1}.b.$$

(λ) (i) $x.(-y) = (-x).y = -(x.y)$; (ii) $(-x).(-y) = x.y$. For to prove (i) $0 = x.0 = x.(y+(-y)) = x.y+x.(-y)$, therefore $x.(-y) = -(x.y)$.
Similarly $$(-x).y = -(x.y).$$
To prove (ii) from (i)

$$(-x).(-y) = -((-x).y) = -(-(x.y)) = (x.y).$$

IV. Ordered Fields

An ordered field is a field T which contains a subset P with the following properties:

(i) $x \in T \Rightarrow$ *either* $x \in P$ *or* $-x \in P$.
(ii) $x \in P$ *and* $-x \in P \Rightarrow x = 0$.
(iii) $x \in P$ *and* $y \in P \Rightarrow x.y \in P$ *and* $x+y \in P$.

We can now define an order relation $\geqslant$ between every two elements of T by saying that $x \geqslant y$ if and only if $x+(-y) \in P$. If $x \geqslant y$ and $x \neq y$ then we write $x > y$. The set P consists of the elements of T for which $x \geqslant 0$.

Properties of ordered fields

(α) $0 \in P$. For $0 = -0$ and either $0 \in P$ or $-0 \in P$.

(β) $x.x \in P$. This follows from (iii) if $x \in P$ and if $x \notin P$ then $-x \in P$ and $x.x = (-x).(-x) \in P$.

(γ) Either $x \geqslant y$ or $y \geqslant x$. For $-(x+(-y)) = y+(-x)$ and by (i) either $x+(-y) \in P$ or $y+(-x) \in P$.

(δ) $x \geqslant x$. For $x+(-x) = 0 \in P$.

(λ) $x \geqslant y$ and $y \geqslant z \Rightarrow x \geqslant z$. For

$$x+(-z) = (x+(-y))+(y+(-z)) \in P \quad \text{by (iii).}$$

(μ) $x \geqslant y \Rightarrow x+z \geqslant y+z$. For

$$(x+z)+(-(y+z)) = x+z+(-y)+(-z)$$
$$= x+(-y) \in P.$$

(ν) $x \geqslant y$ and $z \geqslant w \Rightarrow x+z \geqslant y+w$. For ($\mu$) applied twice gives

$$x+z \geqslant y+z \geqslant y+w.$$

(o) $x > 0, y \geqslant 0 \Rightarrow x+y > 0$. For $x+y \geqslant 0$ since $x+y \in P$ and if $x+y = 0$ then $y = -x$ and we have $x \in P$, $-x \in P$. Therefore $x = 0$.

Extensions of the whole number system

§ 1. *The positive and negative integers*

The whole number system has properties similar to those of a field but where for a field we have a group with respect to both + and . in the case of the whole number system we have hemi-groups. The Distributive Law however holds. Our ultimate aim is to embed the whole number system in a field. If I is a group then we use I_0 to denote the set of elements of I excluding the neutral element.

The first step is to embed the hemi-group over the operation + in a group I and define . between the new elements in such a way that (*a*) I_0 is a hemi-group over . and (*b*) the Distributive Law holds.

We use the general method of embedding a hemi-group in a group. The new elements are classes of ordered pairs $[(a, b)]$ of whole numbers where $[(a, b)] = [(c, d)]$ if and only if $\exists\, k$ such that

$$a = c+k, \quad b = d+k \quad \text{or} \quad c = a+k, \quad d = b+k.$$

Thus each class contains either a pair $(0, p)$ or a pair $(p, 0)$ and only one pair of either of these forms. The classes $[(0, p)]$ are called positive if $p \neq 0$. They are in 1–1 correspondence with the whole numbers excluding 0 and we shall use p to denote the class of pairs $[(0, p)]$. The classes $[(p, 0)]$ $p \neq 0$ are called negative and we use $-p$ instead of $[(p, 0)]$. We write I for the set of these elements. I is a group with respect to +, 0 is the neutral element and $-p$ is the inverse of p. Multiplication has already been defined over the whole numbers and we define $[(0, p)].[(0, q)]$ to be $[(0, pq)]$. We wish to preserve the Distributive Law and thus for $p, q \in I$

$$q(p+(-p)) = q.0 = 0.$$

Hence $(-p).q = -(p.q)$, $(-p).(-q) = p.q$. These are definitions of . and the reader can verify I_0 is a hemi-group over .. I is now a group over +, I_0 a hemi-group over . and the Distributive Law holds. We say $p > q$ if and only if $p+(-q)$ is positive,

§ 2. *The Rational numbers*

We again extend the hemi-group of I_0 over . to a group by considering the classes of pairs (a, b), where $a \in I$, $b \in I$, $a \neq 0$, $b \neq 0$. We put (a, b), (c, d) in the same class if and only if $\exists\, k$ such that either $a = k.c$, $b = k.d$ or $c = k.a$, $d = k.b$. I_0 then becomes a group with respect to . which we denote by R^*. The elements of R^* of the form

$[(a, 1)]$ behave in an exactly similar manner to the elements of I. We write a/b instead of $[a, b]$, a instead of $a/1$ and refer to a as an integer.

We must next define addition in such a way that the distributive property is maintained. We need to have

$$b.d(a/b+c/d) = b.d.(a/b)+b.d.(c/d) = a.d+b.c.$$

Thus we *define* + by $a/b+c/d = (a.d+b.c)/(b.d)$. We also define 0 to be a rational such that

$$a/b+0 = a/b \quad (a/b).0 = 0,$$

then with these definitions it can be seen the R^* is a field called the *rational number field.*

R^* is in fact an ordered field, as may be seen by taking P to be the rational numbers 0, a/b if $a.b > 0$.

In R^* we can form powers: x^y is defined for x rational y an integer by (i) if $y > 0$, x^y is x multiplied by itself y times; (ii) if $y < 0$, x^y is $(1/x)$ multiplied by itself y times.

The properties of powers P(1), P(2), P(3), P(4) all remain valid.

§3. *The Real numbers*

In the rational number system we can add, subtract, multiply and divide. We can also form powers but not take roots, the equation $x^2 = a$ for example may or may not be soluble.

To extend the rational number field further we need a different technique.

One method of performing the extension is that used by Dedekind and called the Dedekind section. By such a section we mean an ordered pair of sets of rational numbers say (A, B) such that

(i) $A \cap B = \phi$.
(ii) $A \cup B = R^*$.
(iii) $A \neq \phi$, $B \neq \phi$.
(iv) If $a \in A$ and $b \in B$ then $a < b$.

It is convenient initially to restrict ourselves to non-negative rational numbers and to use the single sets L instead of the pair (A, B) with the properties

(i) $L \neq \phi$.
(ii) $L \neq$ all positive rationals.
(iii) $r \in L$ and $0 \leqslant s < r \Rightarrow s \in L$.

We also add a fourth property which is a convention introduced to avoid ambiguities later on.

(iv) L has no greatest number.

We define now order, addition and multiplication between sets of the form L which will enable us to use them as numbers. Further those particular sets L of the form $\{x|x < r, x, r \text{ rational}\}$ will have properties similar to those of the rational numbers.

For two sets L, M of the above form we use the following definitions.

Order: $L \leqslant M$ if $L \subset M$.

Addition: $L+M = \{x|x = r+s, \text{ where } r \in L, s \in M\}$.

Multiplication: $L.M = \{x|x = r.s, \text{ where } r \in L, s \in M\}$.

With these definitions the sets $\{x|0 \leqslant x < r, \text{ where } x \in R^*, r \in R^*\}$ behave in a similar manner to the positive rationals. The new numbers are called positive real numbers. If we adjoin the element 0 with the property $0+a = a+0 = a$ for every real positive a then the system is a hemi-group with respect to $+$. As in the case of whole numbers we extend this hemi-group: defining multiplication so that

$$(-p).q = -(p.q), \quad (-p).(-q) = p.q$$

and the Distributive Law holds. But now in contradistinction with the case of whole numbers our new set of entities is a field. To establish this we need only show that the set with 0 omitted is a group with respect to .. Now if a is a positive real number $1.a = a$ follows from the definition of multiplication and if a is not positive then

$$a = -(-a) = -(1.(-a)) = -(-(1.a)) = 1.a.$$

Thus 1 is the neutral element. Again if a is a positive real number there is an inverse a^{-1}, in fact a^{-1} is the set of all rationals of the form $1/s$, where s in a positive rational not in a. If a is not positive or zero then a^{-1} is equal to $-((-a)^{-1})$. The other group axioms are easily verified and thus our new system is a Field. In fact it is ordered as may be seen by taking as P the set of real numbers which we called positive above together with 0. The ordering based on this set agrees with that defined above, (i.e. $L \leqslant M$ if $L \subset M$).

The new system is called the field of real numbers; it contains a subset which behaves like the rational numbers namely elements of the form a or $-a$, where $a = \{x|0 < x < r, x \in R^*, r \in R^*\}$; this in turn contains a subset which behaves like the integers; and this contains a subset which behaves like the whole number system. We refer to these various subsets by the same names as the systems from which we originally obtained them. Strictly speaking we are using an ambiguous nomenclature here but since all the relevant properties are the same in both cases the ambiguity does not have any erroneous consequences.

We could of course repeat the process by which the real field was obtained from the rational field. But this does not lead to anything essentially new. For if we define L as a set of positive real numbers such that if $s \in L$ then $r \in L$, where $0 < r < s$ and r a real number; then each r is a set of rational numbers and if we denote by K the union of all these rational numbers for all $r \in L$ then it is easy to see that K is a real number and that L is precisely the set of positive real numbers less than K. This implies a 1–1 correspondence between new sets of real numbers and real numbers themselves. Thus the new field will be in 1–1 correspondence with the old and will be of no greater utility.

The real number field is *complete*. If S is a set of real numbers and $\exists\, k$ such that $s \leqslant k$ for every $s \in S$ then $\exists\, t$ a real number such that

$$(a)\ s \leqslant t \text{ for every } s \in S$$

and $$(b)\ \text{if } t' < t \text{ then } \exists\, s \in S \text{ such that } s > t'.$$

t is called the *least upper bound of* S. Suppose first, that S contains positive numbers. Denote by t the real number which is the union of all the rationals which belong to the sets of rationals which define the positive numbers in S. Clearly if $s \in S$, $s \leqslant t$ since this follows immediately from the definitions if $s > 0$ whilst if $s \leqslant 0$ we have $s \leqslant 0$ and $0 < t$ implies $s \leqslant t$. But if $t' < t$ $\exists$ a rational contained in t and not in t', i.e. $\exists\, s \in S$ such that $s > t'$. If S contains only negative numbers let one of them be p and consider the set S_1, where

$$S_1 = \{x \mid x = s + p + 1 \quad s \in S\}.$$

$\exists$ the least upper bound q of S_1 and $q - p - 1$ is the least upper bound of S.

Archimedes Axiom holds, i.e. if $a > 0$, b are given $\exists$ an integer n such that $na > b$. If $b \leqslant 0$, $n = 1$ would do. If $b > 0\, \exists$ positive rationals p, q such that $p < a$, $q > b$. Suppose $p = p_1/p_2$, $q = q_1/q_2$, where p_1, p_2, q_1, q_2 are all positive integers and $n = p_2 q_1$ then

$$np = p_1 q_1 \geqslant q_1 \geqslant q.$$

Therefore $$na > np \geqslant q > b.$$

If (L, R) is a Dedekind section of the real number field and t is the least upper bound of L then L and R are defined either by

$$L = \{x \mid x \leqslant t\},\ R = \{x \mid x > t\} \quad \text{or} \quad L = \{x \mid x < t\},\ R = \{x \mid x \geqslant t\}.$$

Thus either L has a greatest number or R has a least number.

To define powers of real numbers we proceed as follows:

p an integer > 0 $\quad a^p$ is a multiplied by itself p times,

$p = 0$ $\quad a^p = 1$ if $a \neq 0$ (0^0 is not defined),

p an integer < 0 $\quad a^p = 1/a^{-p}$.

p a rational $= p_1/p_2$, where $p_2 > 0$. If $a > 0$ then $a^p =$ set of rationals r for which $0 < r$ and $r^{p_2} < a^{p_1}$.

The properties of powers given in P (1)–P (4) may be verified in so far as the appropriate functions have been defined.

We cannot define rational powers of negative numbers. For this we need to use the complex number system. The real numbers have one property namely that of *order* which we cannot hope to preserve if we are to solve all equations of the form $x.x = b$. For if an ordered field had this property then in the notation of IV we should have $P = T$.

It is for this reason that both real numbers and complex numbers are used extensively in analysis.

§ 4. *Complex numbers*

The method of extension from the real number field to the complex number field is of a general type. If we have a field K_1 in which we are unable to solve a certain polynomial equation say

$$a_0x^n + \ldots + a_n = 0,$$

where $a_i \in K_1$ and n is a positive integer then we can construct another field K_2 by forming all the polynomials whose coefficients belong to K_1 defining additions and multiplications in the natural manner and then identifying any two polynomials if they can be reduced to a common form by replacing $a_0x^n + \ldots + a_n$ by 0.

In our case we wish to solve $x^2 + 1 = 0$ and the entities will be polynomials in x with real coefficients in which we can replace x^2 wherever it occurs by -1. Thus all our polynomials are represented by linear forms. We define addition and multiplications by

$$(a+bx)+(c+dx) = (a+c)+(b+d)\,x,$$

$$(a+bx).(c+dx) = ac+(bc+ad)x+bdx^2$$

$$= ac-bd+(bc+ad)\,x.$$

In many respects the symbol $a+bx$ is misleading in that it appears to involve the (undefined) addition of a to the (undefined) product of b with a mysterious (and undefined) x. For this reason we prefer to use an ordered pair of real numbers $[a, b]$. Then we have as our definitions of $+$ and $.$

$$[a,b]+[c,d] = [a+c, b+d],$$
$$[a,b].[c,d] = [ac-bd, ad+bc].$$

The additive neutral element is $[0, 0]$ and the multiplicative neutral element is $[1, 0]$. The additive inverse of $[a, b]$ is $[-a, -b]$ and the multiplicative inverse is $[a.(a^2+b^2)^{-\frac{1}{2}}, -b.(a^2+b^2)^{-\frac{1}{2}}]$.

In this field we can solve the equation $[x,y].[x,y] = [a,b]$ but in fact much more than this is true. It is possible to solve any polynomial equation in complex numbers but the proof of this depends upon complex analysis and is beyond the scope of this book.

§5. *An axiomatic approach to real numbers*

Instead of attempting to define real numbers constructively we can simply lay down axioms which they must satisfy and leave on one side the question as to whether there are any entities satisfying these axioms or not. One set of axioms is as follows.

A set $\mathbf{R}$ is the set of real numbers if and only if it satisfies the following axioms.

(i) $\mathbf{R}$ is an ordered field (so that axioms III, IV hold).

(ii) If $T \subset \mathbf{R}$ is such that $\exists\, r \in \mathbf{R}$ satisfying $t \leqslant r$ all $t \in T$ then $\exists\, p \in \mathbf{R}$ such that (a) $t \leqslant p$ all $t \in T$, (b) if $q < p$ $\exists\, t \in T$ such that $t > q$. p is called the *least upper bound* of T.

We could then deduce all the properties of real numbers from these axioms. Many of these properties are given in III and IV of Algebraic Systems. The positive integers are defined to be 1, $1+1$, $1+1+1, \ldots$ and denoted by $1, 2, 3, \ldots$. Archimedes axiom can easily be deduced. If $\exists\, a > 0$ such that $na \leqslant b$ for every positive integer n then let p be the least upper bound of the set of elements $na, n = 1, 2, \ldots$. We have $a > 0 \Rightarrow p-a < p \Rightarrow \exists\, N$ such that $Na > p-a \Rightarrow (N+1)\,a > p$ which contradicts one of the properties defining p. Hence Archimedes Axiom is true.

Integral powers are easily defined and for rational powers of particular numbers $a^{p/q}$, where p is an integer, q a positive integer and $a > 0$ we consider the set K of positive real numbers x for which $x^q < a^p$. Now either $x < 1$ or $x < x^q < a^p$ thus K is bounded above and by (iii) above $\exists$ a least upper bound say τ. We define τ to be $a^{p/q}$. It may be verified that the properties of powers hold in particular $(a^{p/q})^q = a^p$.

To see this we should first establish the Binomial expansion for positive integers

$$(*)\ (x+y)^q = \sum_{r=0}^{r=q} \binom{q}{r} x^r y^{q-r},$$

where
$$\binom{q}{r} = \frac{q!}{r!(q-r)!},$$

by an inductive argument.

In this the symbols $q!$, $\sum\limits_{r=0}^{r=q}$ can be shown to have a meaning by an inductive argument. When this has been established suppose that $0 < y < 1$ then

$$(x+y)^q < x^q + \sum_{r=0}^{q-1} \binom{q}{r} x^r y$$

and if $x > 0$

$$(**)\ (x+y)^q < x^q + y \sum_{r=0}^{q} \binom{q}{r} x^r = x^q + y.(1+x)^q.$$

Reverting to the properties of powers $(a^{p/q})^q = a$ we see that if $\tau^q < a^p$ then $\exists\, \epsilon > 0$ such that $(\tau+\epsilon)^q < a^p$. For in (**) take $x = \tau$, $y = \epsilon$ and choose $\epsilon < (a^p - \tau^q)(1+\tau)^{-q}$. This is impossible by the definition of τ, therefore $\tau^q \geqslant a^p$. Similarly we can show that $\tau^q \leqslant a^p$, therefore $\tau^q = a^p$.

All other properties of real numbers can be deduced from these axioms.

INDEX